SPIG2018

SPIG2018

Special Issue Editors

Goran Poparić
Bratislav Obradović
Duško Borka
Milan Rajković

MDPI • Basel • Beijing • Wuhan • Barcelona • Belgrade

MDPI

Special Issue Editors

Goran Poparić
University of Belgrade
Serbia

Bratislav Obradović
University of Belgrade
Serbia

Duško Borka
University of Belgrade
Serbia

Milan Rajković
University of Belgrade
Serbia

Editorial Office
MDPI
St. Alban-Anlage 66
4052 Basel, Switzerland

This is a reprint of articles from the Special Issue published online in the open access journal *Atoms* (ISSN 2218-2004) from 2018 to 2019 (available at: https://www.mdpi.com/journal/atoms/special_issues/SPIG2018)

For citation purposes, cite each article independently as indicated on the article page online and as indicated below:

LastName, A.A.; LastName, B.B.; LastName, C.C. Article Title. *Journal Name* **Year**, *Article Number*, Page Range.

ISBN 978-3-03897-850-3 (Pbk)
ISBN 978-3-03897-851-0 (PDF)

Contents

About the Special Issue Editors

Goran Poparić is a Professor at the Faculty of Physics, University of Belgrade. His research interests include physics of atomic collision processes, physics of atoms and molecules, computer physics, and computer simulations of physical processes.

Bratislav Obradović is a Professor at the Faculty of Physics, University of Belgrade. His research interests include spectroscopy diagnostics of electrical discharges and plasmas, and application of non-thermal atmospheric pressure discharges. He is a referee for the following journals: *Plasma Sources Science and Technology*, *Journal of Physics D: Applied Physics*, *Journal of Applied Physics*, *Plasma Chemistry* and *Plasma Processes*, *Plasma Processes and Polymers*, and *Journal of Hazardous Materials*.

Duško Borka received his B.Sc. (1999), M.Sc. (2002), and Ph.D. (2006) degrees from the Faculty of Physics, University of Belgrade. He has been working for twenty years as a scientist at the Institute of Nuclear Sciences Vinča in Belgrade, Serbia. He completed postdoctoral training in the Department of Applied Mathematics at the University of Waterloo, Canada, in 2007. He has published 65 articles in internationally reviewed journals, which have been cited over 500 times. His research interests include gravitation and the structure of the Universe, particle interaction with solids, mathematical modeling, and computer simulations of physical processes. He is a member of Serbian Astronomical Society, Serbian Physical Society, SEENET-MTP (Southeastern European Network in Mathematical and Theoretical Physics), AIS3 (Association of Italian and Serbian Scientists and Scholars). Dr. Borka received the Annual Award of the Institute of Nuclear Sciences "Vinča", for his scientific contributions to basic research in 2012. He is referee in the following journals: *Carbon, Chaos, Applied Mathematical Modelling, International Journal of Computational Methods, Nuclear Instruments and Methods in Physics Research B, European Physical Journal D, Universe, Publications of the Astronomical Observatory of Belgrade.*

Milan Rajković is a Senior Research Scientist at the Institute of Nuclear Sciences Vinca, University of Belgrade, Serbia. His research interests include physics of fusion plasmas, nonlinear dynamical systems, complex networks, and computational topology.

Preface to "SPIG2018"

The 29th International Symposium on the Physics of Ionized Gases was held in Belgrade, Serbia, from 28 August to 1 September 2018, at the Serbian Academy of Sciences and Arts (SASA). Over 150 attendees were welcomed. The conference was organized by the Vinča Institute of Nuclear Sciences, in cooperation with SASA, and under the auspices of the Ministry of Education, Science and Technological Development of the Republic of Serbia. This conference is a series of biennial meetings which began in 1962 in Belgrade. The 29th in this series of events reflected upon the progress in this challenging field of science, with the aim of presenting new results in fundamental and frontier theories as well as technologies in the areas of general plasma physics, atomic collision processes, and particle and laser beam interactions with solids. Within the framework of the conference, the 3rd Workshop on X-ray and VUV Interaction with Biomolecules in Gas Phase–XiBiGP was also organized.

The topic of our conference included the following fields:

- Atomic Collision Processes
  Electron and Photon Interactions with Atomic Particles, Heavy Particle Collisions, Swarms, and Transport Phenomena

- Particle and Laser Beam Interactions with Solids
  Atomic Collisions in Solids, Sputtering and Deposition, and Laser and Plasma Interactions with Surfaces

- Low Temperature Plasmas
  Plasma Spectroscopy and other Diagnostic Methods, Gas Discharges, and Plasma Applications and Devices

- General Plasmas
  Fusion Plasmas, Astrophysical Plasmas, and Collective Phenomena

The present issue of Atoms represents a collection of papers presented as Invited Lectures, Topical Invited Lectures, Progress Reports, and Workshop Lectures. Regarding the handling of these lectures, we would like to thank the Scientific Committee for their expertise and support: G. Poparić (Co-chair, Serbia), B. Obradović (Co-chair, Serbia), D. Borka (Serbia), S. Buckman (Australia), J. Burgdörfer (Austria), J. Cvetić (Serbia), E. Danezis (Greece), Z. Donko (Hungary), V. Guerra (Portugal), D. Ilić (Serbia), M. Ivković (Serbia), I. Mančev (Serbia), D. Marić (Serbia), N. J. Mason (UK), A. Milosavljević (France), K. Mima (Japan), Z. Mišković (Canada), L. Nahon (France), P. Roncin (France), I. Savić (Serbia), Y. Serruys (France), N. Simonović (Serbia), S. Tošić (Serbia), M. Škorić (Japan) and M. Trtica (Serbia). We would also like to thank the following members of the Local Organizing Committee for their exemplary teamwork and excellent planning: D. Borka (Co-chair), M. Rajković (Co-chair), V. Borka Jovanović (Co-secretary), N. Potkonjak (Co-secretary), N. Konjević, N. Cvetanović, A. Hreljac, B. Grozdanić, J. Aleksić, M. Nešić, S. Živković, M. S. Dimitrijević and J. Ciganović.

Goran Poparić, Bratislav Obradović, Duško Borka, Milan Rajković
Special Issue Editors

atoms

MDPI

Article

Atomic and Molecular Processes in a Strong Bicircular Laser Field

Dejan B. Milošević [1,2]

1. Faculty of Science, University of Sarajevo, Zmaja od Bosne 35, 71000 Sarajevo, Bosnia and Herzegovina; milo@bih.net.ba; Tel.: +387-33-610-157
2. Academy of Sciences and Arts of Bosnia and Herzegovina, Bistrik 7, 71000 Sarajevo, Bosnia and Herzegovina

Received: 23 September 2018; Accepted: 5 November 2018; Published: 8 November 2018

check for updates

Abstract: With the development of intense femtosecond laser sources it has become possible to study atomic and molecular processes on their own subfemtosecond time scale. Table-top setups are available that generate intense coherent radiation in the extreme ultraviolet and soft-X-ray regime which have various applications in strong-field physics and attoscience. More recently, the emphasis is moving from the generation of linearly polarized pulses using a linearly polarized driving field to the generation of more complicated elliptically polarized polychromatic ultrashort pulses. The transverse electromagnetic field oscillates in a plane perpendicular to its propagation direction. Therefore, the two dimensions of field polarization plane are available for manipulation and tailoring of these ultrashort pulses. We present a field that allows such a tailoring, the so-called bicircular field. This field is the superposition of two circularly polarized fields with different frequencies that rotate in the same plane in opposite directions. We present results for two processes in a bicircular field: High-order harmonic generation and above-threshold ionization. For a wide range of laser field intensities, we compare high-order harmonic spectra generated by bicircular fields with the spectra generated by a linearly polarized laser field. We also investigate a possibility of introducing spin into attoscience with spin-polarized electrons produced in high-order above-threshold ionization by a bicircular field.

Keywords: strong-field physics; attoscience; bicircular field; high-order harmonic generation; above-threshold ionization; spin-polarized electrons

1. Introduction: Three-Step Model and Bicircular Laser Field

Available strong laser fields allow the study of new laser-field-induced atomic and molecular processes such as high-order harmonic generation (HHG) [1] and above-threshold ionization (ATI) [2]. These processes are commonly considered for a laser field which is linearly polarized and explained by semiclassical three-step model [3–5]. According to this model the electron, liberated in tunnel ionization, moves driven by the laser field and returns to the parent core where it recombines emitting a high-harmonic photon in the HHG process or rescatters and is detected having much larger energy in the high-order ATI (HATI) process.

Let us explain this three-step model in more detail using Figure 1. Initially, the electron is bound with the energy $-I_p$. When the linearly polarized laser field $E_{\rm lin}(t)$ approaches an extremum at the time t_0, the electron can tunnel through the potential barrier, created by an instantaneous laser electric field and the atomic potential, and is "born" in the continuum with zero velocity $v(t_0)$. This is the first step of this model. After that, the field strength decreases, goes through zero and then reaches its next maximum value. Since the field and the corresponding force at the time t' change their signs, the electron velocity changes its direction at the time t'' and the electron starts moving back to its parent core. The corresponding electron velocity is related to the quantity $A(t)$, defined by

$E_{lin}(t) = -dA(t)/dt$. Since the field $E_{lin}(t)$ is extremal at the times t'' and t_0 we have that $A(t) = 0$ at these moments. The electron returns to the parent core at the time t_1 having the velocity $v_{ret}(t_1)$. This is the second step. It can be shown using momentum conservation [6], that the kinetic energy of the returned electron has the maximum value $3.17U_p$, with $U_p = E^2_{lin,max}/(4\omega^2)$ the electron ponderomotive energy in a linearly polarized field having frequency ω (we use atomic units). In the case of HHG process, in the third step, the electron recombines to the ground atomic state and the energy equal to ionization potential energy I_p plus the electron kinetic energy is released in the form of an energetic photon. Maximum high-harmonic photon energy is $I_p + 3.17U_p$, as denoted in Figure 1. The efficiency of the HHG process is approximately the same for all harmonic photons with energies larger than I_p and the HHG spectrum has a shape of a plateau. Since the third step happens during the time interval which is a small part of the laser field optical cycle, it is clear that the described high-order atomic and molecular processes develop during few tens of attoseconds if one uses femtosecond lasers. Therefore, this third step "opens the doors" for attoscience [7–12] which investigates electron dynamics of strong-field processes on the time scale of few attoseconds, a natural scale for electronic motion in atoms and molecules (one atomic unit of time is 24.19 as).

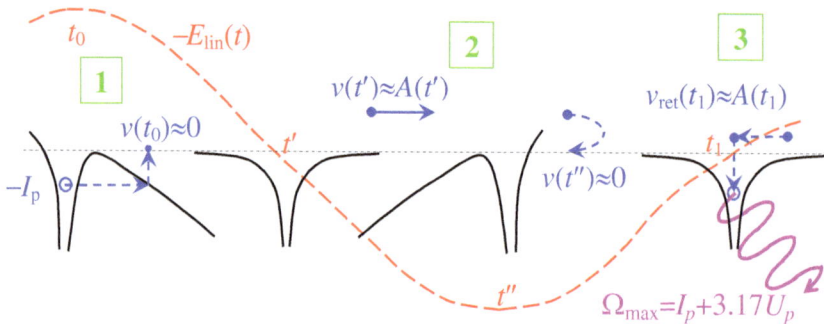

Figure 1. Graphical sketch of the three-step model for high-order harmonic generation. The combined atomic and laser field potential is presented by black lines, while the electron and its velocity are shown in blue. The temporal evolution of a linearly polarized laser field $E_{lin}(t)$ is depicted by the red long-dashed line. The emitted high-harmonic photon is illustrated by a pink wavy line with an arrow at the end.

It is well known that HATI and HHG processes are not possible with a circularly polarized laser field since the liberated electron, driven by such a field, cannot return to the parent core to recombine or rescatter. However, these processes become possible if one uses the (so-called) bicircular laser field consisting of two counter-rotating circularly polarized laser fields which are coplanar and have different frequencies. This was first confirmed experimentally for HHG in 1995 [13,14] (for more references see recent articles [15,16]). ATI process in a bicircular field was first investigated theoretically in [17,18] (see also [19]) and confirmed experimentally in [20]. Theoretical analysis of HATI was performed in [21–23], while the relevant experimental results were published in [24,25].

In 2000 bicircular-field-induced HHG was explained using the quantum-orbit theory [26]. More information about this theory is given in [6,27]. In the present context it is important that, using this theory, two-dimensional trajectories of the electrons which come back to the parent core were identified. In addition, it was found that the emitted higher harmonics are circularly polarized with alternating ellipticities equal to ±1. This was confirmed experimentally in 2014 [28]. For application it is crucial to generate circularly polarized high-order harmonics which can serve as a source of soft X-ray photons. Such photons have application for analysis of various chirality sensitive processes in organic molecules [29,30], magnetic materials [31,32] etc.

Combining a group of circularly polarized high-order harmonics having ellipticities which alternate between +1 and −1, rather than obtaining a circularly polarized pulse, we obtain a pulse having unusual polarization properties. This was first shown in [33] where, for a bicircular field with frequencies ω and 2ω, a star-like form with 3 linearly polarized pulses rotated by 120° was obtained. This theoretical prediction has recently been confirmed in experiment [34]. It was suggested in 2001 [35] that circularly polarized attosecond pulse trains can be generated if the harmonics having helicity +1 are stronger than that of helicity −1 (and vice versa), i.e., if we, by some means, achieve helicity asymmetry in an interval of high-harmonic photon energies. In [35] such asymmetry was noticed for He atom, which has s ground state, for the intensity of the 2ω bicircular field component two times higher than that of the ω component. Later on, in 2015, it was found that the helicity asymmetry for much higher photon energies exists for HHG by noble gases with the p ground state [36–38].

We study an $r\omega$–$s\omega$ bicircular field, with r and s integers, defined by

$$
\begin{aligned}
E_x(t) &= [E_1 \sin(r\omega t) + E_2 \sin(s\omega t + \varphi)] / \sqrt{2}, \\
E_y(t) &= [-E_1 \cos(r\omega t) + E_2 \cos(s\omega t + \varphi)] / \sqrt{2}.
\end{aligned}
\tag{1}
$$

Here $I_1 = E_1^2$ and $I_2 = E_2^2$ are the intensities of the components and φ is the relative phase. Examples of such fields are presented in Figure 2 for various combinations of r and s and the phase $\varphi = 0$ (for different phases the field is rotated but does not change the shape [39]). We see that this field satisfies $(r + s)$-fold rotational symmetry. Furthermore, this field obeys particular dynamical symmetry: simultaneous rotation about the z axis by the angle $r \cdot 360° / (r + s)$ and translation in time by $T / (r + s)$ leaves the field unchanged (see the Appendix A in [38]). For example, ω–2ω bicircular field is invariant with respect to simultaneous rotation by 120° and translation in time by 1/3 optical cycle.

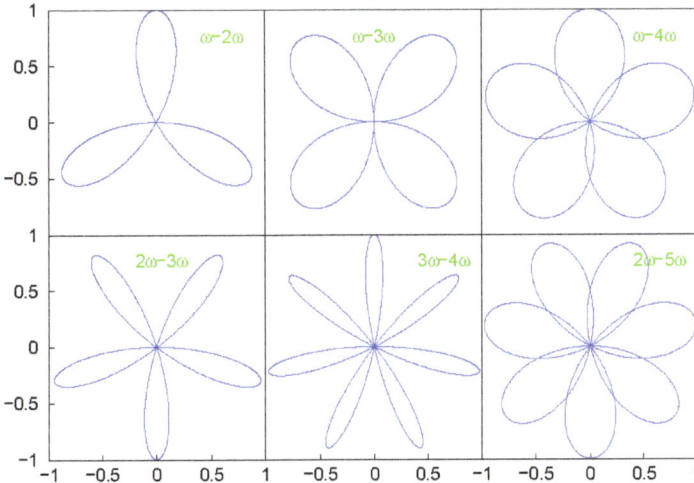

Figure 2. The electric-field-vector polar diagram for the $r\omega$–$s\omega$ bicircular field, having equal component intensities, plotted for $0 \leq t \leq T = 2\pi/\omega$, with the fundamental frequency ω. The six presented panels depict the field for various combinations of the values of r and s, as denoted.

2. Results for High-Order Harmonic Generation by Bicircular Field

According to our theory of HHG by bicircular field, presented in [38], the nth harmonic emission rate is given by

$$w_n = \frac{(n\omega)^3}{2\pi c^3} |\mathbf{T}_n|^2, \quad \mathbf{T}_n = \int_0^T \frac{dt}{T} \mathbf{d}(t) e^{in\omega t}. \tag{2}$$

Here $\mathbf{d}(t)$ is the time-dependent dipole and the nth harmonic and its ellipticity ε_n satisfy the following selection rule [38]

$$\varepsilon_n = \pm 1 \quad \text{for} \quad n = q(r+s) \pm r. \tag{3}$$

These relations can be derived using the dynamical symmetry of the bicircular field.

There are qualitative differences between the HHG spectrum generated by a linearly polarized laser field and the spectrum generated by bicircular field with equal component intensities. We illustrate this in Figure 3 by showing the HHG spectra, generated by Ar atoms subjected to a laser field having the fundamental photon energy $\omega = 1.6$ eV, as a function of the harmonic order and the laser field intensity in atomic units. For HHG by a linearly polarized laser field (upper panel (a)), emitted harmonics are linearly polarized and the spectrum forms a plateau which finishes by a cutoff. The cutoff position, i.e., the maximum harmonic order, is proportional to the laser intensity. The spectrum in the plateau exhibits fast oscillations. These oscillations are caused by the interference of the quantum-orbit contributions [6,40,41]. In the cutoff region there are no such oscillations since only one quantum orbit contributes to the HHG spectrum. The spectrum for ω–2ω bicircular field, presented in the bottom panel (b), also exhibits a plateau with a cutoff. However, the plateau is different. First, the plateau is more inclined and its height decreases with the increase of the harmonic order. Second, the plateau is flat and there are no oscillations as in the linear polarization case. The reason is that the contribution of only one quantum orbit is dominant. However, in the cutoff region there are such oscillations, again contrary to the linear polarization case. The reason is that in the cutoff region more orbits contribute to the HHG spectrum generated by bicircular laser field and the oscillatory structure is due to their interference. The relevant quantum orbits are analyzed in [26]. In the present paper we have shown, using three-dimensional graphs of Figure 3, that this behavior is valid for a wide range of laser intensities and harmonic orders. It is also clear from Figure 3 that, for a range of harmonic orders and laser intensities, the harmonic emission rate is higher for HHG by bicircular field than for HHG by linearly polarized field.

It should be mentioned that for a bicircular field with higher intensity of the second field component the plateau becomes more similar to that of the linearly polarized case, i.e., it is not inclined and oscillatory structures appear due to the interference of contributions of more quantum orbits. This is recently explored in detail in [16,42].

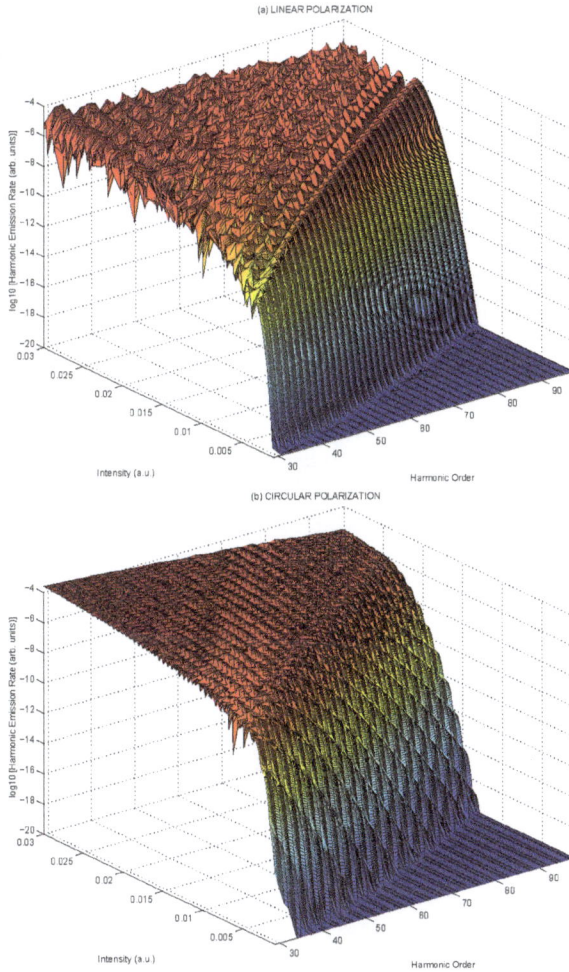

Figure 3. Three-dimensional graphs of the logarithm of the harmonic emission rate as a function of the harmonic order and laser intensity (in atomic units). Upper panel (**a**): Linearly polarized laser field. Lower panel (**b**): ω–2ω bicircular laser field. The fundamental photon energy is $\omega = 1.6$ eV and Ar atoms are modeled by *s* ground state. The same arbitrary units for HHG rates are used in both panels.

3. Spin Asymmetry in Above-Threshold Ionization by Bicircular Field

In [43] (see also more recent references [44,45]) it was suggested to introduce the concept of attospin using spin-polarized electrons emitted in ionization by bicircular laser field. In this paper the differential ionization rate, $w_{\mathbf{p}\ell m}(n) = 2\pi p \left| T_{\mathbf{p}\ell m}(n) \right|^2$, of atoms having initial bound state $\psi_{\ell m}$, was calculated applying the saddle point-method as described in [22]. In this process the energy $n\omega$ is absorbed and an electron with momentum $\mathbf{p}$ and energy $E_{\mathbf{p}} = \mathbf{p}^2/2$ is emitted. The T-matrix element $T_{\mathbf{p}\ell m}(n)$ is presented as a sum of the direct and rescattering T-matrix elements and ℓ and m are, respectively, the orbital and magnetic quantum number. For Xe atoms the ground state is p state ($\ell = 1$ and $m = \pm 1$ (matrix elements are zero for $m = 0$)), and we have two continua corresponding to two ground states of Xe$^+$ ion ($^2P_{3/2}$ and $^2P_{1/2}$). Therefore, there are two ionization potentials $I_p^{3/2} = 12.13$ eV and $I_p^{1/2} = 13.44$ eV. We denote the corresponding differential ionization rates by

$w_{\mathbf{p},m,j}$, where $m = \pm 1$ and $j = 3/2$ for $I_p^{3/2}$ and $j = 1/2$ for $I_p^{1/2}$. For the differential ionization rate for electrons with the spin up ($W_{\mathbf{p}\uparrow}$) and down ($W_{\mathbf{p}\downarrow}$) we get

$$W_{\mathbf{p}\uparrow} = \left(2w_{\mathbf{p},-1,1/2} + w_{\mathbf{p},-1,3/2}\right)/3 + w_{\mathbf{p},1,3/2}, \quad W_{\mathbf{p}\downarrow} = \left(2w_{\mathbf{p},1,1/2} + w_{\mathbf{p},1,3/2}\right)/3 + w_{\mathbf{p},-1,3/2}. \quad (4)$$

We use the rates $W_{\mathbf{p}\uparrow}$ and $W_{\mathbf{p}\downarrow}$ to define the spin asymmetry parameter $A_{\mathbf{p}}$ and the normalized spin asymmetry parameter $\tilde{A}_{\mathbf{p}}$ by the relations:

$$A_{\mathbf{p}} = \frac{W_{\mathbf{p}\uparrow} - W_{\mathbf{p}\downarrow}}{W_{\mathbf{p}\uparrow} + W_{\mathbf{p}\downarrow}}, \quad \tilde{A}_{\mathbf{p}} = A_{\mathbf{p}}\frac{W_{\mathbf{p}\uparrow} + W_{\mathbf{p}\downarrow}}{\max_{\mathbf{p}}(W_{\mathbf{p}\uparrow} + W_{\mathbf{p}\downarrow})}. \quad (5)$$

For $I_p^{1/2} = I_p^{3/2}$, i.e., if we neglect the spin-orbit coupling, Formulas (4) and (5) give $A_{\mathbf{p}} = 0$. In the case of Xe atoms the fine structure splitting is $I_p^{1/2} - I_p^{3/2} = 1.31$ eV and one expects a substantial spin asymmetry. If the rates are equal for $m = 1$ and $m = -1$ then the asymmetry parameter $A_{\mathbf{p}}$ is also zero. However, for ATI of noble gases having p ground state by a circularly polarized laser field the ionization rate exhibits strong $m = \pm 1$ asymmetry, so that for Xe we expect large values of $A_{\mathbf{p}}$. Furthermore, for a bicircular field the electron rescattering is possible, which opens up access to attosecond spin effects, since the rescattering process develops on attosecond time scale [43].

In this paper we evaluate the ionization rate and the spin asymmetry parameter in a different way than in [43]. Namely, we calculate the differential ionization rate using numerical integration instead of the saddle-point method [46]. In addition, the results are averaged over the laser intensity distribution in the focus [47]. The used spatio-temporal averaging is applicable for long pulses, while for few-cycle pulses [8] the dynamical symmetry of the bicircular field is violated and the problem should be explored separately. From the upper left panel (a) of Figure 4 it follows that the direct differential ionization yield exhibits rotational symmetry by the angle $360°/(r+s) = 120°$ and the reflection symmetry corresponding to the axes with angles $60°$, $180°$, and $300°$ with respect to the x axis. The spin asymmetry parameters, shown in the left panels (middle panel (c) and bottom panel (e)), obey the same symmetry. For the rescattered electrons (right panels (b), (d), and (f)) the reflection symmetry is broken, but the rotational symmetry is maintained. The presented yields are normalized to the maximum value which is 1.237×10^{-4} for direct electrons and 7.707×10^{-5} for rescattered electrons (in arbitrary units since we present the results for the focal-averaged spectra). The results are normalized so that the maximum yield is $w_{\max} = 1$ and $\log_{10}(w_{\max}) = 0$. For the direct-electrons yields we show 6 orders, while for the rescattered-electron yields we present 4 orders of magnitude. The asymmetry parameter for direct electrons emitted in a fixed direction (for example at $60°$) exhibits fast oscillations with the increase of the photoelectron energy. This was explained in [43] as the interference of two dominant electron trajectories obtained by the saddle-point method. We now see that this behaviour is preserved in the spectra obtained by numerical integration. In addition, these fast oscillations survive averaging over the laser intensity distribution in the focus. Spin asymmetry parameters change from -0.4996 to 0.98 for direct electrons and from -0.5812 to 0.8581 for rescattered electrons. The most important result is that the spin asymmetry parameter for high-energy electrons can take large values. These electrons come from rescattering and they are characterized by the attosecond time scale so that, measuring the spectra and spin-polarization of these electrons, one can explore spin-dependent effects in atoms and molecules with unprecedented time resolution.

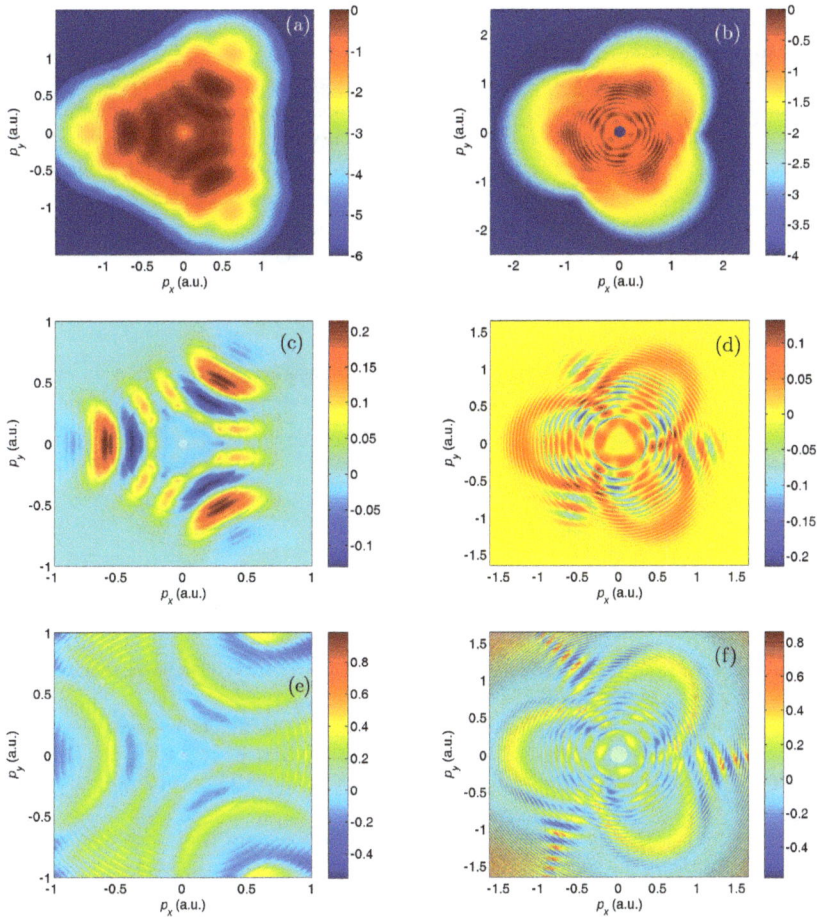

Figure 4. Focal-averaged results for Xe atoms ionized by ω–2ω bicircular field with the fundamental wavelength 800 nm and the same component peak intensities $I_1 = I_2 = 1.1 \times 10^{14}$ W/cm^2, depicted in false colors in the photoelectron momentum plane. Top panels (**a,b**): The logarithm of the summed photoelectron yield $W_{\mathbf{p}\uparrow} + W_{\mathbf{p}\downarrow}$. Middle panels (**c,d**): Normalized spin asymmetry parameter $\tilde{A}_\mathbf{p}$. Bottom panels (**e,f**): Spin asymmetry parameter $A_\mathbf{p}$. Left panels (**a,c,e**): Only the direct electrons are taken into account. Right panels (**b,d,f**): Only the rescattered electrons are accounted for.

4. Conclusions

We have explored two high-order atomic processes induced by bicircular fields. First, we have explicitly compared the HHG spectra generated by a bicircular laser field with the spectra generated by a linearly polarized laser field. In spite of that both spectra exhibit plateau and cutoff features, we observed important differences. Contrary to the case of linear polarization, the plateau is rather smooth for HHG by bicircular field. This is important for obtaining a high-harmonic attosecond pulse train by combining a group of high harmonics. Namely, as in the mode-locking laser technique, the relative phase between combined field components should be constant in order to generate ultrashort pulses. This condition is much better fulfilled for HHG by bicircular field [33].

We expect that bicircular field will have a bright future in application to molecular processes. The reason is that the rotational symmetry of the $r\omega$–$s\omega$ bicircular field (compare Figure 2) can be combined with analog C_{r+s} symmetry of polyatomic molecules. For example, planar molecule BF_3 and nonplanar molecules CF_3I and NH_3 obey the C_3 symmetry, as well as the ω–2ω bicircular field. Examples can be found in recent references [48–53].

Another interesting possibility of application of bicircular fields is to explore the electron spin on the ultrashort time scale. Spin-polarized electrons have important applications [54,55]. Spin asymmetry in above-threshold ionization by a circularly polarized laser field was investigated theoretically in [56–58] and in more recent experiments [59–61]. In our paper we have shown that the spin asymmetry in HATI by bicircular field survives focal-averaging and thus should be observed in future experiments which will open access to attospin. It should also be mentioned that, without the focal averaging, it would not be possible to explore quantum-mechanical effects such as experimentally observed intensity-dependent enhancements in HATI spectra [2]. Such enhancements are caused by the channel-closing effect. We have recently shown that this effect is important not only for linearly polarized laser fields but also for bicircular fields [62].

In addition, the effect of the bicircular field on ATI is especially important since it can be applied to study complicated molecules and materials where the spin dependence plays an important role. For diatomic molecules it was predicted in [63] and confirmed in experiment [64] that two-source double-slit interference effects in angle-resolved HATI spectra survive both the molecular orientation averaging and focal averaging.

Funding: This research received no external funding.

Conflicts of Interest: The authors declare no conflict of interest.

References

1. Kohler, M.; Pfeifer, T.; Hatsagortsyan, K.; Keitel, C. Frontiers of Atomic High-Harmonic Generation. *Adv. At. Mol. Opt. Phys.* **2012**, *61*, 159–208. [CrossRef]
2. Becker, W.; Goreslavski, S.P.; Milošević, D.B.; Paulus, G.G. The plateau in above-threshold ionization: The keystone of rescattering physics. *J. Phys. B* **2018**, *51*, 162002. [CrossRef]
3. Corkum, P.B. Plasma perspective on strong field multiphoton ionization. *Phys. Rev. Lett.* **1993**, *71*, 1994–1997. [CrossRef] [PubMed]
4. Schafer, K.J.; Yang, B.; DiMauro, L.F.; Kulander, K.C. Above threshold ionization beyond the high harmonic cutoff. *Phys. Rev. Lett.* **1993**, *70*, 1599–1602. [CrossRef] [PubMed]
5. Paulus, G.G.; Becker, W.; Nicklich, W.; Walther, H. Rescattering effects in above-threshold ionization: A classical model. *J. Phys. B* **1994**, *27*, L703–L708. [CrossRef]
6. Becker, W.; Grasbon, F.; Kopold, R.; Milošević, D.B.; Paulus, G.G.; Walther, H. Above-threshold ionization: From classical features to quantum effects. *Adv. At. Mol. Opt. Phys.* **2002**, *48*, 35–98.
7. Scrinzi, A.; Ivanov, M.Y.; Kienberger, R.; Villeneuve, D.M. Attosecond physics. *J. Phys. B* **2006**, *39*, R1–R37. [CrossRef]
8. Milošević, D.B.; Paulus, G.G.; Bauer, D.; Becker, W. Above-threshold ionization by few-cycle pulses. *J. Phys. B* **2006**, *39*, R203–R262. [CrossRef]
9. Kling, M.F.; Vrakking, M.J.J. Attosecond electron dynamics. *Annu. Rev. Phys. Chem.* **2008**, *59*, 463–492. [CrossRef] [PubMed]
10. Krausz, F.; Ivanov, M. Attosecond physics. *Rev. Mod. Phys.* **2009**, *81*, 163–234. [CrossRef]
11. Ueda, K.; Ishikawa, K.L. Attosecond science: Attoclocks play devil's advocate. *Nat. Phys.* **2011**, *7*, 371–372. [CrossRef]
12. Calegari, F.; Sansone, G.; Stagira, S.; Vozzi, C.; Nisoli, M. Advances in attosecond science. *J. Phys. B* **2016**, *49*, 062001. [CrossRef]
13. Eichmann, H.; Egbert, A.; Nolte, S.; Momma, C.; Wellegehausen, B.; Becker, W.; Long, S.; McIver, J.K. Polarization-dependent high-order two-color mixing. *Phys. Rev. A* **1995**, *51*, R3414–R3417. [CrossRef] [PubMed]
14. Long, S.; Becker, W.; McIver, J.K. Model calculations of polarization-dependent two-color high-harmonic generation. *Phys. Rev. A* **1995**, *52*, 2262–2278. [CrossRef] [PubMed]

15. Odžak, S.; Hasović, E.; Becker, W.; Milošević, D.B. Atomic processes in bicircular fields. *J. Mod. Opt.* **2017**, *64*, 971–980. [CrossRef]

16. Milošević, D.B. Quantum-orbit analysis of high-order harmonic generation by bicircular field. *J. Mod. Opt.* **2018**, *66*, 47–58. [CrossRef]

17. Kramo, A.; Hasović, E.; Milošević, D.B.; Becker, W. Above-threshold detachment by a two-color bicircular laser field. *Laser Phys. Lett.* **2007**, *4*, 279–286. [CrossRef]

18. Hasović, E.; Kramo, A.; Milošević, D.B. Energy- and angle-resolved photoelectron spectra of above-threshold ionizationand detachment. *Eur. Phys. J. Spec. Top.* **2008**, *160*, 205–216. [CrossRef]

19. Hasović, E.; Milošević, D.B.; Becker, W. A method of carrier-envelope phase control for few-cycle laser pulses. *Laser Phys. Lett.* **2006**, *3*, 200–204. [CrossRef]

20. Mancuso, C.A.; Hickstein, D.D.; Grychtol, P.; Knut, R.; Kfir, O.; Tong, X.M.; Dollar, F.; Zusin, D.; Gopalakrishnan, M.; Gentry, C.; et al. Strong-field ionization with two-color circularly polarized laser fields. *Phys. Rev. A* **2015**, *91*, 031402. [CrossRef]

21. Hasović, E.; Becker, W.; Milošević, D.B. Electron rescattering in a bicircular laser field. *Opt. Express* **2016**, *24*, 6413–6424. [CrossRef] [PubMed]

22. Milošević, D.B.; Becker, W. Improved strong-field approximation and quantum-orbit theory: Application to ionization by a bicircular laser field. *Phys. Rev. A* **2016**, *93*, 063418. [CrossRef]

23. Hoang, V.-H.; Le, V.-H.; Lin, C.D.; Le, A.-T. Retrieval of target structure information from laser-induced photoelectrons by few-cycle bicircular laser fields. *Phys. Rev. A* **2017**, *95*, 031402. [CrossRef]

24. Mancuso, C.A.; Hickstein, D.D.; Dorney, K.M.; Ellis, J.L.; Hasović, E.; Knut, R.; Grychtol, P.; Gentry, C.; Gopalakrishnan, M.; Zusin, D.; et al. Controlling electron-ion rescattering in two-color circularly polarized femtosecond laser fields. *Phys. Rev. A* **2016**, *93*, 053406. [CrossRef]

25. Mancuso, C.A.; Dorney, K.M.; Hickstein, D.D.; Chaloupka, J.I.; Tong, X.-M.; Ellis, J.L.; Kapteyn, H.C.; Murnane, M.M. Observation of ionization enhancement in two-color circularly polarized laser fields. *Phys. Rev. A* **2017**, *96*, 023402. [CrossRef]

26. Milošević, D.B.; Becker, W.; Kopold, R. Generation of circularly polarized high-order harmonics by two-color coplanar field mixing. *Phys. Rev. A* **2000**, *61*, 063403. [CrossRef]

27. Milošević, D.B.; Bauer, D.; Becker, W. Quantum-orbit theory of high-order atomic processes in intense laser fields. *J. Mod. Opt.* **2006**, *53*, 125–134. [CrossRef]

28. Fleischer, A.; Kfir, O.; Diskin, T.; Sidorenko, P.; Cohen, O. Spin angular momentum and tunable polarization in high-harmonic generation. *Nat. Photonics* **2014**, *8*, 543–549. [CrossRef]

29. Cireasa, R.; Boguslavskiy, A.E.; Pons, B.; Wong, M.C.H.; Descamps, D.; Petit, S.; Ruf, H.; Thiré, N.; Ferré, A.; Suarez, J.; et al. Probing molecular chirality on a sub-femtosecond timescale. *Nat. Phys.* **2015**, *11*, 654–658. [CrossRef]

30. Nahon, L.; Nag, L.; Garcia, G.A.; Myrgorodska, I.; Meierhenrich, U.; Beaulieu, S.; Wanie, V.; Blanchet, V.; Geneaux, R.; Powis, I. Determination of accurate electron chiral asymmetries in fenchone and camphor in the VUV range: Sensitivity to isomerism and enantiomeric purity. *Phys. Chem. Chem. Phys.* **2016**, *18*, 12696–12706. [CrossRef] [PubMed]

31. Fan, T.; Grychtol, P.; Knut, R.; Hernández-García, C.; Hickstein, D.D.; Zusin, D.; Gentry, C.; Dollar, F.J.; Mancuso, C.A.; Hogle, C.; et al. Bright circularly polarized soft X-ray high harmonics for X-ray magnetic circular dichroism. *Proc. Natl. Acad. Sci. USA* **2015**, *112*, 14206–14211. [CrossRef] [PubMed]

32. Kfir, O.; Zayko, S.; Nolte, C.; Sivis, M.; Möller, M.; Hebler, B.; Arekapudi, S.S.P.K.; Steil, D.; Schäfer, S.; Albrecht, M.; et al. Nanoscale magnetic imaging using circularly polarized high-harmonic radiation. *Sci. Adv.* **2017**, *3*, eaao4641. [CrossRef] [PubMed]

33. Milošević, D.B.; Becker, W. Attosecond pulse trains with unusual nonlinear polarization. *Phys. Rev. A* **2000**, *62*, 011403. [CrossRef]

34. Chen, C.; Tao, Z.; Hernández-García, C.; Matyba, P.; Carr, A.; Knut, R.; Kfir, O.; Zusin, D.; Gentry, C.; Grychtol, P.; et al. Tomographic reconstruction of circularly polarized high-harmonic fields: 3D attosecond metrology. *Sci. Adv.* **2016**, *2*, e1501333. [CrossRef] [PubMed]

35. Milošević, D.B.; Becker, W.; Kopold, R.; Sandner, W. High-Harmonic Generation by a Bichromatic Bicircular Laser Field. *Laser Phys.* **2001**, *11*, 165–168.

36. Milošević, D.B. Generation of elliptically polarized attosecond pulse trains. *Opt. Lett.* **2015**, *40*, 2381–2384. [CrossRef] [PubMed]

37. Medišauskas, L.; Wragg, J.; van der Hart, H.; Ivanov, M.Y. Generating isolated elliptically polarized attosecond pulses using bichromatic counterrotating circularly polarized laser fields. *Phys. Rev. Lett.* **2015**, *115*, 153001. [CrossRef] [PubMed]

38. Milošević, D.B. Circularly polarized high harmonics generated by a bicircular field from inert atomic gases in the *p* state: A tool for exploring chirality-sensitive processes. *Phys. Rev. A* **2015**, *92*, 043827. [CrossRef]

39. Odžak, S.; Milošević, D.B. Bicircular-laser-field-assisted electron-ion radiative recombination. *Phys. Rev. A* **2015**, *92*, 053416. [CrossRef]

40. Milošević, D.B.; Becker, W. Role of long quantum orbits in high-order harmonic generation. *Phys. Rev. A* **2002**, *66*, 063417. [CrossRef]

41. Milošević, D.B.; Ehlotzky, F. Scattering and reaction processes in powerful laser field. *Adv. At. Mol. Opt. Phys.* **2003**, *49*, 373–532.

42. Milošević, D.B. Control of the helicity of high-order harmonics generated by bicircular laser fields. *Phys. Rev. A* **2018**, *98*, 033405. [CrossRef]

43. Milošević, D.B. Possibility of introducing spin into attoscience with spin-polarized electrons produced by a bichromatic circularly polarized laser field. *Phys. Rev. A* **2016**, *93*, 051402. [CrossRef]

44. Milošević, D.B. Spin-dependent effects in high-order above-threshold ionization: Spin-orbit interaction and exchange effects. *J. Phys. B* **2017**, *50*, 164003. [CrossRef]

45. Ayuso, D.; Jiménez-Galán, A.; Morales, F.; Ivanov, M.; Smirnova, O. Attosecond control of spin polarization in electron-ion recollision driven by intense tailored fields. *New J. Phys.* **2017**, *19*, 073007. [CrossRef]

46. Hasović, E.; Busuladžić, M.; Gazibegović-Busuladžić, A.; Milošević, D.B.; Becker, W. Simulation of Above-Threshold Ionization Experiments Using the Strong-Field Approximation. *Laser Phys.* **2007**, *17*, 376–389. [CrossRef]

47. Milošević, D.B.; Becker, W.; Okunishi, M.; Prümper, G.; Shimada, K.; Ueda, K. Strong-field electron spectra of rare-gas atoms in the rescattering regime: Enhanced spectral regions and a simulation of the experiment. *J. Phys. B* **2010**, *43*, 015401. [CrossRef]

48. Baykusheva, D.; Ahsan, M.S.; Lin, N.; Wörner, H.J. Bicircular High-Harmonic Spectroscopy Reveals Dynamical Symmetries of Atoms and Molecules. *Phys. Rev. Lett.* **2016**, *116*, 123001. [CrossRef] [PubMed]

49. Mauger, F.; Bandrauk, A.D.; Uzer, T. Circularly polarized molecular high harmonic generation using a bicircular laser. *J. Phys. B* **2016**, *49*, 10LT01. [CrossRef]

50. Reich, D.M.; Madsen, L.B. Illuminating Molecular Symmetries with Bicircular High-Order-Harmonic Generation. *Phys. Rev. Lett.* **2016**, *117*, 133902. [CrossRef] [PubMed]

51. Liu, X.; Zhu, X.; Li, L.; Li, Y.; Zhang, Q.; Lan, P.; Lu, P. Selection rules of high-order-harmonic generation: Symmetries of molecules and laser fields. *Phys. Rev. A* **2016**, *94*, 033410. [CrossRef]

52. Odžak, S.; Hasović, E.; Milošević, D.B. High-order harmonic generation in polyatomic molecules induced by a bicircular laser field. *Phys. Rev. A* **2016**, *94*, 033419. [CrossRef]

53. Hasović, E.; Odžak, S.; Becker, W.; Milošević, D.B. High-order harmonic generation in non-planar molecules driven by a bicircular field. *Mol. Phys.* **2017**, *115*, 1750–1757. [CrossRef]

54. Gay, T.J. Physics and technology of polarized electron scattering from atoms and molecules. *Adv. At. Mol. Opt. Phys.* **2009**, *57*, 157–247. [CrossRef]

55. Žutić, I.; Fabian, J.; Das Sarma, S. Spintronics: Fundamentals and applications. *Rev. Mod. Phys.* **2004**, *76*, 323–410. [CrossRef]

56. Barth, I.; Smirnova, O. Spin-polarized electrons produced by strong-field ionization. *Phys. Rev. A* **2013**, *88*, 013401. [CrossRef]

57. Herath, T.; Yan, L.; Lee, S.K.; Li, W. Strong-field ionization rate depends on the sign of the magnetic quantum number. *Phys. Rev. Lett.* **2012**, *109*, 043004. [CrossRef] [PubMed]

58. Barth, I.; Smirnova, O. Comparison of theory and experiment for nonadiabatic tunneling in circularly polarized fields. *Phys. Rev. A* **2013**, *87*, 065401. [CrossRef]

59. Hartung, A.; Morales, F.; Kunitski, M.; Henrichs, K.; Laucke, A.; Richter, M.; Jahnke, T.; Kalinin, A.; Schöffler, M.; Schmidt, L.P.H.; et al. Electron spin polarization in strong-field ionization of xenon atoms. *Nat. Photonics* **2016**, *10*, 526–528. [CrossRef]

60. Liu, M.-M.; Shao, Y.; Han, M.; Ge, P.; Deng, Y.; Wu, C.; Gong, Q.; Liu, Y. Energy- and Momentum-Resolved Photoelectron Spin Polarization in Multiphoton Ionization of Xe by Circularly Polarized Fields. *Phys. Rev. Lett.* **2018**, *120*, 043201. [CrossRef] [PubMed]

61. Trabert, D.; Hartung, A.; Eckart, S.; Trinter, F.; Kalinin, A.; Schöffler, M.; Schmidt, L.P.H.; Jahnke, T.; Kunitski, M.; Dörner, R. Spin and Angular Momentum in Strong-Field Ionization. *Phys. Rev. Lett.* **2018**, *120*, 043202. [CrossRef] [PubMed]

62. Milošević, D.B.; Becker, W. Channel-closing effects in strong-field ionization by a bicircular field. *J. Phys. B* **2018**, *51*, 054001. [CrossRef]

63. Busuladžić, M.; Gazibegović-Busuladžić, A.; Milošević, D.B.; Becker, W. Angle-Resolved High-Order Above-Threshold Ionization of a Molecule: Sensitive Tool for Molecular Characterization. *Phys. Rev. Lett.* **2008**, *100*, 203003. [CrossRef] [PubMed]

64. Okunishi, M.; Itaya, R.; Shimada, K.; Prümper, G.; Ueda, K.; Busuladžić, M.; Gazibegović-Busuladžić, A.; Milošević, D.B.; Becker, W. Two-Source Double-Slit Interference in Angle-Resolved High-Energy Above-Threshold Ionization Spectra of Diatoms. *Phys. Rev. Lett.* **2009**, *103*, 043001. [CrossRef] [PubMed]

Article

The Influence of Secondary Electron Emission and Electron Reflection on a Capacitively Coupled Oxygen Discharge

Andrea Proto [1] and Jon Tomas Gudmundsson [1,2,*]

[1] Science Institute, University of Iceland, Dunhaga 3, IS-107 Reykjavik, Iceland; proto.andrea@yahoo.com
[2] Department of Space and Plasma Physics, School of Electrical Engineering and Computer Science, KTH Royal Institute of Technology, SE-100 44 Stockholm, Sweden
* Correspondence: tumi@hi.is; Tel.: +354-525-4946

Received: 22 September 2018; Accepted: 23 November 2018; Published: 28 November 2018

check for
updates

Abstract: The one-dimensional object-oriented particle-in-cell Monte Carlo collision code oopd1 is applied to explore the role of secondary electron emission and electron reflection on the properties of the capacitively-coupled oxygen discharge. At low pressure (10 mTorr), drift-ambipolar heating of the electrons dominates within the plasma bulk, while at higher pressure (50 mTorr), stochastic electron heating in the sheath region dominates. Electron reflection has negligible influence on the electron energy probability function and only a slight influence on the electron heating profile and electron density. Including ion-induced secondary electron emission in the discharge model introduces a high energy tail to the electron energy probability function, enhances the electron density, lowers the electronegativity, and increases the effective electron temperature in the plasma bulk.

Keywords: capacitively-coupled discharge; oxygen; particle-in-cell/Monte Carlo collision; electron heating; secondary electron emission

1. Introduction

Low pressure radio frequency (rf)-driven capacitively-coupled discharges have a range of material processing applications such as plasma etching and plasma enhanced chemical vapor deposition within the microelectronics industry. These discharges have been explored extensively over the past few decades. However, a few issues remain to be fully understood, including the electron heating mechanism, in particular when driven by multiple frequencies [1], and the role of surfaces regarding recombination and quenching of various species and phenomena such as secondary electron emission and electron reflection [2,3]. The modern capacitively-coupled discharge consists of two parallel electrodes separated by a few cm and is driven by a radio-frequency power generator. The plasma forms when rf voltage is applied between the electrodes. The electrons that gain enough energy from the resulting electric field produce positive ions, negative ions, and electrons through electron impact ionization of neutral atoms and molecules and electron impact dissociative attachment of molecules, which forms the plasma. The plasma is separated from the electrodes by space charge sheaths. Multiple frequencies are commonly applied in order to achieve separate control of ion flux and ion energy, as the ion flux dictates the throughput of the process and the ion energy determines the etching and deposition parameters on the wafer surface.

The particle-in-cell (PIC) method, when combined with Monte Carlo (MC) treatment of collision processes, is a self-consistent kinetic approach that has become a predominant numerical approach to investigate the properties of the low pressure capacitively-coupled discharge. This approach is commonly referred to as particle-in-cell Monte Carlo collision (PIC/MCC) method. The basic idea of the PIC method is to allow typically a few hundred thousand computer-simulated particles (superparticles) to represent a significantly higher number of real particles (density in the range of 10^{14}–10^{18} m^{-3}) [4–6]. In a PIC simulation, the motion of each particle is simulated and the various macro-quantities are calculated from the position and velocity of these particles. The particle interaction is handled through a macro-force acting on the particles, which is calculated from the field equations at points on a computational grid. This method allows us to follow the spatio-temporal evolution of the various plasma parameters such as particle density, particle energy, particle fluxes, and particle heating rates.

The kinetics of the capacitively-coupled oxygen discharge have been studied for over two decades starting with the seminal work of Vahedi and Surendra [7] using the 1D xpdp1 PIC/MCC code. Since then, a number of PIC/MCC studies have been reported on oxygen and Ar/O$_2$ discharges using the xpdx1 series of codes, in both symmetrical and asymmetrical geometry, performed over a range of pressures and compared to experimental findings [8] and to analytical density profiles [9], showing good agreement, to explore the formation of the ion energy distribution function in an O$_2$/Ar mixture in an asymmetric capacitively-coupled discharge [10], and the influence of the secondary electron emission on the density profiles and the electron energy distribution function (EEDF) [11]. Other 1D PIC/MCC codes have been developed to explore the oxygen discharge. A 1D PIC/MCC model developed in Greifswald, that includes the metastable oxygen molecule O$_2(a^1\Delta_g)$ as a fraction of the ground state molecule, was used to determine the ion energy distribution function (IEDF) in oxygen CCP [12,13]. Furthermore, they found by comparison with experiments that one sixth of the oxygen molecules are in the metastable singlet delta state. A 1D PIC/MCC code, developed in Dalian [14,15], was applied to explore the electrical asymmetry effect in a dual-frequency capacitively-coupled oxygen discharge. Similar to Bronold et al. [12], this work assumed a constant density for the singlet metastable molecule O$_2(a^1\Delta_g)$. More recently, a 1D PIC/MCC code that was developed in Budapest was used to explore the heating mechanism in a capacitively-coupled oxygen discharge driven by tailored waveforms (composed of N harmonics in addition to a fundamental frequency f_1) [16,17]. Furthermore, a PIC/MCC fluid hybrid model was applied to explore the electron power absorption and the influence of pressure on the energetics and particle densities [18,19]. In all of these works, only electrons, the positive ion O$_2^+$, and the negative ion O$^-$ were treated kinetically, and the positive ion O$^+$ was neglected. Furthermore, none of the metastable states were treated kinetically. The one-dimensional object-oriented plasma device one (oopd1) code allows having the simulated particles of different weights, which allows for tracking both charged and neutral particles in the simulation. Earlier, we benchmarked the basic reaction set for the oxygen discharge in oopd1 to the xpdp1 code [20].

In recent years, the oxygen reaction set in the oopd1 code was improved significantly [20–22]. Using this improved discharge model, we showed that the singlet metastable molecular states have a significant influence on the electron heating mechanism in the capacitively-coupled oxygen discharge [21–24] as well as the ion energy distribution [25]. We demonstrated that, when operating at low pressure (10 mTorr), the electron heating is mainly located within the plasma bulk (the electronegative core), while, when operating at higher pressures (50–500 mTorr), the electron heating appears almost solely within the sheath regions [22,23]. Furthermore, when operating at low pressure, the electron heating within the discharge is due to a hybrid drift-ambipolar-mode (DA-mode) and α-mode, and while operating at higher pressures, the discharge is operated in a pure α-mode [26,27]. We have also shown that detachment by the singlet molecular metastable states is the process that has the most influence on the electron heating process in the higher pressure regime, while it has almost negligible influence at lower pressures [22–24].

Secondary electron emission and electron reflection from the electrodes have often been neglected in PIC/MCC simulations. When it is included, it is common to assume the secondary electron emission to have a constant value (independent of the discharge conditions such as the energy of the bombarding ions), while only the ion-induced secondary electron emission is taken into account, and thus, the contributions of other species are neglected [2,3]. The effects of including a constant secondary electron emission yield are increased electron density, enhancement of the density profiles and the electron energy distribution functions (EEDFs), decreased sheath width, and the electron heating rate profiles changing significantly in both argon [28] and oxygen [11] discharges. Furthermore, it has been demonstrated that an asymmetry can be introduced by having electrodes with different secondary electron emission properties in a capacitively-coupled discharge [29], which was later extended to also include the electrical asymmetry effect in a dual-frequency capacitively-coupled discharge driven by two consecutive harmonics with different electrode materials [30]. In these studies, the secondary electron emission yield was set to be a constant. A few recent studies have emphasized using realistic secondary electron emission yields for both fast neutrals and ions bombarding the electrodes [2,3,22,28,31–33].

In an earlier study, we explored the role of including an energy-dependent secondary electron emission yield for both O^+ and O_2^+-ions and O and O_2 neutrals in an oxygen discharge [22]. We noted that this had a significant influence on the discharge properties, including increased electron and ion densities and decreased sheath width. Here, we study systematically how the secondary electron emission and the electron reflection from the electrodes influence the charged particle profiles, the electron heating processes, the electron energy probability function (EEPF), and the effective electron temperature, in a single frequency voltage-driven capacitively-coupled oxygen discharge by means of numerical simulation, for a fixed discharge voltage, while the discharge pressure is varied from 10–50 mTorr. The simulation parameters and the cases explored are defined in Section 2, and the simulation results found by including and excluding the ion-induced secondary electron emission and electron reflection are compared in Section 3. We give a summary and concluding remarks in Section 4.

2. The Simulation

The one-dimensional (1d-3v) object-oriented particle-in-cell Monte Carlo collision (PIC/MCC) code oopd1 [34,35] is herein applied to a capacitively-coupled oxygen discharge. The oopd1 code, like the well-known xpdp1 code [7], is a general plasma device simulation tool capable of simulating various types of plasmas, including breakdown, accelerators, beams, as well as processing discharges [20].

The oxygen reaction set included in the oopd1 code is rather extensive. Like xpdp1, it includes the ground state oxygen molecule $O_2(X^3\Sigma_g^-)$, the negative ion O^-, the positive ion O_2^+, and electrons [7,20]. In addition, oxygen atoms in the ground state $O(^3P)$ and ions of the oxygen atom O^+ [20], the singlet metastable molecule $O_2(a^1\Delta_g)$, and the metastable oxygen atom $O(^1D)$ [21], and the singlet metastable molecule $O_2(b^1\Sigma_g^+)$ [22] were added along with the relevant reactions and cross-sections. The full oxygen reaction set was discussed in our earlier works where the cross-sections used were also given [20–22]. Furthermore, oopd1 has energy-dependent secondary electron emission coefficients for oxygen ions and neutrals as they bombard both clean and dirty metal electrodes [22]. Thus, for this current work, the discharge model contains nine species: electrons, the ground state neutrals $O(^3P)$ and $O_2(X^3\Sigma_g^-)$, the negative ions O^-, the positive ions O^+ and O_2^+, and the metastables $O(^1D)$, $O_2(a^1\Delta_g)$, and $O_2(b^1\Sigma_g^+)$. We herein use the secondary electron emission yield for a dirty surface as given in our earlier work [22].

We assume a geometrically-symmetric capacitively-coupled discharge where one of the electrodes is driven by an rf voltage at a single frequency:

$$V(t) = V_0 \sin(2\pi f t) \tag{1}$$

while the other electrode is grounded. Here, V_0 is the voltage amplitude, f the driving frequency, and t the time. The discharge operating parameters assumed are the voltage amplitude of $V_0 = 222$ V, an electrode separation of 4.5 cm, and a capacitor of 1 F connected in series with the voltage source. The driving frequency is assumed to be 13.56 MHz. These are the parameters used in our earlier works using oopd1 [20–22,24,27] and in the work of Lichtenberg et al. [9] using the xpdp1 code. The discharge electrode separation is assumed to be small compared to the electrode diameter so that the discharge can be treated as one dimensional. We assume the electrode diameter to be 10.25 cm, which is needed in order to determine the absorbed power, and set the discharge volume for the global model calculations applied to determine the partial pressure of the neutral species. The time step Δt and the grid spacing Δx are set to resolve the electron plasma frequency and the electron Debye length of the low-energy electrons, respectively, according to $\omega_{pe}\Delta t < 0.2$, where ω_{pe} is the electron plasma frequency, and the simulation grid is taken to be uniform and consists of 1000 cells. The electron time step is set to 3.68×10^{-11} s. The simulation was run for 5.5×10^6 time steps, which corresponds to 2750 rf cycles. It takes roughly 1700 rf cycles to reach equilibrium for all particles, and the time averaged plasma parameters shown, such as the densities, the electron heating rate, and the effective electron temperature, are averages over 1000 rf cycles. All particle interactions are treated by the Monte Carlo method with a null-collision scheme [4]. For the heavy particles, we use sub-cycling, and the heavy particles are advanced every 16 electron time steps [36]. Furthermore, we assume that the initial density profiles are parabolic [36].

The kinetics of the charged particles (electrons, O_2^+-ions, O^+-ions, and O^--ions) was followed for all energies. Since the neutral gas density is much higher than the densities of charged species, the neutral species at thermal energies (below a certain cut-off energy) are treated as a background with fixed density and temperature and maintained uniformly in space. These neutral background species are assumed to have a Maxwellian velocity distribution at the gas temperature (here, $T_n = 26$ mV). The kinetics of the neutrals are followed when their energy exceeds a preset energy threshold value. The energy threshold values and the particle weights used here for the various neutral species included in the simulation are listed in Table 1. The partial pressures of the background thermal neutral species were calculated using a global (volume averaged) model of the oxygen discharge, as discussed in Proto and Gudmundsson [37]. The fractional densities for the neutrals $O_2(X^3\Sigma_g^-)$, $O(^3P)$, $O_2(a^1\Delta_g)$, and $O_2(b^1\Sigma_g)$, estimated using the global model calculations at 10, 25, and 50 mTorr, are listed in Table 2. These values are used as input for the PIC/MCC simulation as the partial pressures of the neutral background gas. Note that not all the neutrals considered in the global model calculations are shown in Table 2. Due to recombination of atomic oxygen and quenching of metastable atoms and molecules on the electrode surfaces, discussed below, there is a drop in the high energy (energy above the threshold value) atomic oxygen density and an increase in the high energy oxygen molecule densities next to the electrodes, as shown in our earlier work [22]. Thus, assuming uniformity of the background gas is thus somewhat an unrealistic assumption.

Table 1. The parameters of the simulation, the particle weight, and the energy threshold above which kinetics of the neutral particles are followed.

Species	Particle Weight	Energy Threshold (meV)
$O_2(X^3\Sigma_g^-)$	5×10^7	500
$O_2(a^1\Delta_g)$	5×10^6	100
$O_2(b^1\Sigma_g)$	5×10^6	100
$O(^3P)$	5×10^7	500
$O(^1D)$	5×10^7	50
O_2^+	10^7	-
O^+	10^6	-
O^-	5×10^7	-
e	1×10^7	-

The electrode surfaces have significant influence on the discharge properties. There are a few parameters regarding the surface interaction of the neutral species that have to be set in the discharge model. For a neutral species that hits the electrode, we assume it returns as a thermal particle with a given probability. Similarly atoms can recombine on the electrode surfaces to form a thermal molecule with a given probability. As the oxygen atom $O(^3P)$ hits the electrode, we assume that half of the atoms are reflected as $O(^3P)$ at room temperature, and the other half recombines to form the ground state oxygen molecule $O_2(X^3\Sigma_g^-)$ at room temperature. Thus, for a neutral oxygen atom in the ground state $O(^3P)$, we use a wall recombination coefficient of 0.5, as measured by Booth and Sadeghi [38], for a pure oxygen discharge in a stainless steel reactor at 2 mTorr. Similarly, as the metastable oxygen atom $O(^1D)$ hits the electrode, we assume that half of the atoms are quenched to form $O(^3P)$ and that the other half recombines to form the ground state oxygen molecule $O_2(X^3\Sigma_g^-)$ at room temperature. For the surface quenching coefficients of the singlet metastables molecules on the electrode surfaces, we assume for the singlet metastable $O_2(a^1\Delta_g)$ a value of $\gamma_{wqa} = 0.0001$, and for the singlet metastable $O_2(b^1\Sigma_g^+)$, we assume a value of $\gamma_{wqb} = 0.1$, based on the suggestion by O'Brien and Myers [39] that the surface quenching coefficient for the $b^1\Sigma_g^+$ state is significantly larger than for the $a^1\Delta_g$ state. We explored the influence of the surface quenching coefficients of the singlet metastable molecule $O_2(a^1\Delta_g)$ on the discharge properties in an earlier work [37]. There, we demonstrated that the influence of γ_{wqa} on the discharge properties and the electron heating mechanism can be significant indeed. The partial pressures listed in Table 2 were calculated by a global model using these surface quenching and recombination parameters as discussed in our earlier study [37].

Table 2. The partial pressures of the thermal neutrals at 10, 25, and 50 mTorr for the wall quenching coefficient for the singlet metastable molecule $O_2(a^1\Delta_g)$ of $\gamma_{wqa} = 0.0001$ calculated by a global (volume averaged) model.

Pressure	$O_2(X^3\Sigma_g^-)$	$O_2(a^1\Delta_g)$	$O_2(b^1\Sigma_g)$	$O(^3P)$
10 mTorr	0.9684	0.0265	0.0018	0.0015
25 mTorr	0.9607	0.0350	0.0019	0.0007
50 mTorr	0.9739	0.0215	0.0022	0.0004

In the simulations, we either neglect electron reflection from the electrode or assume that electrons are reflected from the electrodes with a probability of 0.2, which is the number of elastically-reflected electrons per incoming electron, independent of their energy and angle of incidence. This value is based on the summary of values presented by Kollath [40] for various materials. This value has been used by others in PIC/MCC simulations of capacitively-coupled discharges [2,3]. However, in reality, the reflection of electrons is known to depend on the electrode material, incident electron energy and the angle of incidence [40,41]. Furthermore, for all the cases explored here, we neglect secondary electron emission due to electron impact of the electrodes. The four cases explored for each pressure are listed in Table 3.

Table 3. The four cases explored for each pressure.

Case	γ_{see}	Electron Reflection
1	$\gamma_{see}(\mathcal{E})$ [22]	none
2	$\gamma_{see}(\mathcal{E})$ [22]	20%
3	$\gamma_{see} = 0.0$	none
4	$\gamma_{see} = 0.0$	20%

3. Results and Discussion

The choice of the surface quenching coefficient for the singlet metastable $O_2(a^1\Delta_g)$ of $\gamma_{wqa} = 0.0001$ was based on our earlier study of the time averaged electron heating profile between the electrodes $\langle \mathbf{J}_e \cdot \mathbf{E} \rangle$ [37]. In this study, we found that at 10 mTorr, almost all the electron heating occurred within the plasma bulk (the electronegative core), and the electron heating profile was almost independent of the surface quenching coefficient for the singlet metastable molecule $O_2(a^1\Delta_g)$, while the DA-heating mode dominated the time averaged electron heating over one rf cycle. At 25 mTorr, the time averaged electron heating occurred both in the bulk (the electronegative core) and in the sheath regions, and a hybrid DA- and α-mode heating was observed. When operating at 50 mTorr, electron heating in the sheath region dominated, and the discharge was operated in a pure α-mode. Thus, this choice of pressure values and $\gamma_{wqa} = 0.0001$ gave us three distinct operating regimes to analyze further.

The electron energy probability function (EEPF) in the discharge center is shown in Figure 1, for the various combinations of pressures, including and excluding secondary electron emission and electron reflection from the electrodes, for a total of four cases for each pressure, as shown in Table 3. Figure 1a shows the electron probability function (EEPF) at 10 mTorr. At low electron energy, the EEPF curved outwards, and a high energy tail was apparent when secondary electron emission was excluded from the simulation. We see that adding secondary electron emission to the discharge model enhanced the EEPF. When including the ion-induced energy-dependent secondary electron emission yield, more electrons were created at the electrodes, which were subsequently accelerated to the plasma bulk across the sheath. Thus, more high energy electrons were created in the discharge, and the EEPF exhibits a high energy tail when secondary electrons were emitted from the electrodes. This high energy tail extended up to roughly 240 eV. At 10 mTorr, both cases (including and excluding electron reflection) including secondary electron emission overlapped, and both cases (including and excluding electron reflection) neglecting the secondary electron emission overlapped. Thus, including electron reflection from the electrodes had negligible effects on the EEPF. Figure 1b shows the EEPF at 25 mTorr. We see that the shape changed for all four cases as the pressure increased. The electron reflection had a negligible effect on the EEPF. This can be seen from the overlap of the green dashed line on the black one, when secondary electron emission was excluded, and from the overlap of the red line on the blue one, where secondary electron emission was included. Furthermore, an overall reduction of the high energy part of the curve, compared to the 10 mTorr case, was observed when secondary electron emission was neglected. This means that, when the pressure was raised and the secondary electron emission was neglected, there were fewer hot electrons within the bulk. Figure 1c shows the EEPF at 50 mTorr. Here, the transition, which already started at 25 mTorr, was fully accomplished and the shape of the EEPF now curved inwards or was bi-Maxwellian for all four cases. As before including secondary electron emission led to a high energy tail. Furthermore, now, the black dashed line with the green one and the blue dashed line with the red one overlapped almost perfectly, which indicates that the electron reflection from the electrodes had negligible effects.

Figure 2 shows the profile of the time averaged power absorption by the electrons over an rf cycle $\langle \mathbf{J}_e \cdot \mathbf{E} \rangle$. A predominance of the electron heating within plasma bulk was observed in all four cases at 10 mTorr. We see that, when including the ion-induced secondary electron emission, the difference between including and excluding the electron reflection at the electrodes was very small within the plasma bulk. The same occurred when the secondary electron emission was excluded. A maximum in the power absorption in the bulk and a minimum in the sheath region were observed when the secondary electron emission was excluded and the electron reflection was included in the simulation (black line in Figure 2a). On the contrary, a maximum in the power absorption in the sheath edge and a minimum in the bulk were seen when the secondary electron emission was included and the electron reflection was excluded (red line in Figure 2a). At the transition pressure of 25 mTorr, the situation was drastically changed. A combination of the electron heating in the plasma bulk and in the sheath region was observed in all four cases. Indeed Figure 2b shows that there was a maximum in the power absorption in the bulk and a minimum in the sheath edge when both secondary electron emission and electron reflection were excluded (green line in Figure 2b), while a minimum in the bulk and a maximum in the sheath edge were observed when the secondary electron emission and the electron reflection were included (blue line in Figure 2b). At 50 mTorr, the transition was fully accomplished and the electron heating was almost solely in the sheath region or stochastic electron heating. This is clearly seen in Figure 2c when averaged over the rf cycle. Indeed, the electron reflection did not play much of a role. The maximum in the power absorption was observed when secondary electron emission was included in the simulation and the sheath was slightly narrower.

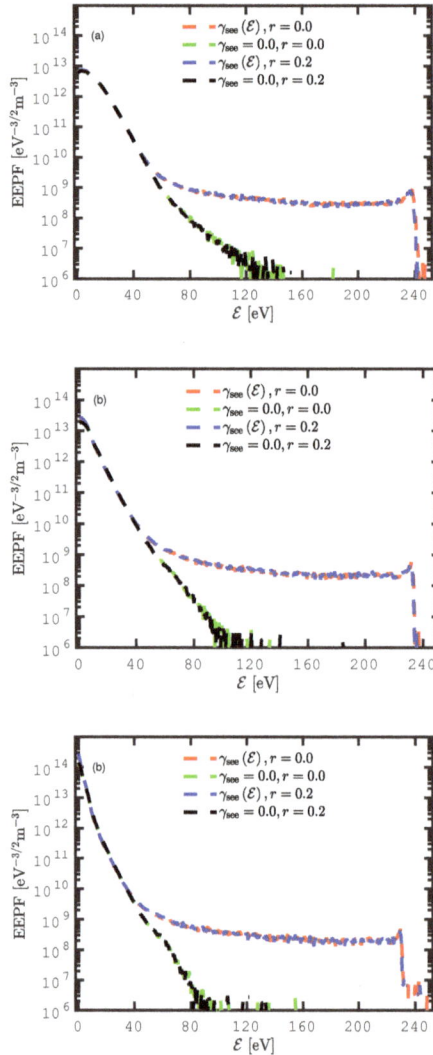

Figure 1. The electron energy probability function (EEPF) in the discharge center for a parallel plate capacitively-coupled oxygen discharge at (**a**) 10 mTorr, (**b**) 25 mTorr, and (**c**) 50 mTorr with a surface quenching coefficient for the singlet metastable molecule $O_2(a^1\Delta_g)$ as $\gamma_{wqa} = 0.0001$ and a gap separation of 4.5 cm driven by a 222 V voltage source at a driving frequency of 13.56 MHz.

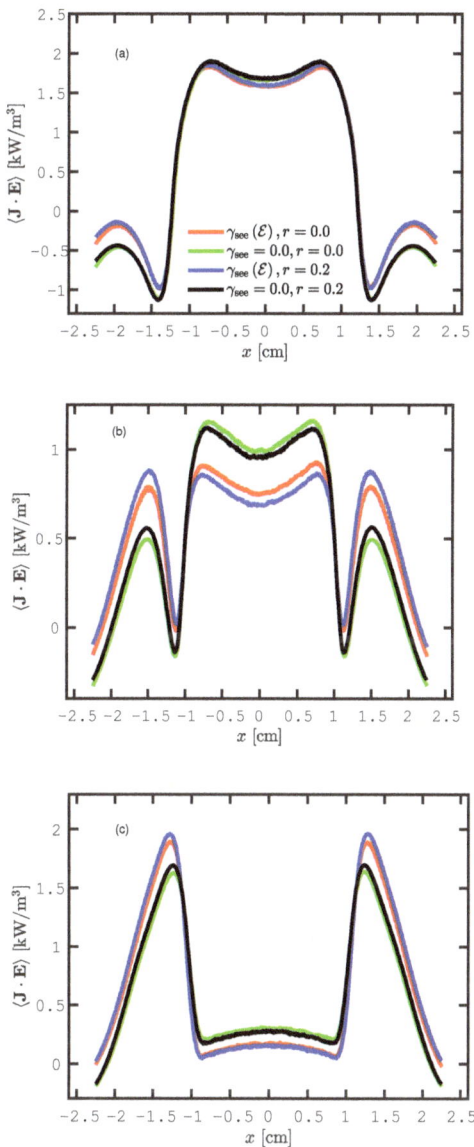

Figure 2. The time averaged electron heating profile for a parallel plate capacitively-coupled oxygen discharge at (**a**) 10 mTorr, (**b**) 25 mTorr, and (**c**) 50 mTorr with a surface quenching coefficient for the singlet metastable molecule $O_2(a^1\Delta_g)$ as $\gamma_{wqa} = 0.0001$ and a gap separation of 4.5 cm driven by a 222 V voltage source at a driving frequency of 13.56 MHz.

In order to explore the observed transition further, we plot the center electron density as a function of pressure in Figure 3a. The electron density increased with increased pressure. At 10 mTorr, all four cases exhibited a similar electron density, and the electron density was slightly enhanced when the electron reflection was included in the simulation. At 25 mTorr, we see that including both the secondary electron emission and the electron reflection gave the highest center electron density, while excluding both processes led to the lowest center electron density. The differences in electron density between including and excluding both secondary electron emission and electron reflection were bigger than at 10 mTorr. At 50 mTorr, we see that including the ion induced secondary electron emission increased the center electron density and that including electron reflection increased the electron density even further.

Further insights about the observed transition are shown in the plot of the center electronegativity as a function of pressure in Figure 3b. At 10 mTorr, the discharge was the most strongly electronegative. The electronegativity decreased from ~110 at 10 mTorr to ~20 at 50 mTorr. The electronegativity was higher (lower) when electron reflection was excluded (included) in the simulation; however, all four cases were very close to each other. The maximum (minimum) value of the electronegativity was reached when both secondary electron emission and electron reflection were excluded (included). At 25 mTorr, we observe that the gap between including and excluding both secondary electron emission and electron reflection was bigger than at 10 mTorr. We observe that, when secondary electron emission was included, excluding the electron reflection enhanced the electronegativity. The same occurred when secondary electron emission was included. Indeed, in this case, excluding both secondary electron emission and electron reflection gave the highest electronegativity. At 50 mTorr, the electronegativity was drastically reduced. We observed that electronegativity was lowest when secondary electron emission was included in the simulation and that electron reflection did not play much of a role. On the other hand, the electronegativity was highest when secondary electron emission and electron reflection were excluded from the simulation.

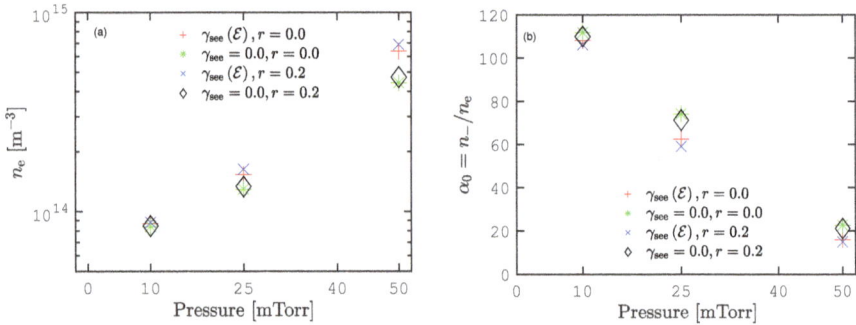

Figure 3. The (**a**) electron density and the (**b**) electronegativity in the discharge center as a function of pressure for a parallel plate capacitively-coupled oxygen discharge with a surface quenching coefficient for the singlet metastable molecule $O_2(a^1\Delta_g)$ as $\gamma_{wqa} = 0.0001$ and a gap separation of 4.5 cm driven by a 222 V voltage source at a driving frequency of 13.56 MHz.

Figure 4 shows the spatio-temporal behavior of the electron power absorption $J_e \cdot E$, where J_e and E are the spatially and temporally-varying electron current density and electric field, respectively. The figures show the electron power absorption for the various combinations of pressures, including and excluding secondary electron emission, while excluding electron reflection from the electrodes. For each of the figures, the abscissa covers the whole inter-electrode gap, from the powered electrode on the left-hand size to the grounded electrode on the right-hand size. Similarly, the ordinate covers the full rf cycle. Note that each of the six figures may have different magnitude scales, represented by the color scales on the right-hand side of each figure. Therefore, there can be differences in the six figures, not only qualitative, but also quantitative. Figure 4a,b shows the spatio-temporal behavior of the electron power absorption including and excluding $\gamma_{see}(\mathcal{E})$ at 10 mTorr, respectively. Figure 4c shows the difference between including and excluding the ion-induced secondary electron emission from the electrodes. In Figure 4a,b, the most significant heating is observed in the sheath region, during the sheath expansion, and the most significant cooling is observed during the sheath collapse. Here, significant energy gain (red and yellow areas) and small energy loss (dark blue areas) were evident within the plasma bulk region. We observe electron heating during the sheath collapse on the bulk side of the edge of the collapsing sheath (next to the instantaneous anode), while there was cooling (electrons loose energy) on the electrode side (the lower left-hand corner and upper center on the right-hand side). This kind of electron heating structure was observed experimentally in a capacitively-coupled SF_6/N_2 discharge [42] and SiH_4 discharge [43] using spatiotemporal optical emission spectroscopy. This heating mechanism was explored further using the relaxation continuum model [44], where electron heating due to three processes was identified: sheath expansion (α-mode), high electric field within the bulk, and ionization due to formation of a double layer on the instantaneous anode side, which resulted in acceleration of electrons. Indeed, in electronegative discharges, this electron heating within the plasma bulk can be the dominating electron heating mechanism [42,44]. This heating mechanism, which is due to electrons that are accelerated by strong drift and ambipolar electric fields within the plasma bulk and at the sheath edges in strongly electronegative discharges, was later coined as drift ambipolar (DA) electron heating [45]. In highly electronegative discharges, these electrons are often found to dominate the ionization processes. As seen in Figure 3b, the electronegativity was high $\sim$110 at 10 mTorr, which is essential for the DA-heating to be effective. The electron heating occurred both within the bulk and in the sheath regions, and a hybrid DA- and α-mode heating was observed. By looking at the time averaged electron heating profile in Figure 2a, we see that there was electron cooling in the sheath region and all the electron heating occurred in the bulk region averaged over one rf cycle. Figure 4c shows that there were no significant differences in the power absorption between the two cases, and in fact, there was a slightly higher electron heating within the discharge when secondary electron emission was excluded. Figure 4d,e shows the spatio-temporal behavior of the electron power absorption for $\gamma_{see}(\mathcal{E})$ and $\gamma_{see} = 0.0$, respectively, at 25 mTorr. Figure 4f shows the difference between including and excluding secondary electron emission from the electrodes. At 25 mTorr, a transition process was observed. Indeed, Figure 4e shows that the heating and the cooling in the sheath regions were reduced, while Figure 4d shows that a significant contribution to the electron heating in the bulk region was observed. Therefore, a hybrid DA- and α-mode heating was observed, where the DA-heating was more important when secondary electron emission was excluded (Figure 4e) than when it was included (Figure 4d). This is clearly seen in the difference plot shown in Figure 4f when cooling was seen as the difference. We see in Figure 2b that in this case, the time averaged power absorption was observed both in the plasma bulk, as well as in the sheath regions. Figure 4g,h shows the spatio-temporal behavior of the electron power absorption for $\gamma_{see}(\mathcal{E})$ and $\gamma_{see} = 0.0$ at 50 mTorr, respectively. Figure 4i shows the difference between including secondary electron emission (Figure 4g) and excluding secondary electron emission (Figure 4h). Here, the electron heating rate in the sheath regions reduced again, and there was almost no heating in the plasma bulk, as seen in Figure 4g,h; a pure α-mode was observed for both plots. This is clearly seen in Figure 2a when averaged over the rf cycle. As seen from the difference plot shown in Figure 4i, the electron heating in the sheath region was quantitatively more important for $\gamma_{see}(E)$ than for $\gamma_{see} = 0.0$. There was clearly higher electron heating at sheath expansion when secondary electron emission was included. However, including ion-induced secondary electron emission from the electrodes decreased the overall electron power absorption.

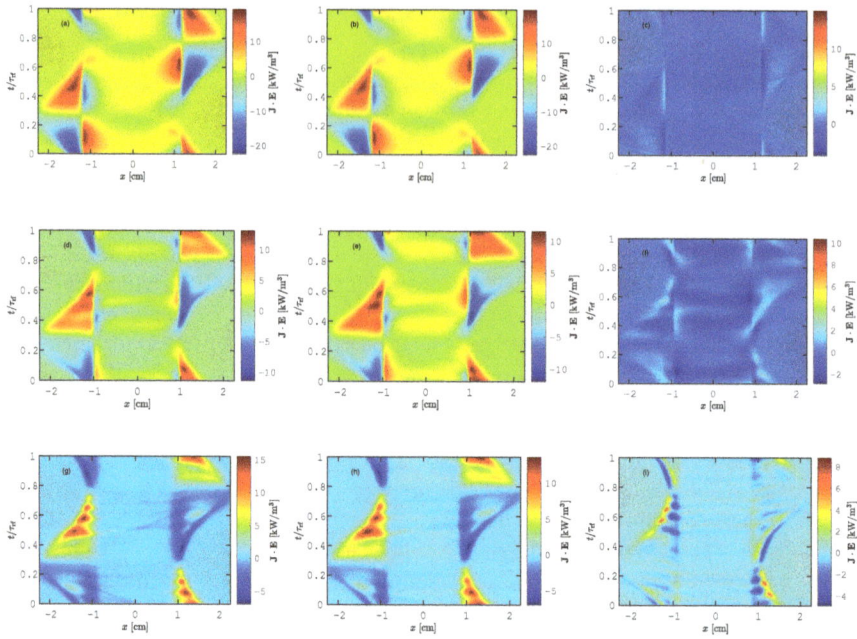

Figure 4. The spatio-temporal behavior of the electron power absorption for a parallel plate capacitively-coupled oxygen discharge at 10 mTorr for (**a**) $\gamma_{see}(\mathcal{E})$ and (**b**) $\gamma_{see} = 0.0$, (**c**) the difference between $\gamma_{see}(\mathcal{E})$ and $\gamma_{see} = 0.0$, at 25 mTorr, for (**d**) $\gamma_{see}(\mathcal{E})$ and (**e**) $\gamma_{see} = 0.0$, (**f**) the difference between $\gamma_{see}(\mathcal{E})$ and $\gamma_{see} = 0.0$, at 50 mTorr for, (**g**) $\gamma_{see}(\mathcal{E})$ and (**h**) $\gamma_{see} = 0.0$, and (**i**) the difference between $\gamma_{see}(\mathcal{E})$ and $\gamma_{see} = 0.0$ with a surface quenching coefficient for the singlet metastable molecule $O_2(a^1\Delta_g)$ as $\gamma_{wqa} = 0.0001$, $r = 0.0$, and a gap separation of 4.5 cm driven by a 222 V voltage source at a driving frequency of 13.56 MHz.

Figure 5 shows the spatio-temporal behavior of the effective electron temperature. It shows the effective electron temperature as a function of position between the electrodes within one rf cycle, for the various combinations of pressures including and excluding secondary electron emission from the electrodes. At 10 mTorr, the effective electron temperature was high within the plasma bulk, and no important difference was observed in the spatio-temporal behavior of the effective electron temperature between $\gamma_{see}(\mathcal{E})$ and $\gamma_{see} = 0.0$, as seen in Figure 5a,b, respectively. A peak in the effective electron temperature was observed within the bulk region on the instantaneous anode side and agrees with the region of peak electron heating seen in Figure 4a,b. The difference in the effective electron temperature calculated excluding and including the secondary electron emission is seen in Figure 5c. We see that the effective electron temperature within the plasma bulk was slightly higher when secondary electron emission was included. The peaks in the effective electron temperature were higher when secondary electron emission was included. We also observe that the effective electron temperature had a peak within the plasma bulk at the instantaneous anode side and at the sheath expansion at 25 mTorr for both $\gamma_{see}(\mathcal{E})$ and $\gamma_{see} = 0.0$, as seen in Figure 5d,e, respectively. This is clearly seen in the difference plot in Figure 5f. The electron effective temperature was higher for $\gamma_{see}(\mathcal{E})$ than for $\gamma_{see} = 0.0$ in the bulk region. In particular, the peak in the bulk region on the instantaneous anode side increased when secondary electron emission was included. At 50 mTorr, we observe a peak in the effective electron temperature during the sheath expansion. We also see that there was an increase in the effective electron temperature within the bulk when secondary electron emission was included. For $\gamma_{see} = 0.0$, the effective electron temperature in the bulk and in the sheath region was lower than when secondary electron emission was included, as seen in Figure 5g,h. This behavior is clearly manifest in the difference plot shown in Figure 5i. At all pressures, we found that including secondary electron emission in the discharge model increased the electron energy.

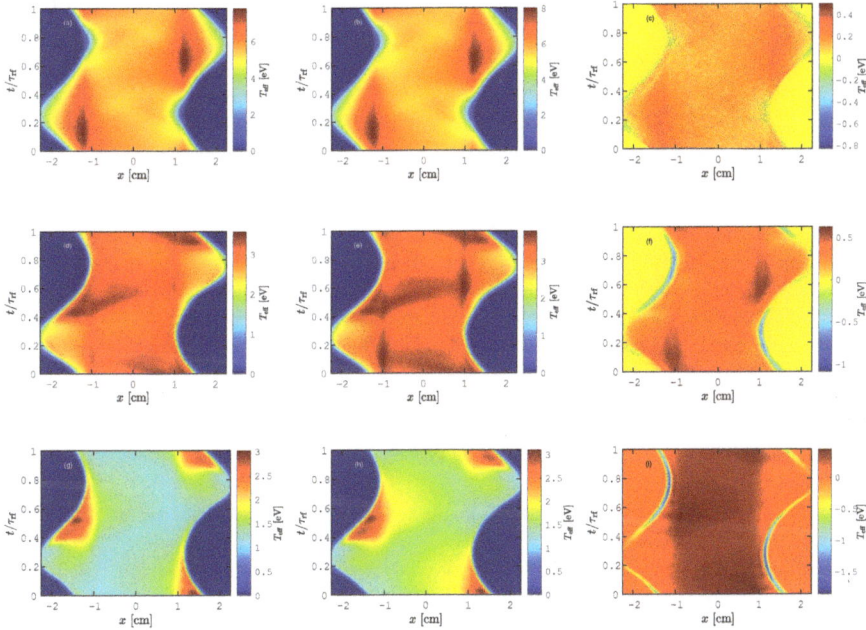

Figure 5. The spatio-temporal behavior of the effective electron temperature for a parallel plate capacitively-coupled oxygen discharge at 10 mTorr for (**a**) $\gamma_{see}(\mathcal{E})$ and (**b**) $\gamma_{see} = 0.0$, (**c**) the difference between $\gamma_{see} = 0.0$ and $\gamma_{see}(\mathcal{E})$, at 25 mTorr, for (**d**) $\gamma_{see}(\mathcal{E})$, and (**e**) $\gamma_{see} = 0.0$, (**f**) the difference between $\gamma_{see} = 0.0$ and $\gamma_{see}(\mathcal{E})$, at 50 mTorr for, (**g**) $\gamma_{see}(\mathcal{E})$ and (**h**) $\gamma_{see} = 0.0$, and (**i**) the difference between $\gamma_{see} = 0.0$ and $\gamma_{see}(\mathcal{E})$ with a surface quenching coefficient for the singlet metastable molecule $O_2(a^1\Delta_g)$ as $\gamma_{wqa} = 0.0001$, $r = 0.0$, and a gap separation of 4.5 cm driven by a 222 V voltage source at a driving frequency of 13.56 MHz.

4. Conclusions

The one-dimensional object-oriented PIC/MCC code oopd1 was applied to explore the evolution of the EEPF and of the electron heating mechanism in a capacitively-coupled oxygen discharge while including and excluding the ion-induced secondary electron emission and electron reflection. Adding secondary electron emission enhances the EEPF with a high energy tail for all the pressures. At 10 mTorr, the EEPF curves outwards. The electron heating at 10 mTorr is a hybrid DA- and α-mode heating, and no significant difference is observed including and excluding secondary electron emission from the electrodes. Averaged over one rf cycle, a predominance of the electron heating in the plasma bulk was observed for all the cases. At 25 mTorr, the shape of the EEPF starts to develop an inward curving behavior and a hybrid DA- and α-mode heating is observed. The role of sheath heating increases when secondary electron emission from the electrodes is included in the simulation. At 50 mTorr, the transition, which had already started at 25 mTorr, is fully accomplished, and the shape of the EEPF is now bi-Maxwellian, while no electron heating is observed in the plasma bulk.

Author Contributions: Conceptualization, J.T.G. and A.P.; Formal Analysis, A.P. and J.T.G.; Investigation, A.P. and J.T.G.; Resources, J.T.G.; Data Curation, A.P.; Writing—Original Draft Preparation, J.T.G. and A.P.; Writing—Review & Editing, J.T.G. and A.P.; Supervision, J.T.G.; Project Administration, J.T.G.; Funding Acquisition, J.T.G.

Funding: This work was partially supported by the Icelandic Research Fund Grant No. 163086, the University of Iceland Research Fund, and the Swedish Government Agency for Innovation Systems (VINNOVA) Contract No. 2014-04876.

Conflicts of Interest: The authors declare no conflict of interest.

References

1. Donkó, Z.; Schulze, J.; Czarnetzki, U.; Derzsi, A.; Hartmann, P.; Korolov, I.; Schüngel, E. Fundamental investigations of capacitive radio frequency plasmas: Simulations and experiments. *Plasma Phys. Control. Fusion* **2012**, *54*, 124003. [CrossRef]

2. Derzsi, A.; Korolov, I.; Schüngel, E.; Donkó, Z.; Schulze, J. Effects of fast atoms and energy-dependent secondary electron emission yields in PIC/MCC simulations of capacitively coupled plasmas. *Plasma Sources Sci. Technol.* **2015**, *24*, 034002. [CrossRef]

3. Daksha, M.; Derzsi, A.; Wilczek, S.; Trieschmann, J.; Mussenbrock, T.; Awakowicz, P.; Donkó, Z.; Schulze, J. The effect of realistic heavy particle induced secondary electron emission coeffcients on the electron power absorption dynamics in single- and dual-frequency capacitively coupled plasma. *Plasma Sources Sci. Technol.* **2017**, *26*, 085006. [CrossRef]

4. Birdsall, C.K. Particle-in-cell charged-particle simulations, plus Monte Carlo collisions with neutral atoms, PIC-MCC. *IEEE Trans. Plasma Sci.* **1991**, *19*, 65–85. [CrossRef]

5. Verboncoeur, J.P. Particle simulation of plasmas: Review and advances. *Plasma Phys. Control. Fusion* **2005**, *47*, A231–A260. [CrossRef]

6. Tskhakaya, D.; Matyash, K.; Schneider, R.; Taccogna, F. The Particle-In-Cell Method. *Contrib. Plasma Phys.* **2007**, *47*, 563–594. [CrossRef]

7. Vahedi, V.; Surendra, M. A Monte Carlo collision model for the particle-in-cell method: Applications to argon and oxygen discharges. *Comput. Phys. Commun.* **1995**, *87*, 179–198. [CrossRef]

8. Lee, S.H.; Iza, F.; Lee, J.K. Particle-in-cell Monte Carlo and fluid simulations of argon-oxygen plasma: Comparisons with experiments and validations. *Phys. Plasmas* **2006**, *13*, 057102. [CrossRef]

9. Lichtenberg, A.J.; Vahedi, V.; Lieberman, M.A.; Rognlien, T. Modeling electronegative plasma discharges. *J. Appl. Phys.* **1994**, *75*, 2339–2347. [CrossRef]

10. Babaeva, N.Y.; Lee, J.K.; Shon, J.W.; Hudson, E.A. Oxygen ion energy distribution: Role of ionization, resonant, and nonresonant charge-exchange collisions. *J. Vac. Sci. Technol. A* **2005**, *23*, 699–704. [CrossRef]

11. Roberto, M.; Verboncoeur, J.; Verdonck, P.; Cizzoto, E. Effects of the Secondary Electron Emission Coeffcient on the Generation of Charged Particles in RF Oxygen Discharge. *ECS Trans.* **2006**, *4*, 563–571.

12. Bronold, F.X.; Matyash, K.; Schneider, D.T.R.; Fehske, H. Radio-frequency discharges in oxygen: I. Particle-based modelling. *J. Phys. D Appl. Phys.* **2007**, *40*, 6583–6592. [CrossRef]

13. Matyash, K.; Schneider, R.; Dittmann, K.; Meichsner, J.; Bronold, F.X.; Tskhakaya, D. Radio-frequency discharges in oxygen: III. Comparison of modelling and experiment. *J. Phys. D Appl. Phys.* **2007**, *40*, 6601–6607. [CrossRef]

14. Schüngel, E.; Zhang, Q.Z.; Iwashita, S.; Schulze, J.; Hou, L.J.; Wang, Y.N.; Czarnetzki, U. Control of plasma properties in capacitively coupled oxygen discharges via the electrical asymmetry effect. *J. Phys. D Appl. Phys.* **2011**, *44*, 285205. [CrossRef]

15. Zhang, Q.Z.; Jiang, W.; Hou, L.J.; Wang, Y.N. Numerical simulations of electrical asymmetry effect on electronegative plasmas in capacitively coupled rf discharge. *J. Appl. Phys.* **2011**, *109*, 013308. [CrossRef]

16. Derzsi, A.; Lafleur, T.; Booth, J.P.; Korolov, I.; Donkó, Z. Experimental and simulation study of a capacitively coupled oxygen discharge driven by tailored voltage waveforms. *Plasma Sources Sci. Technol.* **2016**, *25*, 015004. [CrossRef]

17. Derzsi, A.; Bruneau, B.; Gibson, A.; Johnson, E.; O'Connell, D.; Gans, T.; Booth, J.P.; Donkó, Z. Power coupling mode transitions induced by tailored voltage waveforms in capacitive oxygen discharges. *Plasma Sources Sci. Technol.* **2017**, *26*, 034002. [CrossRef]

18. Bera, K.; Rauf, S.; Collins, K. PIC-MCC/Fluid Hybrid Model for Low Pressure Capacitively Coupled O$_2$ Plasma. *AIP Conf. Proc.* **2011**, *1333*, 1027–1032.

19. Bera, K.; Rauf, S.; Collins, K. Plasma Dynamics in Low-Pressure Capacitively Coupled Oxygen Plasma Using PIC–MCC/Fluid Hybrid Model. *IEEE Trans. Plasma Sci.* **2011**, *39*, 2576–2577. [CrossRef]

20. Gudmundsson, J.T.; Kawamura, E.; Lieberman, M.A. A benchmark study of a capacitively coupled oxygen discharge of the oopd1 particle-in-cell Monte Carlo code. *Plasma Sources Sci. Technol.* **2013**, *22*, 035011. [CrossRef]

21. Gudmundsson, J.T.; Lieberman, M.A. On the role of metastables in capacitively coupled oxygen discharges. *Plasma Sources Sci. Technol.* **2015**, *24*, 035016. [CrossRef]

22. Hannesdottir, H.; Gudmundsson, J.T. The role of the metastable O$_2$(b$^1\Sigma_g^+$) and energy-dependent secondary electron emission yields in capacitively coupled oxygen discharges. *Plasma Sources Sci. Technol.* **2016**, *25*, 055002. [CrossRef]

23. Gudmundsson, J.T.; Ventéjou, B. The pressure dependence of the discharge properties in a capacitively coupled oxygen discharge. *J. Appl. Phys.* **2015**, *118*, 153302. [CrossRef]

24. Gudmundsson, J.T.; Hannesdottir, H. On the role of metastable states in low pressure oxygen discharges. *AIP Conf. Proc.* **2017**, *1811*, 120001.

25. Hannesdottir, H.; Gudmundsson, J.T. On singlet metastable states, ion flux and ion energy in single and dual frequency capacitively coupled oxygen discharges. *J. Phys. D Appl. Phys.* **2017**, *50*, 175201. [CrossRef]

26. Gudmundsson, J.T.; Snorrason, D.I. On electron heating in a low pressure capacitively coupled oxygen discharge. *J. Appl. Phys.* **2017**, *122*, 193302. [CrossRef]

27. Gudmundsson, J.T.; Snorrason, D.I.; Hannesdottir, H. The frequency dependence of the discharge properties in a capacitively coupled oxygen discharge. *Plasma Sources Sci. Technol.* **2018**, *27*, 025009. [CrossRef]

28. Bojarov, A.; Radmilović-Radjenović, M.; Petrović, Z.L. The influence of the ion induced secondary electron emission on the characteristics of rf plasmas. *Publ. Astron. Obs. Belgrade* **2010**, *89*, 131–134.

29. Lafleur, T.; Chabert, P.; Booth, J.P. Secondary electron induced asymmetry in capacitively coupled plasmas. *J. Phys. D Appl. Phys.* **2013**, *46*, 135201. [CrossRef]

30. Korolov, I.; Derzsi, A.; Donkó, Z.; Schulze, J. The influence of the secondary electron induced asymmetry on the electrical asymmetry effect in capacitively coupled plasmas. *Appl. Phys. Lett.* **2013**, *103*, 064102. [CrossRef]

31. Radmilović-Radjenović, M.; Petrović, Z.L. Influence of the surface conditions on rf plasma characteristics. *Eur. Phys. J. D* **2009**, *54*, 445–449. [CrossRef]

32. Bojarov, A.; Radmilović-Radjenović, M.; Petrović, Z.L. Modeling the effects of the secondary electron emission in a dual-frequency capacitively coupled plasma reactor. In Proceedings of the 20th Europhysics Sectional Conference on Atomic and Molecular Physics of Ionized Gases (ESCAMPIG XX), Novi Sad, Serbia, 13–17 July 2010; p. P2.38.

33. Bojarov, A.; Radmilović-Radjenović, M.; Petrović, Z.L. Particle in cell simulation of the electrical asymmetric effect with a realistic model of the ion induced secondary electron emission. In Proceedings of the 27th Summer School and International Symposium on the Physics of Ionized Gases, Belgrade, Serbia, 26–29 August 2014; pp. 407–410.

34. Hammel, J.; Verboncoeur, J.P. DC Discharge Studies Using PIC-MCC. *Bull. Am. Phys. Soc.* **2003**, *48*, 66.

35. Verboncoeur, J.P.; Langdon, A.B.; Gladd, N.T. An object-oriented electromagnetic PIC code. *Comput. Phys. Commun.* **1995**, *87*, 199–211. [CrossRef]

36. Kawamura, E.; Birdsall, C.K.; Vahedi, V. Physical and numerical methods of speeding up particle codes and paralleling as applied to RF discharges. *Plasma Sources Sci. Technol.* **2000**, *9*, 413–428. [CrossRef]

37. Proto, A.; Gudmundsson, J.T. The role of surface quenching of the singlet delta molecule in a capacitively coupled oxygen discharge. *Plasma Sources Sci. Technol.* **2018**, *27*, 074002. [CrossRef]

38. Booth, J.P.; Sadeghi, N. Oxygen and fluorine kinetics in electron cyclotron resonance plasmas by time-resolved actinometry. *J. Appl. Phys.* **1991**, *70*, 611–620. [CrossRef]

39. O'Brien, R.J.; Myers, G.H. Direct flow measurement of O$_2$(b$^1\Sigma_g^+$) quenching rates. *J. Chem. Phys.* **1970**, *53*, 3832–3835. [CrossRef]

40. Kollath, R. Sekundärelektronen-Emission fester Körper bei Bestrahlung mit Elektronen. In *Elektronen-Emission Gasentladungen I*; Flügge, S., Ed.; Handbuch der Physik; Springer: Berlin, Germany, 1956; Volume 21, pp. 232–303.

41. Braginsky, O.; Kovalev, A.; Lopaev, D.; Proshina, O.; Rakhimova, T.; Vasilieva, A.; Voloshin, D.; Zyryanov, S. Experimental and theoretical study of dynamic effects in low-frequency capacitively coupled discharges. *J. Phys. D Appl. Phys.* **2012**, *45*, 015201. [CrossRef]

42. Petrović, Z.L.; Tochikubo, F.; Kakuta, S.; Makabe, T. Spatiotemporal optical emission spectroscopy of rf discharges in SF_6. *J. Appl. Phys.* **1993**, *73*, 2163–2172. [CrossRef]

43. Makabe, T.; Tochikubo, F.; Nishimura, M. Influence of negative ions in rf glow discharges in SiH_4 at 13.56 MHz. *Phys. Rev. A* **1990**, *42*, 3674–3677. [CrossRef] [PubMed]

44. Nakano, N.; Shimura, N.; Petrović, Z.L.; Makabe, T. Simulations of rf glow discharges in SF_6 by the relaxation continuum model: Physical structure and function of the narrow-gap reactive-ion etcher. *Phys. Rev. E* **1994**, *49*, 4455–4465. [CrossRef]

45. Schulze, J.; Derzsi, A.; Dittmann, K.; Hemke, T.; Meichsner, J.; Donkó, Z. Ionization by drift and ambipolar electric fields in electronegative capacitive radio frequency plasmas. *Phys. Rev. Lett.* **2011**, *107*, 275001. [CrossRef] [PubMed]

atoms

MDPI

Review

Advanced Helical Plasma Research towards a Steady-State Fusion Reactor by Deuterium Experiments in Large Helical Device

Yasuhiko Takeiri [1,2]

1   National Institute for Fusion Science, National Institutes of Natural Sciences, 322-6 Oroshi,
     Toki 509-5292, Japan; takeiri@nifs.ac.jp; Tel.: +81-572-58-2008
2   SOKENDAI (The Graduate University for Advanced Studies), 322-6 Oroshi, Toki 509-5292, Japan

Received: 3 October 2018; Accepted: 4 December 2018; Published: 8 December 2018

check for
updates

Abstract: The Large Helical Device (LHD) is one of the world's largest superconducting helical system fusion-experiment devices. Since the start of experiments in 1998, it has expanded its parameter regime. It has also demonstrated world-leading steady-state operation. Based on this progress, the LHD has moved on to the advanced research phase, that is, deuterium experiment, which started in March 2017. During the first deuterium experiment campaign, an ion temperature of 10 keV was achieved. This was a milestone in helical systems research: demonstrating one of the conditions for fusion. All of this progress and increased understanding have provided the basis for designing an LHD-type steady-state helical fusion reactor. Moreover, LHD plasmas have been utilized not only for fusion research, but also for diagnostics development and applications in wide-ranging plasma research. A few examples of such contributions of LHD plasmas (spectroscopic study and the development of a new type of interferometer) are introduced in this paper.

Keywords: Large Helical Device (LHD); deuterium experiment; ion temperature of 10 keV; plasma research; spectroscopic study; dispersion interferometer

1. Introduction

The Large Helical Device (LHD) [1] is one of the world's largest magnetically-confined fusion-experiment devices and is categorized as a helical system. Experiments started in March 1998, and the LHD has taken part in pioneering research in the worldwide fusion research community since then. The LHD has the critical advantage and engineering capability of steady-state operation. It has played a complementary and alternative role to the tokamak approach. The main goals of the LHD are to establish a scientific basis for a steady-state helical fusion reactor and to promote academic study for a comprehensive physics-based understanding of toroidal plasmas.

It is worth noting that helical fusion research has been performed worldwide, as shown in Figure 1 [2]. Another large-scale superconducting device, Wendelstein 7-X, started operation in 2015 [3]. Costa Rica [4] and China have also commenced helical fusion research. In July 2017, the National Institute for Fusion Science (NIFS) agreed with Southwest Jiaotong University to collaboratively construct, and then conduct fusion research using, the Chinese First Quasi-Axisymmetric Stellarator (CFQS) [5]. The start of its experiments is envisaged for 2021.

This paper is organized as follows. In Section 2, one significant achievement (i.e., an ion temperature record) and a few other wide-ranging physics results obtained in the LHD's first deuterium campaign are briefly described. Note that most of these results are before publication, and details will be given in future individual publications. Section 3 is devoted to introducing the design activity of a steady-state helical fusion reactor—the so-called FFHR. Section 4 emphasizes LHD as a platform for wide-ranging plasma research by describing two examples of its use within such research. These examples are the LHD's use in spectroscopic study, and the development of a new type of interferometer that is applicable to atmospheric pressure plasmas.

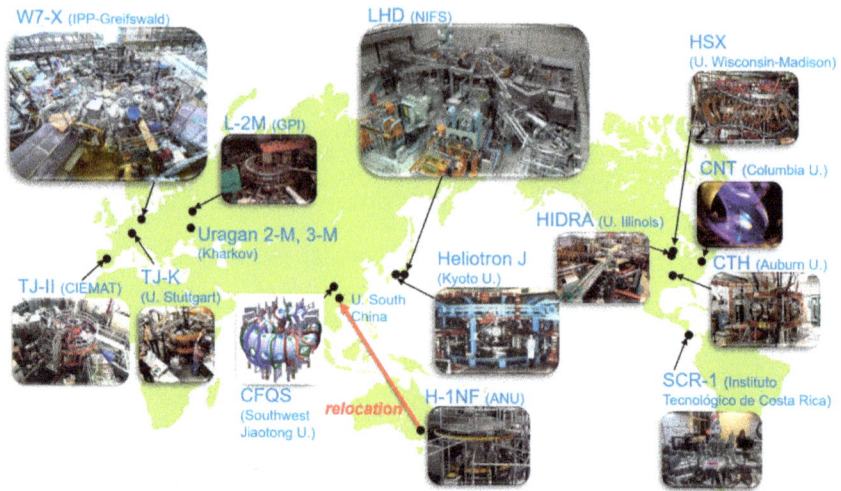

Figure 1. The world-wide helical systems research. This figure has been modified and translated into English from its original version [2].

2. LHD Project Entering Deuterium Experiment Phase

Progress in both the physics and engineering aspects of hydrogen experiment phase of LHD has been summarized in recent reviews [6,7]. More specifically, engineering aspects which were the focus of Ref. [6] include the reliable operation of the LHD's large-scale superconducting magnetic system, progress in heating systems, the closed helical diverter, and the successful development and installation of a tritium removal system (installed for use in deuterium experiments). Conversely, in Ref. [7], plasma parameters (e.g., density, temperature, beta values, and long-pulse operation) and their extension in the hydrogen experiment phase were reviewed along with associated physics findings and understandings.

One of the highlights of the hydrogen experiment phase has been the demonstration of a 47 min 39 s-long discharge, with a few-keV range, achieving the world record in total injected energy (3.36 GJ) [8]. The LHD has explored a world-leading long-pulse operation regime, although its fusion triple product is much lower than those of break-even tokamaks. Complementary research involving helical systems and tokamaks has been envisaged to work towards a steady-state and high-performance fusion reactor regime. Plasma parameters such as temperature, density, and beta value have also steadily developed during the hydrogen experiment phase, as a result of the reliable large-scale superconducting magnet system and physics findings as well as a steady increase in heating power.

Based on this progress, the LHD has entered its deuterium experiment phase (i.e., a more fusion-relevant phase), with the first deuterium plasma on 7 March 2017. The most notable result from the first deuterium campaign was the achievement of an ion temperature of 10 keV [9], as shown in Figure 2. This was a milestone in helical fusion research, in the sense that helical plasma has now reached one of the conditions for fusion. This achievement was made possible by the confinement improvement in deuterium plasmas compared to hydrogen plasmas [9], in addition to an increase in ion heating power via upgraded neutral beam injection (NBI) [10], careful choice of magnetic configuration to retain the heating efficiency of NBI, and the extensive wall conditioning [9]. Experimental observations of the so-called "isotope effect" have taken place not only in high-ion-temperature plasmas but also in pure electron cyclotron heated (ECH) plasmas during LHD deuterium experiments. However, these findings need to be investigated further [11–14] to clarify the mechanism of the isotope effect. A simultaneous increase in ion and electron temperatures should be pursued in subsequent deuterium campaigns.

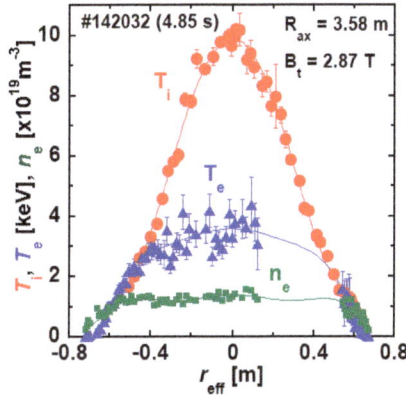

Figure 2. The achievement of an ion temperature of 10 keV in the first deuterium campaign of the Large Helical Device (LHD). The ion temperature profile of 10 keV, along with the electron temperature and density profiles, is depicted. The r_{eff} denotes the effective plasma minor radius (negative values correspond to the inner side of a torus), which is defined as the radius of the equivalent simple torus which encloses the same volume as the flux surface of interest. (This figure is modified from Figure 5g in [9]).

Neutrons produced in deuterium plasmas in conjunction with a well-prepared set of neutron diagnostics [15] (e.g., neutron emission rate and triton burn-up ratio) have provided the capacity for the quantitative assessment of the energetic particles' confinement property [16,17] and their interaction with MHD modes [18,19] in LHD plasmas.

The first deuterium campaign has already provided interesting physics findings such as those on impurity behavior [20,21] and the penetration threshold of resonant magnetic perturbation (RMP) [22]. It has also made progress in engineering aspects, including negative-ion-based (NB) injectors [23] and neutron flux distribution in LHD torus hall [24]. All these findings will be presented in other opportunities.

3. Conceptual Design of the LHD-Type Helical Fusion Reactor FFHR-d1

Based on progress in both the physics and engineering aspects of the LHD project, conceptual design activity has been extensively conducted for an LHD-type helical fusion reactor—the so-called FFHR-d1 [25]. Its expected fusion power is 3 GW. It has the following principal device parameters: major radius of 15.6 m (four times larger than that of LHD), magnetic field strength of 4.7 T (at the helical coil center), plasma volume of 1900 m^3, and stored magnetic energy of 170 GJ. The envisaged plasma parameters are as follows [25]: the central density is ~1.5 × 10^{19} m^{-3}, the central electron temperature ~16.5 keV, and the energy confinement time ~1.5 s. The operation point is explored using a systems code (HELIOSCOPE [26]). An operational point with Q > 10 has been found with a sub-ignition based on LHD data from hydrogen experiments, where Q is the fusion energy gain factor. Confinement improvement that has been identified in LHD deuterium experiments should widen the scope of the operation. Quantitative assessment for start-up scenarios reaching such an identified operation point has also largely been achieved, as reported in Ref. [27]. Scenario developments have been conducted based on the time evolution of plasma radial profiles by solving 1D transport equations.

Since this 1D transport code merely employs a simple empirical transport model deduced from LHD experimental results, consistency with detailed physics criteria such as MHD stability and neoclassical transport should be checked by integrating numerical modules for physics analyses [28] that are being or were already validated by LHD experiments. Using these models, control algorithms of auxiliary heating power and fueling amounts have been examined to reach the identified operation point. In this example, smooth control of fusion power was successfully confirmed. In this way, conceptual design of the LHD-type helical fusion reactor FFHR-d1 has progressed, incorporating the time evolution of plasma profiles. A derivation of FFHR-d1, the so-called FFHR-c1, has also been designed with targeting year-long electric power generation by allowing for auxiliary heating along with innovative ideas for its engineering system [29].

4. LHD as a Platform for Wide-Ranging Plasma Research

LHD plasmas have been utilized not only for fusion research, but also for wide-ranging plasma research by making use of its steady-state and well diagnosed plasma parameters (e.g., temperature). In this section, two examples of such contributions made by the LHD are introduced.

The first example is the LHD's use in spectroscopic study. Spectroscopic studies have been systematically conducted using LHD on a variety of heavy elements relevant to fusion, as well as in other fields, from basic atomic physics to plasma applications. In the periodic table shown in Figure 3, elements which have been injected into LHD plasmas by means of gas puff or tracer encapsulated solid pellet (TESPEL) [30] are labelled by year. The most frequently studied element is tungsten, which will be used as a material for the ITER diverter. Iron has been investigated for its application in solar astrophysics. Tin, xenon, and lanthanide elements are candidate materials for a light source in EUV lithography. Very heavy elements such as platinum and gold may be used for a water-window light source in biological microscopy.

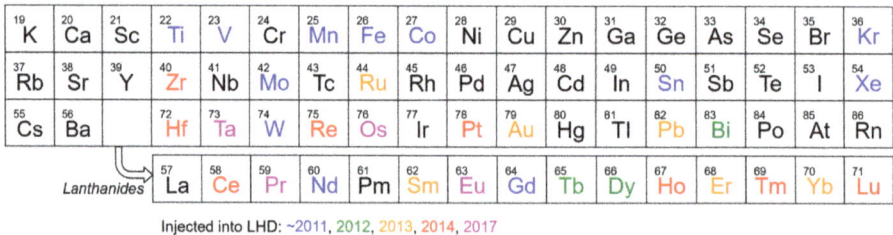

19 K	20 Ca	21 Sc	22 Ti	23 V	24 Cr	25 Mn	26 Fe	27 Co	28 Ni	29 Cu	30 Zn	31 Ga	32 Ge	33 As	34 Se	35 Br	36 Kr
37 Rb	38 Sr	39 Y	40 Zr	41 Nb	42 Mo	43 Tc	44 Ru	45 Rh	46 Pd	47 Ag	48 Cd	49 In	50 Sn	51 Sb	52 Te	53 I	54 Xe
55 Cs	56 Ba		72 Hf	73 Ta	74 W	75 Re	76 Os	77 Ir	78 Pt	79 Au	80 Hg	81 Tl	82 Pb	83 Bi	84 Po	85 At	86 Rn

Lanthanides	57 La	58 Ce	59 Pr	60 Nd	61 Pm	62 Sm	63 Eu	64 Gd	65 Tb	66 Dy	67 Ho	68 Er	69 Tm	70 Yb	71 Lu

Injected into LHD: ~2011, 2012, 2013, 2014, 2017

Figure 3. Elements which have been injected into LHD plasmas are shown in this periodic table, with the year of injection indicated by colors.

With the development of an experimental database for several elements, new spectral lines have been identified for the first time via LHD. By controlling the heating (and thus the electron temperature), the temperature dependence of the EUV spectrum can be systematically obtained. This is the unique advantage of the LHD having excellent spatial and temporal resolutions for electron temperature measurement using a Thomson scattering system. In the case of terbium ions, as described in [30], the spectrum is discrete at electron temperatures above 1 keV and is composed of higher charge states around a Cu-like ion, which has been found for the first time in LHD. As the electron temperature drops below 0.5 keV, the spectrum becomes quasi-continuous because the dominant charge states become lower, eventually producing Ag-like ions. Similarly, several isolated spectral lines have been found experimentally for the first time in the LHD, as has been reported in Refs. [31,32]. Z-dependence of the lanthanide spectra has also been studied, and has been recently discussed in Ref. [33].

A second example of the LHD's use in plasma research is the development of a new type of interferometer for electron density measurement. Interferometers are one of the main types of electron density diagnostic systems. However, interferometers suffer from large measurement errors caused by mechanical vibrations and changes in air conditions. The installation of vibration isolation systems and the control of air are thus required.

A new type of interferometer called a "dispersion interferometer" is insensitive to mechanical vibrations. It is essentially an interferometer, but it can cancel vibration components automatically by using the second harmonic component 2ω and a special interferometer configuration, as shown in Figure 4 (all details can be found in Ref. [34]). The second harmonic components are generated twice from the laser fundamental component ω with nonlinear crystals (i.e., once before and once after the plasma passage) and the interference signal between two second harmonics $I_{DC} + I_{AC} \cos (1.5\varphi_p)$ is detected, where I_{DC}, I_{AC}, and φ_p denote DC and AC values determined by the laser intensity, and the phase shift caused by a plasma, respectively. Since these two wavelength components have almost identical optical paths, the phase shifts caused by vibrations and by the air are the same. By contrast, phase shifts caused by the plasma differ between two wavelength components due to the dispersion of a plasma. Since the phase of the interference signal from which the electron density is calculated is the subtraction of the phases of the two second harmonic components, the phases due to vibrations and air are cancelled and only that due to φ_p remains. This is the reason for the invulnerability of the dispersion interferometer to vibrations and air. We have been developing the dispersion interferometer and have implemented phase modulation for further enhancement of resolutions. The interferometer has also been installed in the LHD to demonstrate its feasibility as an electron density diagnostic system for fusion plasmas. Following its successful demonstration within the LHD, it has been decided that it will be installed as a density measurement system for the first plasma of ITER. It is currently being designed and tested for ITER [35].

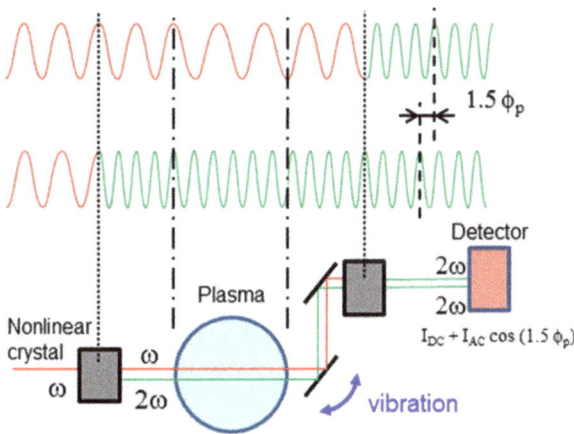

Figure 4. Schematic arrangement of the dispersion interferometer. The dispersion interferometer uses a mixture of the fundamental ω and second harmonic 2ω light as a probe beam. The second harmonic light is generated from the incident fundamental light with a nonlinear crystal. After passing through a plasma, the fundamental light is converted to the second harmonic light, and an interference signal between two second harmonic lights is detected. The light path of the second harmonic light generated by a nonlinear crystal is almost the same as that of the fundamental one. Hence, the variations of the light path length caused by mechanical vibrations are the same between the two lights. On the other hand, the phase shifts caused by the plasma are different due to the dispersion of plasma. As a result, phase shifts caused by vibrations are cancelled and only that belonging to the plasma $1.5\varphi_p$ remains in the interference signal, because the phase of the interference signal is a subtraction of the phases of the two second harmonic lights. In this way, the dispersion interferometer is free from mechanical vibrations.

Moreover, the dispersion interferometer has been proven to be effective for atmospheric pressure plasmas, not merely for fusion plasmas. Conventional interferometry for atmospheric pressure plasma is not straightforward, because changes in air pressure due to heating by plasma cause a 10–100 times larger phase shift than that caused by a plasma. However, the dispersion interferometer can significantly suppress the effect of air, similarly to how mechanical vibrations do. Proof-of-principle experiments for atmospheric pressure plasmas have been conducted with a dispersion interferometer that was developed for the LHD [36]. As shown in Figure 5, the phase shift quickly increases and decreases when the discharge current turns on and off, respectively. These are the phase shifts caused by the plasma, which corresponds to 1.4×10^{20} m^{-3}. Air also has dispersion, although it is minor. The gradual decrease in the phase shift immediately following the plasma ignition is caused by the dispersion of air. Even though effects due to air remain, the dispersion interferometer enables us to distinguish the plasma phase shift and to evaluate the electron density. In this way, new diagnostics developed within the LHD have been able to contribute not only to fusion plasmas but also to atmospheric pressure plasmas.

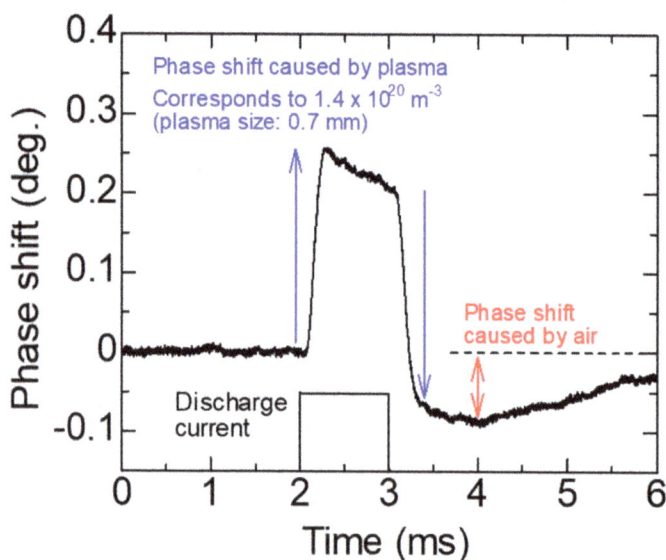

Figure 5. Example of electron density measurements of atmospheric pressure nitrogen plasmas.

5. Conclusions

The LHD has progressed as a large-scale superconducting device since 1998, having demonstrated its advantageous capacity for steady-state operation. It has now entered its advanced deuterium experiment phase. A fusion-relevant ion temperature of 10 keV was successfully achieved during the first deuterium campaign. This was a milestone achievement in helical systems research. The LHD will continue to provide research opportunities for reactor-relevant regimes (including high-performance and steady-state plasmas). Ongoing research in the LHD should provide a firm basis for high-precision predictability towards a steady-state helical fusion reactor. The LHD also acts as a platform for diagnostics development and shows promise for applications in wide-ranging plasma research.

Funding: This research was funded by the Ministry of Education, Culture, Sports, Science and Technology (MEXT), Japan.

Acknowledgments: The author would like to acknowledge all domestic and international collaborators for their continued and extensive cooperation. Technical staff are also greatly appreciated for their devoted efforts in maintaining and running the LHD. The LHD's experiments and all of NIFS' activities have been continuously and strongly supported by the Ministry of Education, Culture, Sports, Science and Technology (MEXT), Japan.

Conflicts of Interest: The author declares no conflicts of interest.

References

1. Takeiri, Y.; Morisaki, T.; Osakabe, M.; Yokoyama, M.; Sakakibara, S.; Takahashi, H.; Nakamura, Y.; Oishi, T.; Motojima, G.; Murakami, S.; et al. Extension of the operational regime of the LHD towards a deuterium experiment. *Nucl. Fusion* **2017**, *57*, 102023. [CrossRef]
2. Yokoyama, M. History and status of helical fusion research. *OHM Magazine*, 5 December 2017, pp. 18–23. (In Japanese)
3. Klinger, T.; Alonso, A.; Bozhenkov, S.; Burhenn, R.; Dinklage, A.; Fuchert, G.; Geiger, G.; Grulke, O.; Langenberg, A.; Hirsch, M.; et al. Performance and properties of the first plasmas of Wendelstein 7-X. *Plasma Phys. Control. Fusion* **2017**, *59*, 014018. [CrossRef]
4. Vargas, V.I.; Mora, J.; Asenjo, J.; Zamora, E.; Otarola, C.; Barillas, L.; Carvajal-Godínez, J.; González-Gómez, J.; Soto-Soto, C.; Piedras, C. Constructing a small modular stellarator in Latin America. *J. Phys. Conf. Ser.* **2015**, *591*, 012016. [CrossRef]
5. Liu, H.; Shimizu, A.; Isobe, M.; Okamura, S.; Nishimura, S.; Suzuki, C.; Xu, Y.; Zhang, X.; Liu, B.; Huang, J.; et al. Magnetic Configuration and Modular Coil Design for the Chinese First Quasi-Axisymmetric Stellarator. *Plasma Fusion Res.* **2018**, *13*, 3405067. [CrossRef]
6. Takeiri, Y. Prospect Toward Steady-State Helical Fusion Reactor Based on Progress of LHD Project Entering the Deuterium Experiment Phase. *IEEE Trans. Plasma Sci.* **2018**, *46*, 1141–1148. [CrossRef]
7. Takeiri, Y. The Large Helical Device—Entering Deuterium Experiment Phase Toward Steady-State Helical Fusion Reactor Based on Achievements in Hydrogen Experiment Phase. *IEEE Trans. Plasma Sci.* **2018**, *46*, 2348–2353. [CrossRef]
8. Seki, T.; Mutoh, T.; Saito, K.; Kasahara, H.; Seki, R.; Kamio, S.; Nomura, G.; Zhao, Y.; Wang, S.; LHD Experiment Group. ICRF Heating Experiment on LHD in Foreseeing a Future Fusion Device. *Plasma Fusion Res.* **2015**, *10*, 3405046. [CrossRef]
9. Takahashi, H.; Nagaoka, K.; Murakami, S.; Osakabe, M.; Nakano, H.; Ida, K.; Tsujimura, T.I.; Kubo, S.; Kobayashi, T.; Tanaka, K.; et al. Realization of high Ti plasmas and confinement characteristics of ITB plasmas in the LHD deuterium experiments. *Nucl. Fusion* **2018**, *58*, 106028. [CrossRef]
10. Osakabe, M.; Isobe, M.; Tanaka, M.; Motojima, G.; Tsumori, K.; Yokoyama, M.; Morisaki, T.; Takeiri, Y.; LHD Experiment Group. Preparation and Commissioning for the LHD Deuterium Experiment. *IEEE Trans. Plasma Sci.* **2018**, *46*, 2324–2331. [CrossRef]
11. Nakata, M.; Nunami, M.; Sugama, H.; Watanabe, T.H. Isotope Effects on Trapped-Electron-Mode Driven Turbulence and Zonal Flows in Helical and Tokamak Plasmas. *Phys. Rev. Lett.* **2017**, *118*, 165002. [CrossRef]
12. Warmer, F.; Takahashi, H.; Tanaka, K.; Yoshimura, Y.; Beidler, C.D.; Peterson, B.; Igami, H.; Ido, T.; Seki, R.; Nakata, M.; et al. Energy confinement of hydrogen and deuterium electron-root plasmas in the Large Helical Device. *Nucl. Fusion* **2018**, *58*, 106025. [CrossRef]
13. Yamada, H.; Tanaka, K.; Tokuzawa, T.; Seki, R.; Suzuki, C.; Yokoyama, M.; Ida, K.; Yoshimura, M.; Fujii, K.; Yamaguchi, H.; et al. Characterization of Isotope Effect on Confinement of Dimensionally Similar NBI-Heated Plasmas in LHD. Presented at the 27th IAEA Fusion Energy Conference, Gandhinagar, India, 22–27 October 2018; paper EX/P3-5.
14. Tanaka, K.; Nakata, M.; Ohtani, Y.; Tsujimura, T.I.; Takahashi, H.; Yokoyama, M.; Warmer, F.; The LHD Experiment Group. Isotope effects on confinement and turbulence in ECRH plasma of LHD. Presented at the 27th IAEA Fusion Energy Conference, Gandhinagar, India, 22–27 October 2018. paper EX/P3-6.
15. Isobe, M.; Ogawa, K.; Nishitani, T.; Miyake, H.; Kobuchi, T.; Pu, N.; Kawase, H.; Takada, E.; Tanaka, T.; Li, S.; et al. Neutron Diagnostics in the Large Helical Device. *IEEE Trans. Plasma Sci.* **2018**, *46*, 2050–2058. [CrossRef]
16. Isobe, M.; Ogawa, K.; Nishitani, T.; Pu, N.; Kawase, H.; Seki, R.; Nuga, H.; Takada, E.; Murakami, S.; Suzuki, Y.; et al. Fusion neutron production with deuterium neutral beam injection and enhancement of energetic-particle physics study in the large helical device. *Nucl. Fusion* **2018**, *58*, 082004. [CrossRef]

17. Ogawa, K.; Isobe, M.; Nishitani, T.; Murakami, S.; Seki, R.; Nuga, H.; Kamio, S.; Fujiwara, Y.; Yamaguchi, H.; Kawase, H.; et al. Energetic-ion Confinement Studies by using Comprehensive Neutron Diagnostics in the Large Helical Device. Presented at the 27th IAEA Fusion Energy Conference, Gandhinagar, India, 22–27 October 2018; paper EX/P3-20.

18. Bando, T.; Ohdachi, S.; Isobe, M.; Suzuki, Y.; Toi, K.; Nagaoka, K.; Takahashi, H.; Seki, R.; Du, X.D.; Ogawa, K.; et al. Excitation of helically-trapped-energetic-ion driven resistive interchange modes with intense deuterium beam injection and enhanced effect on beam ions/bulk plasmas of LHD. *Plasma Phys. Control. Fusion* **2018**, *58*, 082025. [CrossRef]

19. Ohdachi, S.; Bando, T.; Nagaoka, K.; Takahashi, H.; Suzuki, Y.; Watanabe, K.Y.; Du, X.D.; Toi, K.; Osakabe, M.; Morisaki, T. Excitation mechanism of the energetic particle driven resistive interchange mode and strategy to control the mode in Large Helical Device. Presented at the 27th IAEA Fusion Energy Conference, Gandhinagar, India, 22–27 October 2018; paper EX/1-3Rb.

20. Ida, K.; Sakamoto, R.; Yoshinuma, M.; Yamazaki, K.; Kobayashi, T.; The LHD Experiment Group. Isotope effect on impurity and bulk ion particle transport in the Large Helical Device. Presented at the 27th IAEA Fusion Energy Conference, Gandhinagar, India, 22–27 October 2018; paper EX/10-1.

21. Oishi, T.; Morita, S.; Kobayashi, M.; Kawamura, G.; Liu, Y.; Goto, M.; The LHD Experiment Group. Effect of deuterium plasmas on carbon impurity transport in the edge stochastic magnetic field layer of Large Helical Device. Presented at the 27th IAEA Fusion Energy Conference, Gandhinagar, India, 22–27 October 2018; paper EX/P3-11.

22. Watanabe, K.Y.; Sakakibara, S.; Narushima, Y.; Ohdachi, S.; Suzuki, Y.; Takemura, Y.; The LHD Experiment Group. Dependence of RMP penetration threshold on plasma parameters and ion species in helical plasmas. Presented at the 27th IAEA Fusion Energy Conference, Gandhinagar, India, 22–27 October 2018. paper EX/P3-15.

23. Ikeda, K. Exploring Deuterium Beam Operation and Behavior of Co-Extracted Electron in Negative-Ion-Based Neutral Beam Injector. Presented at the 27th IAEA Fusion Energy Conference, Gandhinagar, India, 22–27 October 2018; paper FIP/P1-54.

24. Kobayashi, M.; Tanaka, T.; Nishitani, T.; Ogawa, K.; Isobe, M.; Motojima, G.; Kato, A.; Saze, T.; Yoshihashi, S.; Osakabe, M.; The LHD Experiment Group. Neutron flux distributions in the LHD torus hall evaluated by an imaging plate technique in the first campaign of deuterium plasma experiment. Presented at the 27th IAEA Fusion Energy Conference, Gandhinagar, India, 22–27 October 2018; paper FIP/P3-4.

25. Sagara, A.; Miyazawa, J.; Tamura, H.; Tanaka, T.; Goto, T.; Yanagi, N.; Sakamoto, R.; Masuzaki, S.; Ohtani, H.; The FFHR Design Group. Two conceptual designs of helical fusion reactor FFHR-d1A based on ITER technologies and challenging ideas. *Nucl. Fusion* **2017**, 086046. [CrossRef]

26. Goto, T.; Suzuki, Y.; Yanagi, N.; Watanabe, K.Y.; Imagawa, S.; Sagara, A. Importance of helical pitch parameter in LHD-type heliotron reactor designs. *Nucl. Fusion* **2011**, *51*, 083045. [CrossRef]

27. Goto, T.; Miyazawa, J.; Sakamoto, R.; Suzuki, Y.; Suzuki, C.; Seki, R.; Satake, S.; Huang, B.; Nunami, M.; Yokoyama, M.; et al. Development of a real-time simulation tool towards self-consistent scenario of plasma start-up and sustainment on helical fusion reactor FFHR-d1. *Nucl. Fusion* **2017**, *57*, 066011. [CrossRef]

28. Yokoyama, M.; Seki, R.; Suzuki, C.; Sato, M.; Emoto, M.; Murakami, S.; Osakabe, M.; Tsujimura, T.I.; Yoshimura, Y.; Ido, T.; et al. Extended capability of the integrated transport analysis suite, TASK3D-a, for LHD experiment. *Nucl. Fusion* **2017**, *57*, 126016. [CrossRef]

29. Miyazawa, J.; Goto, T.; Tamura, H.; Tanaka, T.; Yanagi, N.; Murase, T.; Sakamoto, R.; Masuzaki, S.; Ohgo, T.; Sagara, A.; et al. Maintainability of the helical reactor FFHR-c1 equipped with the liquid metal divertor and cartridge-type blankets. *Fusion Eng. Des.* **2018**, *136 Pt B*, 1278–1285. [CrossRef]

30. Sudo, S.; Tamura, N. Tracer-encapsulated solid pellet injection system. *Rev. Sci. Instrum.* **2012**, *83*, 023503. [CrossRef]

31. Suzuki, C.; Koike, F.; Murakami, I.; Tamura, N.; Sudo, S. Temperature dependent EUV spectra of Gd, Tb and Dy ions observed in the Large Helical Device. *J. Phys. B At. Mol. Opt. Phys.* **2015**, *48*, 144012. [CrossRef]

32. Suzuki, C.; Murakami, I.; Koike, F.; Tamura, N.; Sakaue, H.A.; Morita, S.; Goto, M.; Kato, D.; Ohashi, H.; Higashiguchi, T.; et al. Extreme ultraviolet spectroscopy and atomic models of highly charged heavy ions in the Large Helical Device. *Plasma Phys. Control. Fusion* **2017**, *59*, 014009. [CrossRef]

33. Suzuki, C.; Koike, F.; Murakami, I.; Tamura, N.; Sudo, S. Systematic Observation of EUV Spectra from Highly Charged Lanthanide Ions in the Large Helical Device. *Atoms* **2018**, *6*, 24. [CrossRef]

34. Akiyama, T.; Yasuhara, R.; Kawahata, K.; Okajima, S.; Nakayama, K. Dispersion interferometer using modulation amplitudes on LHD. *Rev. Sci. Instrum.* **2014**, *85*, 11D301. [CrossRef] [PubMed]

35. Akiyama, T.; Sirinelli, A.; Watts, C.; Shigin, P.; Vayakls, G.; Walsh, M. Design of a dispersion interferometer combined with a polarimeter to increase the electron density measurement reliability on ITER. *Rev. Sci. Instrum.* **2016**, *87*, 11E133. [CrossRef] [PubMed]

36. Urabe, K.; Akiyama, T.; Terashima, K. Application of phase-modulated dispersion interferometry to electron-density diagnostics of high-pressure plasma. *J. Phys. D Appl. Phys.* **2014**, *47*, 262001. [CrossRef]

Article

Fundamental Plane of Elliptical Galaxies in $f(R)$ Gravity: The Role of Luminosity

Vesna Borka Jovanović [1],*, **Predrag Jovanović [2]**, **Duško Borka [1]** and **Salvatore Capozziello [3,4,5]**

[1]   Atomic Physics Laboratory (040), Vinča Institute of Nuclear Sciences, University of Belgrade, P.O. Box 522, 11001 Belgrade, Serbia; dusborka@vinca.rs
[2]   Astronomical Observatory, Volgina 7, P.O. Box 74, 11060 Belgrade, Serbia; pjovanovic@aob.rs
[3]   Dipartimento di Fisica "E. Pancini", Università di Napoli "Federico II", Compl. Univ. di Monte S. Angelo, Edificio G, Via Cinthia, I-80126 Napoli, Italy; capozzie@na.infn.it
[4]   Istituto Nazionale di Fisica Nucleare (INFN) Sez. di Napoli, Compl. Univ. di Monte S. Angelo, Edificio G, Via Cinthia, I-80126 Napoli, Italy
[5]   Gran Sasso Science Institute, Viale F. Crispi, 7, I-67100 L'Aquila, Italy
*   Correspondence: vborka@vinca.rs; Tel.: +381-11-630-8425

Received: 22 November 2018; Accepted: 28 December 2018; Published: 31 December 2018

Abstract: The global properties of elliptical galaxies are connected through the so-called fundamental plane of ellipticals, which is an empirical relation between their parameters: effective radius, central velocity dispersion and mean surface brightness within the effective radius. We investigated the relation between the parameters of the fundamental plane equation and the parameters of modified gravity potential $f(R)$. With that aim, we compared theoretical predictions for circular velocity in $f(R)$ gravity with the corresponding values from a large sample of observed elliptical galaxies. Besides, we consistently reproduced the values of coefficients of the fundamental plane equation as deduced from observations, showing that the photometric quantities like mean surface brightness are related to gravitational parameters. We show that this type of modified gravity, especially its power-law version—R^n, is able to reproduce the stellar dynamics in elliptical galaxies. In addition, it is shown that R^n gravity fits the observations very well, without the need for a dark matter.

Keywords: modified theories of gravity; methods: analytical; methods: numerical; galaxies: elliptical; galaxies: fundamental parameters

1. Introduction

It is well established that there are three main global observables of elliptical galaxies: the central projected velocity dispersion σ_0, the effective radius r_e, and the mean effective surface brightness (within r_e) I_e. It is well known that elliptical galaxies do not populate uniformly this three dimensional parameter space; they are rather confined to a narrow logarithmic plane, thus called the fundamental plane (FP) [1,2]. Any of the three parameters may be estimated from the other two. Together they describe a plane in three-dimensional space. Many characteristics of a galaxy might be correlated.

To describe the velocity of populations of stars, one defines a rotational velocity v_c—net rotational velocity of a group of stars, and a dispersion σ—the characteristic random velocity of stars. The relation v_c/σ characterizes the kinematics of the galaxies, and it is the main parameter which differentiates spiral from elliptical galaxies. In this manner, spiral galaxies with $v_c/\sigma \gg 1$, are kinematically cold systems, while ellipticals with $0 < v_c/\sigma < 1$, are kinematically hot systems.

In galactic dynamics, the Virial Theorem (VT) which relates the total mass of the galaxies with mean (rotation or dispersion) velocity of their stars, is commonly used for inferring the mass estimates of the galaxies. Stationarity is a sufficient condition for the validity of the VT, and so for ellipticals the VT holds [2]. In terms of r_e, σ_0 and I_e both VT and FP can be described by similar expressions (Equations (1) and (2) in [3]). However, the values of constants a and b in the case of FP differ from those predicted by VT, causing the tilt of the FP with respect to the VT plain. Namely, the calculated values of tilt angle of the FP from astronomical observations give coefficients a and b different from those predicted by the VT ($a = 2$, $b = -1$). When written in logarithmic form, the two planes appear to be tilted by an angle of $\sim$15° [3]. This tilt can be caused by different structural and dynamical effects in elliptical galaxies [4].

It is now established that the dark matter (DM) fraction is likely a major contributor to the tilt of FP, and that the DM fractions increase for larger galaxies, because the effective radii extend further out to regions dominated by the halo [5]. However, in paper [6], the authors derived accurate total mass-to-light ratios $(M/L)_e$ and DM fractions, within a sphere of radius $r = r_e$ centred on the galaxies. They tested the accuracy of the mass determinations by running models with and without DM, and have found that the enclosed total $(M/L)_e$ is independent of the inclusion of a DM halo, with good accuracy and small bias.

Therefore, in this paper we try to explain tilt of the FP without DM, but using modified gravity instead. The plan of this paper is as follows. In Section 2, we briefly describe elliptical galaxies fundamental plane. We also describe used observations and methods. In Section 3 we give the basics of power-law $f(R)$ extended gravity theories in the case of a point-like source and the generalization to a spherically symmetric system which represents elliptical galaxies. In Section 4 we give a connection between the parameters of FP equation and parameters of the R^n extended gravity potential. Section 5 is devoted to summary of the conclusions.

2. Elliptical Galaxies and Their Fundamental Plane

2.1. Surface Brightness of Ellipticals

Surface brightness I is flux F within angular area Ω^2 on the sky ($\Omega = D/d$, where D is side of a small patch in a galaxy located at a distance d). Let us emphasize here that I is independent of distance d: $I = F/\Omega^2 = L/(4\pi d^2) \times (d/D)^2 = L/(4\pi D^2)$, where L is luminosity (see e.g., Section 1.3.1 in [7]).

Main sources of luminosity in elliptical galaxies would be: stellar plasma, hot gas, accreting black holes in the cores of stellar bulges (see e.g., [7] and references therein). According to luminosity, their classification is the following:

1. Massive/luminous ellipticals ($L > 2 \times 10^{10} L_\odot$).
   These ellipticals have low central surface brightness with flat distribution (cores-regions where the surface brightness flattens). They have lots of hot X-ray emitting gas, very old stars, lots of globular clusters, and are characterized by little rotation.
2. Intermediate mass/luminosity ellipticals ($L > 3 \times 10^9 L_\odot$).
   Their characteristic is power law central brightness distribution. They have little cold gas, and their oblate symmetry is consistent with their moderate rotation.
3. Dwarf ellipticals ($L < 3 \times 10^9 L_\odot$).
   Their surface brightness is exponential. There is no rotation.

In contrast to spirals galaxies, ellipticals show regularity in their global luminosity distributions. Surface brightness of most elliptical galaxies, measured along the major axis of a galaxy's image, can be fit by de Vaucouleurs profile: $I(r) = I_e \times 10^{-3.33\left((r/r_e)^{1/4}-1\right)}$. This empirical model, also known as $r^{1/4}$ law, describes how the surface brightness varies as a function of apparent distance r from the center of the galaxy. The Sersic $r^{1/n}$ profile: $I(r) = I_e \times 10^{-b_n\left((r/r_e)^{1/n}-1\right)}$ (the constant b_n is chosen such that half of the luminosity comes from $r < r_e$), which generalizes the de Vaucouleurs profile, is also well suited to describe the surface brightness distribution of these systems (see more about this profile e.g., in [4,8]). De Vaucouleurs profile is a particularly good description of the surface brightness of giant and midsized elliptical galaxies, while dwarf ellipticals are better fit by Sersic profile for $n = 1$ (exponential profiles).

2.2. Fundamental Plane of Elliptical Galaxies

The global properties of elliptical galaxies are connected, and empirical relation which shows this connection is called fundamental plane [3]:

$$log(r_e) = a \, log(v_c) + b \, log(I_e) + c, \tag{1}$$

with r_e—effective radius (which encloses half of the total luminosity emitted by a galaxy), v_c—central velocity dispersion, I_e—mean surface brightness within r_e, and a, b, c—coefficients.

Some object can be represented as a point in the parameter space (r_e, v_c, I_e), and if we present it in logarithmic form, we obtain a plane [9,10]. The angle between the virial plane and FP is the so-called "tilt". Prediction of the VT (Virial equilibrium and constant mass-to-light M/L ratio) for FP coefficients a and b is: $a = 2$, $b = -1$, and the empirical result (using the Virgo Cluster elliptical galaxies as a sample) gives $a = 1.4$, $b = -0.85$ [11]. So, when presented in logarithmic form, these two planes appear to be tilted by an angle of ~15° [3]. This can be explained by stellar population effects and by spatial non-homology in the dynamical structures of the systems. As VT uses the simplified assumptions, the tilt provides fundamental information about galaxy evolution.

2.3. Observations and Method

The observational data of interest for our study are publicly available (in ASCI format) among the source files of the arxiv version of the paper Burstein et al. (1997) [12]: https://arxiv.org/format/astro-ph/9707037, see 'metaplanetab1'. We use some physical properties of stellar systems: among 1150 observed galaxies, there is a sample of 401 ellipticals. In this study, we use the values from the following columns of Table 1 in [12]: column (5)—circular velocity (observed): $\log v_c$ (km/s); column (6)—central velocity dispersion (derived): $\log \sigma_0$ (km/s); column (7)—effective or half-light radius: $\log r_e$ (kpc); column (8)—mean surface brightness within r_e: $\log I_e$ (L$_\odot$/pc^2). Here, we would like to emphasize that σ_0 is derived in that way to get the consistent values for all stellar systems, and that for elliptical galaxies the circular velocity inside effective radius is $v_c(r_e) = \sigma_0$, while for other stellar systems it is $v_c \neq \sigma_0$.

Then, using the relation for v_c which consists of Newtonian contribution and the correction term due to modified gravity, we also calculate the theoretical values for circular velocity v_c^{theor} and FP coefficients (see more in Section 4).

2.4. Region of Parameter Space of Fundamental Plane

Early-type galaxies are observed to populate a tight plane in the space defined by their effective radii, velocity dispersions and surface brightnesses [13]. If elliptical galaxies were perfectly homologous stellar systems with identical stellar populations, then r_e, v_c, and I_e would be related by the VT. Instead, non-homology and/or stellar population variations tend to place elliptical galaxies on a nearby fundamental plane, with the remarkably small width [14]. It has been shown that these parameters are rather stable to gravitational perturbation. The FP parameters do change during close encounters of galaxies, but within a very short time interval just before their final merger, and furthermore, the amplitudes of these changes are comparable to the scatter of the observed FP [15].

As we stated before, elliptical galaxies are not randomly distributed within the 3D parameter space (I_e, r_e, v_c), and when presented in logarithmic form, they lie in a plane. See Figure 1, showing the parameter space in log scale.

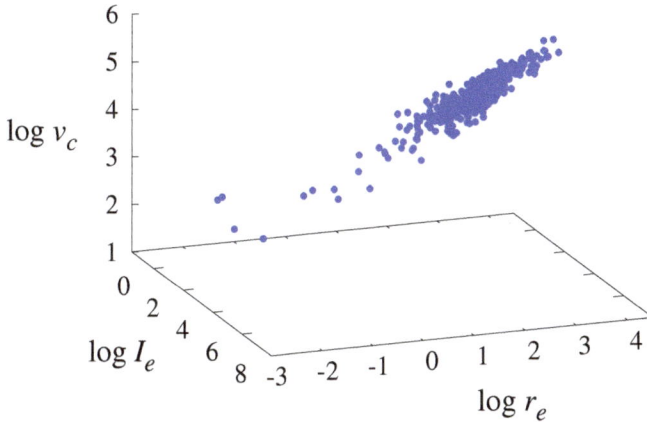

Figure 1. The fundamental plane (FP) parameter space, presented by logarithms of the three parameters: mean surface brightness (within effective radius) $\log I_e$, effective radius $\log r_e$ and circular velocity $\log v_c$, for a sample of elliptical galaxies listed in Table 1 from [12]. Note: in paper [12] only the first page is printed, and we used the whole sample of 401 ellipticals, available among the source files of its arxiv version.

3. $f(R)$ Modified Gravity

Extended Theories of Gravity (ETGs) [16] have been proposed to explain galactic and extragalactic dynamics without introducing DM, and as such they can be used to test if FP of ellipticals could be explained taking into account only their luminous matter content. For that purpose, we adopt $f(R)$ gravity which is the straightforward generalization of Einstein's General Relativity as soon as the function is $f(R) \neq R$, that is, it is not linear in the Ricci scalar R as in the Hilbert-Einstein action. As simple choice, one assumes a generic function $f(R)$ of the Ricci scalar R (in particular, analytic functions) and searches for a theory of gravity having suitable behavior at all scales (at small and large scale lengths).

We start from the action [17]:

$$\mathcal{A} = \int d^4x \sqrt{-g} \left[f(R) + \mathcal{L}_m \right], \tag{2}$$

(with g—metric tensor and $\mathcal{L}_m$—the standard matter Lagrangian), and consider power-law case:

$$f(R) = f_0 R^n, \tag{3}$$

with n the slope of the gravity Lagrangian, and f_0 a dimensional constant (dimensions for f_0 chosen in such a way to give $f(R)$ the right physical dimensions).

R^n gravity is the power-law version of $f(R)$ modified gravity. In the weak field limit, its potential (generated by a pointlike mass m at the distance r) is [17]:

$$\Phi(r) = -\frac{Gm}{2r} \left[1 + \left(\frac{r}{r_c} \right)^{\beta} \right], \tag{4}$$

with r_c—scalelength depending on the gravitating system properties, and β—universal constant:

$$\beta = \frac{12n^2 - 7n - 1 - \sqrt{36n^4 + 12n^3 - 83n^2 + 50n + 1}}{6n^2 - 4n + 2}. \tag{5}$$

here, the case $n = 1 \Rightarrow \beta = 0$ represents Newtonian case. About the power-law fourth-order theories of gravity, as well as about determination of the space parameters of $f(R)$ gravity, see [17–23].

The solution (4) has been obtained in the case of a point-like source, but it can be generalized to the case of extended systems. The generalization of Equation (4) to a spherically symmetric system, gives the correction term of the potential [17]:

$$\Phi_c(r) = -\frac{\pi G \alpha r_c^2}{3} \left[\mathcal{I}_1(r) + \mathcal{I}_2(r) \right] \qquad (6)$$

with parameter $\alpha = 1$ for R^n gravity, and with:

$$\mathcal{I}_1 = 3\pi \int_0^\infty (\xi^2 + \xi'^2)^{(\beta-1)/2} \rho(\xi') \xi'^2 d\xi' \times {}_2F_1 \left[\left\{ \frac{1-\beta}{4}, \frac{3-\beta}{4} \right\}, \{2\}, \frac{4\xi^2 \xi'^2}{(\xi^2 + \xi'^2)^2} \right], \qquad (7)$$

$$\mathcal{I}_2 = 4(1-\beta)\xi \int_0^\infty (\xi^2 + \xi'^2)^{(\beta-3)/2} \rho(\xi') \xi'^2 d\xi' \times {}_3F_2 \left[\left\{ 1, \frac{3-\beta}{4}, \frac{5-\beta}{4} \right\}, \left\{ \frac{3}{2}, \frac{5}{2} \right\}, \frac{4\xi^2 \xi'^2}{(\xi^2 + \xi'^2)^2} \right], \qquad (8)$$

where ξ is generically defined as $\xi = r/r_c$, and the notation for the hypergeometric functions is used: ${}_pF_q[\{a_1, \ldots, a_p\}, \{b_1, \ldots, b_q\}, x]$.

4. Fundamental Plane in $f(R)$ Gravity

4.1. Recovering Fundamental Plane from R^n Gravity

We want to show the connection of the FP of elliptical galaxies with R^n gravity potential, by showing the correlation between the corresponding parameters:

- addend with r_e: correlation between r_e and r_c;
- addend with σ_0: correlation between σ_0 and v_{vir} (v_{vir}—virial velocity);
- addend with I_e: correlation between I_e and r_e (through r_c/r_e ratio).

Here, the reader should note that r_e is the observational gravitational radius (derived from photometry, i.e., its value is determined by the self-gravitating luminous matter content in the inner part of the elliptical galaxy), and r_c is the theoretical gravitational radius (from R^n gravity). We assumed that these two radii were mutually proportional and we tested their different ratios. It is also important here to emphasize that if we introduce the assumption $r_c \sim r_e$, then in point r_e the integrals $\mathcal{I}_1(r_e)$ and $\mathcal{I}_2(r_e)$ (see Equations (7) and (8)) do not depend on r, then $\Phi_c(r_e)$ (see Equation (6)) does not depend on r, so the correction velocity is $v_{c,corr}(r_e) = 0$. In other words, under the condition $r_c \sim r_e$, R^n gravity gives the same σ_0 for elliptical galaxies as in the Newtonian case. For more details about our method, see [10,18].

4.2. Fundamental Plane Coefficients

The empirical result for FP coefficients are given in Bender et al., 1992 [11]: $a = 1.4$, $b = -0.85$. Empirically derived values means that a and b are calculated using observed FP parameters, as coefficients of the FP equation (with the Virgo Cluster elliptical galaxies as a sample). The test for our method is recovering this profile: starting from the gravitational potential derived from $f(R)$ gravity, these values have to be consistently reproduced.

According to Equation (25) for the rotation curve $v_c(r)$ in paper [17], it consists of Newtonian part and the correction term due to R^n gravity:

$$v_c^2(r) = \frac{v_{c,N}^2(r)}{2} + \frac{r}{2} \frac{\partial \Phi_c}{\partial r}, \qquad (9)$$

and therefore we used the above expression to calculate v_c^{theor}—the theoretical prediction of R^n gravity for circular velocity v_c, in the case of extended spherically symmetric systems, taking into account the so-called Hernquist profile for density distribution [24]:

$$\rho(r) = \frac{a_H M}{2\pi r(r + a_H)^3}, \quad a_H = \frac{r_e}{1 + \sqrt{2}}. \tag{10}$$

FP of elliptical galaxies with 3D fit, with the calculated values v_c^{theor}, and the observed values r_e, I_e, presenting the dependence of FP parameters (a, b) on parameters of R^n gravity (r_c, β), we show in Figure 2. In this figure, we presented only the case $\beta = 0.6$, but we tested other values of this parameter in a similar way as well. The phrase "3D fit" denotes a fit of a function z depending on two independent variables (x, y) to the observational data, and in this case Equation (1) is fitted with function $z(x, y) = ax + by + c$, where $x = \log(v_c^{theor})$, $y = \log(I_e)$ and $z = \log(r_e)$, using the least-squares algorithm implemented in "fit" command of Gnuplot (http://www.gnuplot.info/). As a result we obtained the best fit coefficients of FP equation: a, b and c. The procedure is the following (see our Ref. [18] for a detailed explanation): we varied R^n gravity parameters (r_c, β) and for each given pair of the parameters (r_c, β), i.e., for the certain ratios r_c/r_e and certain values β, we calculated the terms x, y and z and finally obtained coefficients (a, b, c). Once this procedure is performed, the obtained values of a and b, are compared with a and b values obtained from observations [11]. As it can be seen from Figure 2, a smaller r_c/r_e ratio results with a larger value for FP parameter a and smaller value of parameter b, obtained by fitting the FP equation through the (I_e, r_e, v_c^{theor}) data points. The best fit (and the smallest scatter of these data points) is obtained for $r_c/r_e \approx 0.05$.

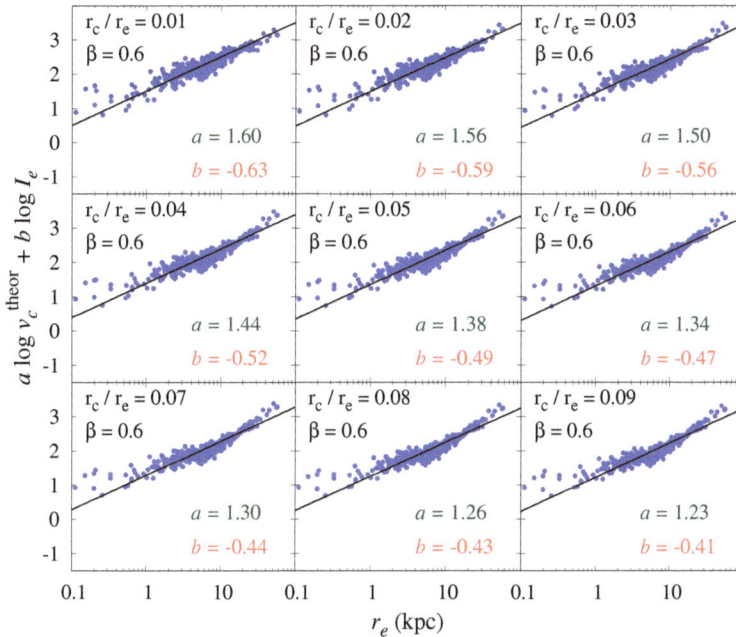

Figure 2. Fundamental plane of elliptical galaxies with calculated circular velocity v_c^{theor}, observed effective radius r_e and observed mean surface brightness (within the effective radius) I_e. For each given pair of R^n gravity parameters (r_c, β), i.e., for the certain cases $r_c/r_e = 0.01, 0.02, 0.03, 0.04, 0.05, 0.06,$ 0.07, 0.08 and 0.09, and $\beta = 0.6$, we present calculated FP coefficients (a, b). Black solid line is result of 3D fit of FP.

4.3. Luminosity and Parameters of R^n Gravity

Correlation of I_e with r_c is reflected through the coefficient b of FP in equation: $r_e \sim I_e^b \times v_c^a$ [11,25]. This means that being r_c related to r_e through $r_c \sim r_e$, also r_c is related to I_e. On the other hand, analytic expression for v_c includes both modified gravity parameters (r_c, β).

Coefficients a, b, c of FP are also correlated with (r_c, β): see Table 3 in our paper [18].

In general, this means that photometric quantities like I_e are related (in a complex way) to the parameters of modified gravitational potential.

5. Discussion and Conclusions

Here we studied the possible connection between the empirical parameters of FP for ellipticals and the theoretical parameters of R^n gravity in order to test if corrections predicted by this type of gravity could explain both photometry and dynamics of ellipticals without DM hypothesis.

Our main conclusions may be summarized as follows:

- We connected fundamental plane of elliptical galaxies with R^n gravity potential, relating together observational and theoretical quantities (i.e., tying the corresponding FP and R^n parameters).
- We reproduced the FP generated by the power law $f(R)$ gravity without considering the presence of DM in galaxies.
- We obtained that the characteristic radius r_c of R^n gravity is proportional to the effective radius r_e: more precisely, $r_c \approx 0.05 r_e$ gives the best fit with data. This fact points out that the gravitational corrections induced by R^n can lead photometry and dynamics of the system.
- We demonstrated that not only stellar kinematics of ellipticals could be affected by modified gravity (as we have already shown in our previous papers), but so too could their most important physical properties, such as their luminosity.

We compared, for the first time, theoretical predictions for circular velocity in $f(R)$ gravity with the corresponding values from the large sample of observed elliptical galaxies. Using gravitational potential derived from $f(R)$ gravity, we consistently reproduced the values of FP parameters. We pointed out that the photometric quantities, like mean surface brightness, are related to gravitational parameters. In addition, we explained that R^n gravity fits the observations very well, taking into account only the luminous matter content of ellipticals, hence not needing DM.

Author Contributions: All the coauthors participated in calculation and discussion of obtained results. These authors contributed equally to this work.

Acknowledgments: This work is supported by Istituto Nazionale di Fisica Nucleare, Sezione di Napoli, Italy, iniziative specifiche TEONGRAV and QGSKY, and by Ministry of Education, Science and Technological Development of the Republic of Serbia, through the project 176003 "Gravitation and the Large Scale Structure of the Universe". The authors also acknowledge the support by Bilateral cooperation between Serbia and Italy 451-03-01231/2015-09/1 "Testing Extended Theories of Gravity at different astrophysical scales" and of the COST Action CA15117 (CANTATA), supported by European Cooperation in Science and Technology (COST). The authors would like to thank Vladimir Reković for improving the English of the paper.

Conflicts of Interest: The authors declare no conflict of interest.

Abbreviations

The following abbreviations are used in this manuscript:

DM	Dark matter
ETGs	Extended Theories of Gravity
FP	Fundamental plane
GC	Galactic Center
GR	General Relativity
VT	Virial Theorem

References

1. Dressler, A.; Lynden-Bell, D.; Burstein, D.; Davies, R.L.; Faber, S.M.; Terlevich, R.; Wegner, G. Spectroscopy and photometry of elliptical galaxies. I—A new distance estimator. *Astrophys. J.* **1987**, *313*, 42–58. [CrossRef]
2. Ciotti, L. The Physical Origin of the Fundamental Plane (of Elliptical Galaxies). In Proceedings of the ESO Astrophysics Symposia "Galaxy Scaling Relations: Origins, Evolution and Applications", Garching, Germany, 18–20 November 1996; pp. 38–43.
3. Busarello, G.; Capaccioli, M.; Capozziello, S.; Longo, G.; Puddu, E. The relation between the virial theorem and the fundamental plane of elliptical galaxies. *Astron. Astrophys.* **1997**, *320*, 415–420.
4. Ciotti, L.; Lanzoni, B.; Renzini, A. The tilt of the fundamental plane of elliptical galaxies—I. Exploring dynamical and structural effects. *Mon. Not. R. Astron. Soc.* **1996**, *282*, 1–12. [CrossRef]
5. Taranu, D.; Dubinski, J.; Yee, H.K.C. Mergers in galaxy groups. II. The fundamental plane of elliptical galaxies. *Astrophys. J.* **2015**, *803*, 1–16. [CrossRef]
6. Cappellari, M.; Scott, N.; Alatalo, K.; Blitz, L.; Bois, M.; Bournaud, F.; Bureau, M.; Crocker, A.F.; Davies, R.L.; Davis, T.A.; et al. The ATLAS^{3D} project —XV. Benchmark for early-type galaxies scaling relations from 260 dynamical models: Mass-to-light ratio, dark matter, Fundamental Plane and Mass Plane. *Mon. Not. R. Astron. Soc.* **2013**, *432*, 1709–1741. [CrossRef]
7. Sparke, L.S.; Gallagher, J.S. *Galaxies in the Universe: An Introduction*; Cambridge University Press: Cambridge, UK, 2007; ISBN 9780521671866.
8. Cardone, V.F. The lensing properties of the Sersic model. *Astron. Astrophys.* **2004**, *415*, 839–848. [CrossRef]
9. Borka Jovanović, V.; Jovanović, P.; Borka, D.; Capozziello, S. Fundamental plane of elliptical galaxies and $f(R)$ gravity. In Proceedings of the 28th Summer School and International Symposium on the Physics of Ionized Gases (SPIG 2016), Belgrade, Serbia, 29 August–2 September 2016; pp. 393–396.
10. Capozziello, S.; Borka, D.; Borka Jovanović, V.; Jovanović, P. Galactic structures from gravitational radii. *Galaxies* **2018**, *6*, 22. [CrossRef]
11. Bender, R.; Burstein, D.; Faber, S.M. Dynamically hot galaxies. I. Structural properties. *Astrophys. J.* **1992**, *399*, 462–477. [CrossRef]
12. Burstein, B.; Bender, R.; Faber, S.M.; Nolthenius, R. Global relationships among the physical properties of stellar systems. *Astron. J.* **1997**, *114*, 1365–1392. [CrossRef]
13. Desroches, L.B.; Quataert, E.; Ma, C.-P.; West, A.A. Luminosity dependence in the Fundamental Plane projections of elliptical galaxies. *Mon. Not. R. Astron. Soc.* **2007**, *377*, 402–414. [CrossRef]
14. Mathews, W.G.; Brighenti, F. Hot gas in and around elliptical galaxies. *Ann. Rev. Astron. Astrophys.* **2003**, *41*, 191–239. [CrossRef]
15. Evstigneeva, E.A.; Reshetnikov, V.P.; Sotnikova, N.Y. Effect of the environment on the fundamental plane of elliptical galaxies. *Astron. Astrophys.* **2002**, *381*, 6–12. [CrossRef]
16. Capozziello, S.; De Laurentis, M. Extended theories of gravity. *Phys. Rep.* **2011**, *509*, 167–321. [CrossRef]
17. Capozziello, S.; Cardone, V.F.; Troisi, A. Low surface brightness galaxy rotation curves in the low energy limit of R^n gravity: No need for dark matter? *Mon. Not. R. Astron. Soc.* **2007**, *375*, 1423–1440. [CrossRef]
18. Borka Jovanović, V.; Capozziello, S.; Jovanović, P.; Borka, D. Recovering the fundamental plane of galaxies by $f(R)$ gravity. *Phys. Dark Univ.* **2016**, *14*, 73–83. [CrossRef]
19. Borka, D.; Jovanović, P.; Borka Jovanović, V.; Zakharov, A.F. Constraints on R^n gravity from precession of orbits of S2-like stars. *Phys. Rev. D* **2012**, *85*, 124004. [CrossRef]
20. Borka, D.; Jovanović, P.; Borka Jovanović, V.; Zakharov, A.F. Orbital precession in R^n gravity: Simulations vs. observations (the S2 star orbit case). *Sveske Fizičkih Nauka (SFIN)* **2013**, *A1*, 61–66.
21. Zakharov, A.F.; Borka, D.; Borka Jovanović, V.; Jovanović, P. Constraints on R^n gravity from precession of orbits of S2-like stars: A case of a bulk distribution of mass. *Adv. Space Res.* **2014**, *54*, 1108–1112. [CrossRef]
22. Borka, D.; Jovanović, P.; Borka Jovanović, V.; Zakharov, A.F. *Advances in General Relativity Research*; Williams, C., Ed.; Nova Science Publishers: Hauppauge, NY, USA, 2015; Chapter 9, pp. 343–362, ISBN 978-1-63483-120-8.
23. Capozziello, S.; Jovanović, P.; Borka Jovanović, V.; Borka, D. Addressing the missing matter problem in galaxies through a new fundamental gravitational radius. *J. Cosmol. Astropart. Phys.* **2017**, *6*, 44. [CrossRef]

24. Hernquist, L. An analytical model for spherical galaxies and bulges. *Astrophys. J.* **1990**, *356*, 359–364. [CrossRef]
25. Bender, R.; Burstein, D.; Faber, S.M. Dynamically hot galaxies. II. Global stellar populations. *Astrophys. J.* **1993**, *411*, 153–169. [CrossRef]

Article

Thermochemical Non-Equilibrium in Thermal Plasmas

Arnaud Bultel [1,*], Vincent Morel [1] and Julien Annaloro [2]

[1]   CORIA, UMR CNRS 6614, Normandie Université, 76801 St-Etienne du Rouvray, France;
       vincent.morel@coria.fr
[2]   CNES, French Spatial Agency, 31400 Toulouse, France; julien.annaloro@cnes.fr
*   Correspondence: arnaud.bultel@coria.fr

Received: 30 November 2018; Accepted: 25 December 2018; Published: 1 January 2019

Abstract: In this paper, we analyze the departure from equilibrium in two specific types of thermal plasmas. The first type deals with the plasma produced during the atmospheric entry of a spatial vehicle in the upper layers of an atmosphere, specifically the one of Mars. The second type concerns the plasma produced during the laser-matter interaction above the breakdown threshold on a metallic sample. We successively describe the situation and give the way along which modeling tools are elaborated by avoiding any assumption on the thermochemical equilibrium. The key of the approach is to consider the excited states of the different species as independent species. Therefore, they obey to conservation equations involving collisional-radiative contributions related to the other excited states. These contributions are in part due to the influence of electrons and heavy particles having a different translation temperature. This 'state-to-state' approach then enables the verification of the excitation equilibrium by analyzing Boltzmann plots. This approach leads finally to a thorough analysis of the progressive coupling until the equilibrium asymptotically observed.

Keywords: non-equilibrium; collisions; radiation; planetary atmospheric entry; laser matter interaction

1. Introduction

The question of the existence of the thermodynamic equilibrium is crucial in plasma physics [1,2]. Indeed, since the energy can be freely distributed over the different excited states, radiation can be easily emitted, which leads to discrepancies in terms of population with respect to the Boltzmann distribution. In addition, ionization or recombination deals with excited states whose population density cannot be easily estimated. Moreover, temporary species, whose density would be negligible in case of thermodynamic equilibrium, can be formed and can have a significant influence on the behavior of the plasma.

Many experimental and theoretical works have been devoted to plasma physics and the main objective of this communication is not to overview the field. This would be necessarily incomplete. Conversely, it is more interesting to focus our attention on unusual situations to enrich the analysis. The CORIA laboratory in France, with the collaboration of the French spatial agency CNES (Centre National d'Etudes Spatiales), works on plasmas produced during the planetary atmospheric entry of spatial vehicles. The CORIA laboratory works also on laser-induced plasmas in the framework of the multi-elemental composition determination based on the laser-induced breakdown spectroscopy (LIBS) technique with the CEA (Commissariat à l'Energie Atomique et aux Energies Alternatives). In these two cases, the plasma can depart significantly from thermodynamic equilibrium, and analyzing this departure helps to enlarge our understanding of the global behavior of the plasmas. This is why we propose in the present communication to focus our attention on these situations.

As a result, the structure of the communication is separated in two main parts. The first part is dedicated to planetary atmospheric entry plasmas and the second part is devoted to the laser-induced plasmas. In each part, the context is given as well as the main features of the related plasmas. Then, tools are presented to characterize the plasmas formed using relevant models.

2. Planetary Atmospheric Entry Plasmas

2.1. Context

Due to gravity, the speed of bodies coming from space can be high. Typically, this speed u_1 reaches several km s^{-1}. When the body approaches a planet having an atmosphere, the penetration of its upper layers generates the formation of a shock layer by compression around the forward part of the body [3,4]. The temperature in this layer increases significantly at levels reaching easily several 10^4 K. A gas to plasma transition takes place in the shock layer and leads to a strong heat transfer to the surface of the body. This heat transfer can be high enough to increase its surface temperature beyond the melting point and to cause the destruction of this surface. In the case of a manned-controlled mission, from the technological point of view, the covering of the spatial vehicle external surface by a thermal protection system (TPS) based on the use of ceramic materials is therefore mandatory.

Literature reports many entries in the Earth [5] or in the Mars [6] atmospheres. They can be controlled in the case of the flight of spatial vehicles, or totally uncontrolled in the case of natural bodies. For Earth, the human missions are well controlled and are particularly illustrated by the supplying of the International Space Station (ISS) at an altitude of $\sim$ 410 km. One of the most impressive events of natural entry took place on 15th February 2013 when a meteorite crossed the sky of the city of Chelyabinsk in Russia before to crash on the ground. Many cameras filmed this event, which clearly put in evidence the strong level of temperature reached in the shock layer through the strong emitted radiance.

2.2. Inside the Shock Layer

Figure 1 illustrates the production of the shock layer (thickness of $\Delta \sim$ 5 cm in order of magnitude) due to the fast external gas motion relative to the surface of a spatial vehicle. In particular, a fluid particle is followed along its trajectory. Entering the shock layer by crossing the detached shock front, its volume is strongly decreased due to the increase in pressure. As a result, the temperature increases and provokes chemical non-equilibrium leading to the global dissociation and ionization of the flow. If the trajectory is close to the stagnation streamline, the fluid particle enters the boundary layer in which the plasma flow gives energy to the body surface (in $x = \Delta$). In this part, the temperature decreases approaching the body, which leads to significant gradients. In this region, whose typical thickness is $\Delta - \delta \sim$ 1 cm, the recombination occurs. Consequently, these recombined species will interact with the TPS.

The thickness of these layers is pretty low. In addition, although the speed is decreased at the crossing of the shock front, the speed remains sufficiently high in the shock layer to prevent local thermodynamic equilibrium. Indeed, the three Damköhler numbers—Da_1, Da_2, and Da_3—defined as

$$Da_1 = \frac{\tau_u}{\tau_{CR}} \quad Da_2 = \frac{\tau_u}{\tau_{MB}} \quad Da_3 = \frac{\tau_u}{\tau_{e-h}} \tag{1}$$

are not much higher than unity. Since the speed is high, the value of the characteristic time scale for convection τ_u is weak. The time scale τ_{CR} for the collisional-radiative source term of the species is of the same order of magnitude, which leads to a Damköhler number Da_1 close to unity. The flow is therefore out of chemical equilibrium (in case of chemical equilibrium, $Da_1 \gg 1$) and not frozen (in case of frozen flow, $Da_1 \ll 1$). The time scale to reach a Maxwell–Boltzmann distribution for the translation (of electrons or heavy species) τ_{MB} is much shorter than τ_u, which leads to $Da_2 \gg 1$. As a result, the translation equilibrium is reached. However, the coupling between electrons and heavy species is difficult: the time scale of coupling τ_{e-h} is then higher than τ_u and the third Damkhöler number Da_3 is therefore lower than unity. Thus, the flow is out of thermal and chemical equilibrium, in other words the flow is in thermochemical non-equilibrium.

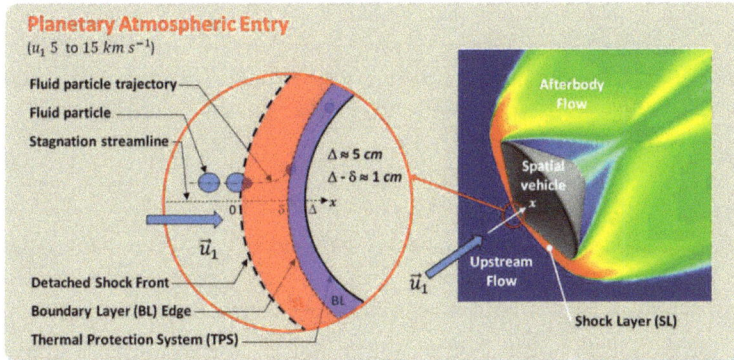

Figure 1. Global situation close to the TPS of a spatial vehicle showing the structure of the shock layer, the motion of fluid particles and the boundary layer. Δ is the typical thickness of the shock layer and $\Delta - \delta$ the one of the boundary layer.

These features have important consequences on the heat transfer to the surface. The presence of species not formed in case of chemical equilibrium and the spectral radiance emitted by the plasma lead to additional contributions. The local parietal heat flux density is then given by

$$
\varphi_w \approx \underset{Modes\ i}{\sum} k_i\ \vec{\nabla}\ T_i \cdot \vec{n}_w + \underset{Species\ j}{\sum} \gamma_j\ \beta_j\ \varphi_j^{(E)}
$$
$$
+ \underset{Frequency\ \nu}{\int} \alpha_\nu \underset{Plasma\ volume}{\iiint} \tau_\nu(\vec{r})\ \varepsilon_\nu(\vec{r})\cos\theta\ \frac{d^3r}{r^2}\ d\nu \tag{2}
$$

and can easily exceed 1 MW m^{-2}. In Equation (2), the first part of the right side refers to the influence of translation and internal modes transfer, the second part to parietal catalysis and the third part to radiation. k_i is the thermal conductivity for the mode i whose temperature is T_i, γ_j is the recombination probability for species j, β_j is the energy accommodation coefficient, α_ν is the spectral absorptivity, τ_ν is the spectral transmittivity along $\vec{r}$, r is the distance with the elementary volume d^3r, θ is the angle with respect to the normal vector $\vec{n}_w$ and ε_ν is the spectral emission coefficient.

The estimation of the wall heat flux density requires a detailed knowledge of the plasma upstream the boundary layer. The CORIA laboratory develops models to go deeper in its understanding. In the upcoming section, we focus our attention on the work dedicated to the EXOMARS mission.

2.3. The EXOMARS Mission

Recently, the European Space Agency (ESA) in cooperation with the Russian Space Agency (ROSCOSMOS) managed the EXOMARS mission whose first step was the landing on the ground of Mars of a rover on board of the Schiaparelli lander on 19th October 2016 [7]. The TPS of the lander was equipped with sensors called ICOTOM embedded in the COMARS+ housing whose role was to provide information on the radiative signature in the infrared part of the spectrum [8]. The related radiation is due to the production of hot CO_2 and CO during the entry in the Martian atmosphere and corresponds to the radiative contribution to φ_w in Equation (2) in the afterbody flow. This atmosphere is indeed mainly composed of CO_2 (95.97%), Ar (1.93%), and N_2 (1.89%). Crossing the shock front, this mixture is then put at high temperature and pressure. The composition then changes since this composition does not correspond to chemical equilibrium. To estimate chemical relaxation time scales

τ_{CR} of this mixture, we have developed models able to show how this relaxation occurs in a typical situation. Since the composition of the upstream atmosphere is well known, we focus our attention on the shock crossing when the heat transfer to the TPS is maximum, therefore over the first centimeters after the shock. This situation of "peak heating" corresponds to an altitude of 45 km [9].

2.4. Modeling of the Shock Front Crossing

Since the excitation equilibrium condition is not systematically fulfilled, the modeling is based on the solution of the excited states population density balance equation written as

$$\frac{dy_{X_m}}{dx} = \frac{m_{X_m} \left[\dot{X}_m \right]}{\rho \, u} \tag{3}$$

assuming negligible the diffusion phenomena within the post-shock flow at steady-state. In Equation (3), y_{X_m} is the mass fraction of the species X on its excited state m whose mass is m_{X_m}. The collisional-radiative term $\left[\dot{X}_m \right]$ results from the influence of all the elementary inelastic/superelastic processes enabling a change in the population density $[X_m]$. The mass flow density is written ρ.

Equation (3) is coupled with the momentum balance equation

$$\frac{d(p + \rho u^2)}{dx} = 0 \tag{4}$$

whose pressure p is calculated by the Dalton law $p = \Sigma_{X,m}[X_m] \, k_B \, T_X$ assuming a perfect gas-like behavior.

Equations (3) and (4) are finally coupled with the energy balance of heavy particles

$$\frac{d}{dx}\left[\frac{e_A}{\rho} + \frac{p_A}{\rho} + \frac{\rho_A}{\rho} \frac{u^2}{2} \right] = \frac{-Q_{A\to e} - \varepsilon_{SE} - Q_{A-RR}}{\rho \, u} \tag{5}$$

and of electrons

$$\frac{d}{dx}\left[\frac{e_e}{\rho} + \frac{p_e}{\rho} + \frac{\rho_e}{\rho} \frac{u^2}{2} \right] = \frac{Q_{A\to e} - \varepsilon_{RR} + Q_{A-RR}}{\rho \, u} \tag{6}$$

These equations must be considered separately because the energy per unit volume e_A and e_e for heavies and electrons depend on the heavy species temperature T_A and on the electron temperature T_e, respectively. The Dalton law $p = p_A + p_e$ is obvious and the mass density results from the summation of the contributions of heavy species ρ_A and electrons ρ_e. Inelastic/superelastic elementary processes are considered through the term $Q_{A\to e}$ while the term Q_{A-RR} results from the influence of the radiative recombination whose emission coefficient is ε_{RR}. The spontaneous emission is related to the emission coefficient ε_{SE}.

The complexity of the upstream flow composed of CO_2, N_2, and Ar induces a very complex chemistry past the shock front. To be relevant, this chemistry must include enough species that can be formed in the post-shock conditions based on C, O, and N atoms. The pressure conditions are insufficient to produce Ar_2^+ dimers. The species taken into account in the resulting CoRaM-MARS collisional-radiative model are listed in Table 1. This list involves 21 species and electrons, 1600 excited vibrational and electronic excited states. All the vibrational states of the molecular electronic ground states are taken into account to reproduce realistically the global dissociation processes.

When the mixture crosses the shock front, a drastic reduction of the mean free path takes place due to the strong increase in mass density ρ. The typical thickness of a shock front is of several mean free paths. In a first approximation, we can consider this increase so rapid that this corresponds to a discontinuity. The Rankine–Hugoniot jump conditions at the shock front can then be used, where the chemistry is frozen. Electron temperature T_e remains unchanged because the rare electrons are not perturbed by the shock front. Indeed, due to their very weak mass, the electrons are in a quite subsonic flow regime. Heavy species temperature T_A is conversely strongly increased at a level incompatible with the upstream chemical conditions. Chemistry then starts, which modifies the composition as a result of the elementary processes, first due to heavy species impact, the electron density n_e being too weak. When n_e is high enough, the elementary processes are driven by the electron-induced collisions. These elementary processes are collisional (vibrational excitation, dissociation, electronic excitation, ionization, excitation transfer, neutral and charge exchanges, and backward processes driven by the detailed balance principle), radiative (spontaneous emission) or mixt (radiative recombination, photo-ionization, self-absorption). This underlying chemistry represents a set of around 10^6 elementary reactions [10].

Table 1. List of the species and their excited states involved in CoRaM-MARS, the CR model developed at the CORIA laboratory for the CO_2-N_2-Ar mixtures.

Species	States
CO_2	$X^1\Sigma_g^+$ (14 states (v_1,v_2,v_3) with $E_v < 0.8$ eV, 106 states (i00,0j0,00k) with $E_v > 0.8$ eV), $^3\Sigma_u^+$, $^3\Delta_u$, $^3\Sigma_u^-$
N_2	$X^1\Sigma_g^+$ ($v = 0 \rightarrow 67$), $A^3\Sigma_u^+$, $B^3\Pi_g$, $W^3\Delta_u$, $B'^3\Sigma_u^-$, $a'^1\Sigma_u^-$, $a^1\Pi_g$, $w^1\Delta_u$, $G^3\Delta_g$, $C^3\Pi_u$, $E^3\Sigma_g^+$
O_2	$X^3\Sigma_g^-$ ($v = 0 \rightarrow 46$), $a^1\Delta_g$, $b^1\Sigma_g^+$, $c^1\Sigma_u^-$, $A'^3\Delta_u$, $A^3\Sigma_u^+$, $B^3\Sigma_u^-$, $f^1\Sigma_u^+$
C_2	$X^1\Sigma_g^+$ ($v = 0 \rightarrow 36$), $a^3\Pi_u$, $b^3\Sigma_g^-$, $A^1\Pi_u$, $c^3\Sigma_u^+$, $d^3\Pi_g$, $C^1\Pi_g$, $e^3\Pi_g$, $D^1\Sigma_u^+$
NO	$X^2\Pi$ ($v = 0 \rightarrow 53$), $a^4\Pi$, $A^2\Sigma^+$, $B^2\Pi$, $b^4\Sigma^-$, $C^2\Pi$, $D^2\Sigma^+$, $B'^2\Delta$, $E^2\Sigma^+$, $F^2\Delta$
CO	$X^1\Sigma^+$ ($v = 0 \rightarrow 76$), $a^3\Pi$, $a'^3\Sigma^+$, $d^3\Delta$, $e^3\Sigma^-$, $A^1\Pi$, $I^1\Sigma^-$, $D^1\Delta^-$, $b^3\Sigma^+$, $B^1\Sigma^+$
CN	$X^2\Sigma^+$ ($v = 0 \rightarrow 41$), $A^2\Pi$, $B^2\Sigma^+$, $D^2\Pi$, $E^2\Sigma^+$, $F^2\Delta$
N_2^+	$X^2\Sigma_g^+$, $A^2\Pi_u$, $B^2\Sigma_u^+$, $a^4\Sigma_u^+$, $D^2\Pi_g$, $C^2\Sigma_u^+$
O_2^+	$X^2\Pi_g$, $a^4\Pi_u$, $A^2\Pi_u$, $b^4\Sigma_g^-$
C_2^+	$X^4\Sigma_g^-$, $1^2\Pi_u$, $^4\Pi_u$, $1^2\Sigma_g^+$, $2^2\Pi_u$, $B^4\Sigma_u^-$, $1^2\Sigma_u^+$
NO^+	$X^1\Sigma^+$, $a^3\Sigma^+$, $b^3\Pi$, $W^3\Delta$, $b'^3\Sigma^-$, $A'^1\Sigma^+$, $W^1\Delta$, $A^1\Pi$
CO^+	$X^2\Sigma^+$, $A^2\Pi$, $B^2\Sigma$, $C^2\Delta$
CN^+	$X^1\Sigma^+$, $a^3\Pi$, $^1\Delta$, $c^1\Sigma^+$
N	$^4S^\circ_{3/2}$, $^2D^\circ_{5/2}$, $^2D^\circ_{3/2}$, $^2P^\circ_{1/2}$, ... (252 states)
O	$^3P_2$, $^3P_1$, $^3P_0$, $^1D_2$ ... (127 states)
C	$^3P_0$, $^3P_1$, $^3P_2$, $^1D_2$ ... (265 states)
Ar	$^1S_0$, $^2[3/2]^\circ_2$, $^2[3/2]^\circ_1$, $^2[1/2]^\circ_0$, ... (379 states)
N^+	$^3P_0$, $^3P_1$, $^3P_2$, $^1D_2$ ... (9 states)
O^+	$^4S^\circ_{3/2}$, $^2D^\circ_{5/2}$, $^2D^\circ_{3/2}$, $^2P^\circ_{3/2}$, ... (8 states)
C^+	$^2P^\circ_{1/2}$, $^2P^\circ_{3/2}$, $^4P_{1/2}$, $^4P_{3/2}$, ... (8 states)
Ar^+	$^2P^\circ_{3/2}$, $^2P^\circ_{1/2}$, $^2S_{1/2}$, $^4D_{7/2}$, ... (7 states)

2.5. Some Results

Due to the upstream conditions in terms of pressure, temperature and speed relative to the spatial vehicle when the peak heating occurs, the crossing of the shock front in $x = 0$ induces high values of temperature and pressure. Table 2 gives the jump conditions. We can see that the temperature is clearly incompatible with an insignificant dissociation degree for CO_2. The collision frequency and the energy available in the collisions then start the chemistry.

Table 2. Jump conditions in $x = 0$ due to the Rankine–Hugoniot assumption at 45 km altitude corresponding to the peak heating.

Variable	Upstream Conditions	Conditions in $x = 0$
Speed (m s^{-1})	5270	690
Mach number	26.4	0.34
Pressure (Pa)	7.6	6000
Temperature (K)	162	16,800

Figure 2 displays the resulting distribution of the aerodynamic variables (pressure, mass density, and speed). Even if Equations (3)–(6) are available only in the region where the diffusion phenomena are negligible (typically before the boundary layer, for $x < \delta \approx 5$ cm), we have displayed all the relaxation as we could observe in shock tubes, to see the final convergence of the flow. Since the boundary layer starts at $\delta \approx 5$ cm, only the first centimeters of the solution displayed on Figure 1 are relevant with respect to the real situation.

Figure 2 clearly shows that the relaxation is still in progress at $\delta \approx 5$ cm. This is also clearly shown by Figure 3 where the Boltzmann plots of the $CO_2(i, 0, 0)$ vibrational states is displayed as a function of the position from the shock front. The vibrational distribution is far from being linear, that reveals a departure from vibrational excitation equilibrium.

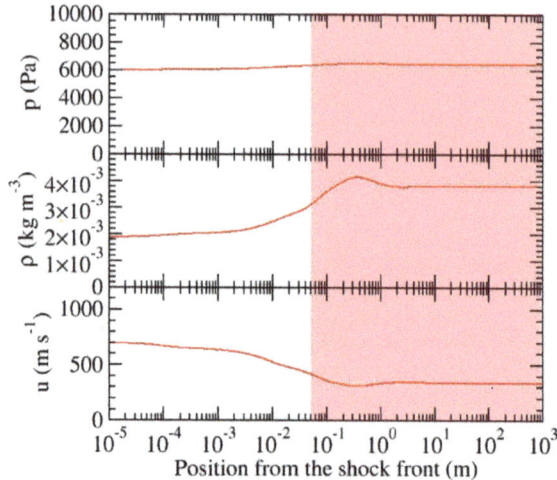

Figure 2. Post-shock relaxation for the pressure, the mass density and the speed resulting from Equations (3)–(6). Since the diffusion phenomena are assumed negligible in these equations, the solution corresponds to the real flow before $x = \delta \approx 5$ cm.

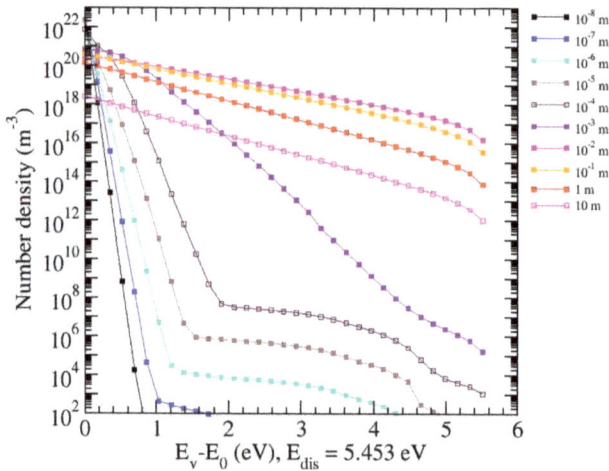

Figure 3. Evolution with the position from the shock front (indicated on the right) of the Boltzmann plot of the $CO_2(i, 0, 0)$ vibrational states. The vibrational excitation energy relative to the ground state is given in abscissa. The dissociation energy of the CO_2 molecule is reminded.

The way to the dissociation is mainly driven by the excited vibrational states close to the dissociation limit [11]. As a result, these distributions prove that the dissociation equilibrium is not reached. Indeed, the distribution of the species number densities displayed on Figure 4 illustrates the current global dissociation process. The formation of CO and O resulting from the dissociation of CO_2 is clearly observed. The dissociation degree is close to 0.01 at 1 cm after the shock front and to 0.35 at $\delta \approx 5$ cm. Molecular and atomic ions start to be produced just after the dissociation of CO_2. A maximum electron density of 1.8×10^{19} m^{-3} is obtained at a location close to 4 cm and corresponds to an ionization degree of $\sim 4 \times 10^{-4}$. Even if this amount seems to be rather small, the electron density is nevertheless high enough to significantly influence the chemistry. Indeed, due to their weak mass, the efficiency of electrons in terms of inelastic/superelastic collisions is much stronger than the one of the heavy particles. However, this influence is reduced since the electron temperature T_e is lower than the heavy particle temperature T_A.

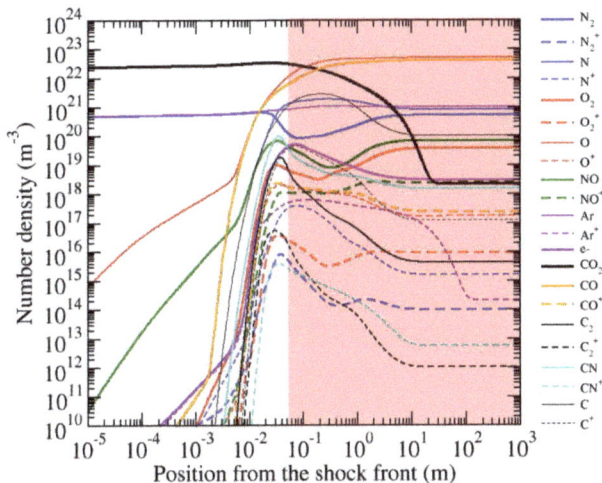

Figure 4. Distribution of the number density of the different species taken into account in the chemistry (see Table 1) behind the shock front. Same as Figure 2: the solution is relevant with respect to the real situation until 5 cm from the shock front.

Figure 5 illustrates the distribution of temperatures. T_A and T_e have been plotted. We have also plotted the distribution of the post-processed values of the energy-dependent vibrational temperature for each vibrational mode of CO_2 resulting from our vibrational state-to-state approach. The total energy-dependent vibrational temperature has been also determined.

This energy-dependent vibrational temperature $T_{vib}^E(CO_2)\_i$ for a mode i is derived from the vibrational energy per unit volume $e_{vib}(CO_2)\_i$ by the equation

$$e_{vib}(CO_2)\_i = \sum_v [CO_2(i,v)]E_v = \frac{\sum_v [CO_2(i,v)] \cdot \sum_v g_v \, E_v \, e^{-\frac{E_v}{k_B \, T_{vib}^E(CO_2)\_i}}}{\sum_v g_v \, e^{-\frac{E_v}{k_B \, T_{vib}^E(CO_2)\_i}}} \tag{7}$$

In the case of the mean vibrational temperature $T_{vib}(CO_2)$, the summations of Equation (7) are extended to the vibrational modes.

On Figure 5, we can see the progressive coupling between the three modes with the distance from the shock front. No one is perfectly coupled with the translation temperature of electrons or heavy particles before the limit of the boundary layer. We can also observe that the temperature departure between the first (symmetric stretching) mode and the second (bending) mode is pretty low. This is mainly due to the Fermi resonance resulting from the energy diagram. The quasi resonance between the related states leads to easy excitation transfer between them. The third (asymmetric stretching) mode is more difficult to excite and remains at temperature rather low. It is particularly interesting to see that, despite the thermal non-equilibrium between these states, they do not contribute in the same way to the mean vibrational temperature. Indeed, Figure 5 shows that this temperature follows the vibrational temperature of the second (bending) mode. This is mainly due to the degeneracy of the states of the second vibrational mode. While the first and third vibrational modes are not degenerated, the second mode presents a degeneracy equal to $v_2 + 1$ resulting from the rotational motion induced by the bending of the molecule. One finds further molecules per unit volume in the related states and the global vibrational temperature follows the one of this mode.

The differences observed on Figure 5 between the vibrational modes of CO_2 are the result of the thermal non-equilibrium $T_e \neq T_A$ and of the efficiency of electrons and heavies in terms of collisions. In addition, the electrons and heavy particles dynamics is deeply different since electrons are produced and heated behind the shock front while the heavies leave their energy along the flow where the global dissociation process takes place. This corresponds therefore to a strong non-equilibrium situation. This situation relaxes over typical length scales longer than the shock layer thickness as illustrated by Figure 5. The thermal non-equilibrium would be resorbed around 1 m from the shock front in case of infinite shock layer thickness. Since $\delta \approx 5$ cm, we conclude to a limit of the boundary layer departing from thermal equilibrium. This conclusion departs from the usual one considered for the case of Earth atmospheric entries [12].

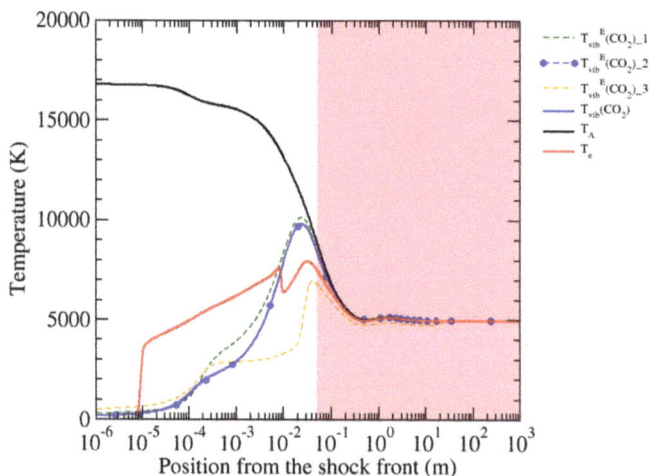

Figure 5. Distribution of the electron temperature T_e, the heavy particle temperature T_A, and the post-processed energy-dependent vibrational temperature of CO_2 for the first (symmetric stretching) mode, second (bending) mode and third (asymmetric stretching) mode. The total energy-dependent vibrational temperature is also displayed. The limit of the boundary layer is located at $\delta \approx 5$ cm: in the red region, feature of the flow in case of infinite shock layer thickness.

3. Laser-Induced Plasmas

3.1. Context

The laser-induced plasmas correspond also to an interesting situation where thermodynamic non-equilibrium plays an important role. These plasmas are produced when a (typically nanosecond) laser pulse reaches a sample at a spectral irradiance of 10^{13}–10^{14} W m^{-2} higher than the breakdown threshold (see Figure 6). The absorption of the laser energy leads to a strong increase in the local temperature of the sample at values (of the order of several 10^4 K) exceeding largely the conditions of a phase change. A plasma is indeed produced whose pressure is initially very high (of the order of several 10^{11} Pa) with respect to the background gas one. The produced plasma then expands according to a hypersonic regime (the Mach number can reach values of the order of 25), produces a shock wave propagating in the background gas and cools mainly owing to the radiative losses. This leads to a physical object having lifetimes of the order of several μs.

The atoms and ions composing the plasma are initially inside the sample. Once the plasma relaxed, a crater is formed where the laser pulse reached the sample. The radiative signature of the plasma can therefore give valuable information about the composition of the sample. This explains why this laser–matter interaction is at the basis of the laser-induced breakdown spectroscopy (LIBS) technique to determine the multi-elemental composition of samples. Measuring the spectral radiance of relevant lines, it is possible to derive the relative population of the different species if thermochemical equilibrium conditions are fulfilled [13].

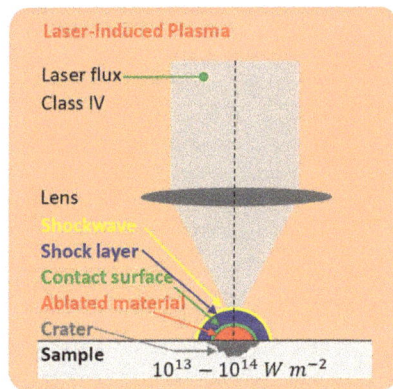

Figure 6. Laser-induced plasma situation. The laser pulse is focused on the sample using a converging lens at an irradiance higher than the breakdown threshold. The ablated material expands according to a hypersonic regime and produces a shock wave propagating in the background gas. As a result, two layers are formed. The first one corresponds to the ablated material and the second one corresponds to the shock layer. These two layers are separated by a contact surface across which the diffusion phenomena can be considered as negligible in a first approximation.

3.2. Possible Non-Equilibrium Situation

The pulse duration plays an important role on the interaction. In the case of ultrashort (fs or ps) pulses, the laser energy is directly deposited within the material and leads to thermal non-equilibrium because electrons and heavies have not enough time to be coupled. In the case of ns laser pulses, the end of the pulse is absorbed by the plasma in expansion with a good coupling between electrons and heavies. In addition, thermal effects due to heat diffusion within the material can be observed for ns laser pulses and can produce micro-droplets, contrary to the case of the ultrashort laser pulses where nanoparticles can be observed. This explains the use of fs laser sources for the micromachining devices.

In terms of ablation precision, it is therefore better to use ultrashort laser pulses. According to the experimental setup used, a nominal ablation rate as low as some 10 nm pulse^{-1} can be reached. This means that the matter forming the laser-induced plasma is in low amount. Typically, experiments are performed by accumulating signals over a tenth of pulses. Then the net minimum ablation rate is of 100 nm. A depth profiling of the sample is therefore possible if the spectral radiance of the relevant lines is high enough to provide significant results. We consider only picosecond laser pulses in the upcoming sections. As a result, the absorption of the laser pulse by the expanding plasma is totally avoided.

In some specific applications, experiments must be performed in low pressure conditions. This corresponds to in situ measurements if the sample cannot be removed for the analysis. If the matter is radioactive or toxic, low-pressure experiments are often better to avoid any dissemination and to keep safe the environment. Then the plasma expands freely. The collision frequency collapses and the analysis of the situation in the light of the Damkhöler numbers performed in Section 2.2 then reveals that the equilibrium conditions are not satisfied. The LIBS determination of the multi-elemental composition of the sample cannot be directly and easily performed. Developing state-to-state approaches may be valuable in this context [14].

3.3. Tokamak and Tungsten

The tungsten tiles of the divertor of a tokamak like WEST from the CEA Cadarache or ITER must be kept inside the machine as much as possible. A possible LIBS analysis of the fuel (hydrogen isotopes) contamination within these tungsten tiles must therefore be performed in low pressure conditions, at a maximum pressure of ~ 10 Pa.

In this context, the CORIA laboratory develops modeling tools similar to those developed for Section 1. The structure of the plasma flow is close to the one developed around the surface of a spatial vehicle. The main difference results from the geometry of the flow. Fundamentally, the entry plasma is a 1D flow regarding the stagnation stream line. Conversely, the laser-induced plasma is a 3D flow but with a hemispherical symmetry and a radial dependency much higher than for the other coordinates. In a first approximation, this flow can be considered as made of two layers separated by a contact surface (see Figure 6) separating the matter ablated from the sample and the shock layer. This shock layer corresponds to the background gas across which the shock front has propagated since the laser pulse. As a result, its external limit corresponds to the shock front. In a second approximation, we can assume these two parts as uniform [15].

For tokamak studies, we are working on tungsten. For comparison with laboratory studies, we focus our attention on the modeling of the behavior of tungsten laser-induced plasmas in rare gases. We have therefore elaborated collisional-radiative models based on state-to-state descriptions of tungsten and of the retained rare gas. The rare gas is denoted as Rg in the upcoming sections. In the shock layer, electrons and the species Rg, Rg^+, and Rg_2^+ can be found on their different excited states. The pressure in the shock layer can be high enough to promote the formation of the dimer molecule Rg_2^+. In the central plasma, electrons, W, W^+, and W^{2+} have been considered.

The knowledge of the electronic excited states structure of W is satisfactory. This is not the case for the ions. The last known W^+ excited state corresponds to an excitation energy of 9.23 eV in the NIST database while the ionization limit is 16.37 eV [16]. We have therefore assumed a hydrogen-like behavior up to the ionization limit. Moreover, to reduce the total number of excited states considered in the conservation equations, the classical lumping procedure has been performed. It consists in the grouping of states sufficiently close in terms of energy. The statistical weight of the grouped levels is taken as the summation of those of the individual levels [17].

Table 3 lists the species and the excited states finally accounted for tungsten and rare gas in the case the rare gas is argon. 230 different excited states are considered.

Table 3. List of the species and their excited states involved in the CR model CoRaM-Ar and CoRaM-W developed at the CORIA laboratory for the laser-induced plasmas on W in a rare gas (here for argon).

Plasma Layer	Species	States
(1) shock layer	Ar	$^1S_0, {}^2[3/2]^\circ_2, {}^2[3/2]^\circ_1, {}^2[1/2]^\circ_0, \ldots$ (90 states)
	Ar^+	$^2P^\circ_{3/2}, {}^2P^\circ_{1/2}, {}^2S_{1/2}, {}^4D_{7/2}, \ldots$ (7 states)
	Ar_2^+	$X^2\Sigma_u^+$
(2) central plasma	W	$^5D_0, {}^5D_1, {}^5D_2, {}^5D_3, \ldots$ (60 states)
	W^+	$^6D_{1/2}, {}^6D_{3/2}, {}^6D_{5/2}, {}^6D_{7/2}, \ldots$ (74 states)
	W^{2+}	$^5D_{0\ldots4}, {}^3P2_{0\ldots2}, {}^5F_{1\ldots5}, {}^3H_{4\ldots6}, {}^3F2_{2\ldots4}$ (5 states)

3.4. Collisional and Radiative Processes in the State-to-State Approach

Since the layers are assumed uniform, equations similar to Equations (3)–(6) written in hemispherical symmetry can be spatially integrated to obtain pure temporal equations. Coupled with the shock propagation from the sample at a speed driven by the Rankine–Hugoniot assumptions, they lead to a system of non-linear ordinary differential equations whose solution can be derived.

In the source terms of the related equations, the spontaneous emission is accounted for. As for the energy diagram of W^+, a lack of elementary data (Einstein coefficients) can be observed. This induces an underestimate of the radiative losses. The radiative recombination is also accounted for. In each layer, electrons and heavies are assumed Maxwellian, but at a different temperature. These particles collide with the different species on their excited states, which leads to their excitation, deexcitation, ionization, and recombination. Each elementary process is considered with its backward process using the detailed balance principle. The derived collisional-radiative model involves almost 550,000 elementary reactions, therefore an order of magnitude similar to atmospheric entry calculations. This number is lower than for atmospheric entry because a lower number of species is involved. In addition, except Rg_2^+ for which the chemistry is simple since no vibrational state is considered, no molecule is concerned.

3.5. Results

We consider the classical laser conditions 10 mJ, 30 ps with a wavelength of 532 nm in argon at atmospheric pressure. The ablated mass is then of the order of 10^{-10} kg by pulse. The laser pulse duration is shorter than the typical time scale of expansion of the plasma and the deposited energy does not diffuse significantly within the sample. As a result, the pulse energy is totally given to the ablated mass. Its initial temperature and pressure are then quite high. They induce the subsequent evolution of the plasma.

Figure 7 illustrates the pressure evolution in the central plasma and in the shock layer. Due to the initial pressure of the central tungsten plasma, the expansion starts around 1 ns and induces the compression of the external background gas whose pressure increases in the shock layer. The expansion leads to the decrease in the pressure of the central plasma until sufficient inversion with respect to the shock layer. Then a recompression of the central plasma due to the shock layer takes place before a coupling between the two layers from the pressure point of view observed along the remaining part of the evolution.

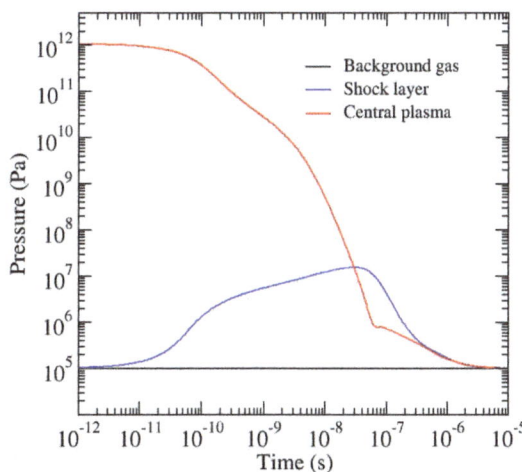

Figure 7. Evolution of the pressure inside the central tungsten plasma and inside the shock layer (case of argon) for a classical laser (10 mJ, 30 ps, 532 nm)-induced plasma experiment at atmospheric pressure.

In the framework of the present assumptions in terms of flow continuity, a minimum pressure of 10 Pa can be considered for argon. Figure 8 illustrates the pressure evolution of the layers in this case. We see that the recompression does not take place. The outside pressure is too low to ensure the confinement of the plasma. Its lifetime is therefore considerably shortened.

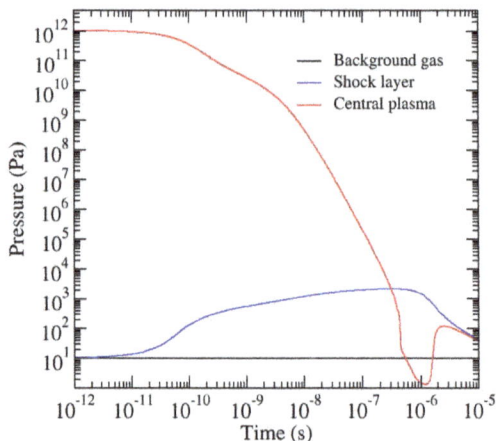

Figure 8. Same as Figure 7, but with an argon pressure of 10 Pa.

These trends can be also observed on the temperature evolutions of the different layers. Figure 9 illustrates the results for an argon gas at atmospheric pressure and Figure 10 those obtained at 10 Pa. It is interesting to see on Figure 9 that the thermal coupling resulting from the elastic collisions is efficient in the central plasma due to the high level of pressure. This is not the case for the shock layer where $T_e \neq T_A$ along almost the complete evolution. In the case of a 10 Pa pressure for the background gas, the thermal coupling is satisfactory in the central plasma until a characteristic time of the order of 100 ns. Before this time, the evolution is quite the same as the one obtained at atmospheric pressure until $t < 30$ ns. The plasma evolves independently from the presence of the background gas. Then, the pressure has sufficiently decreased, and the collision frequency is too weak to ensure the coupling between T_e and T_A. Electron density is very weak and the energy of the plasma is mainly stored in the kinetic energy due to expansion. Internal energy collapses: temperature T_A rapidly decreases.

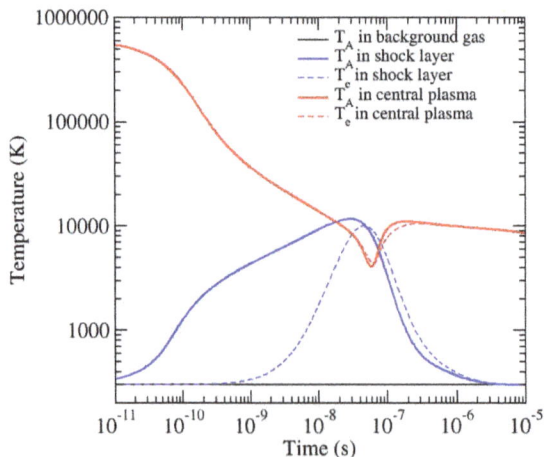

Figure 9. Same as Figure 7, but for the temperatures.

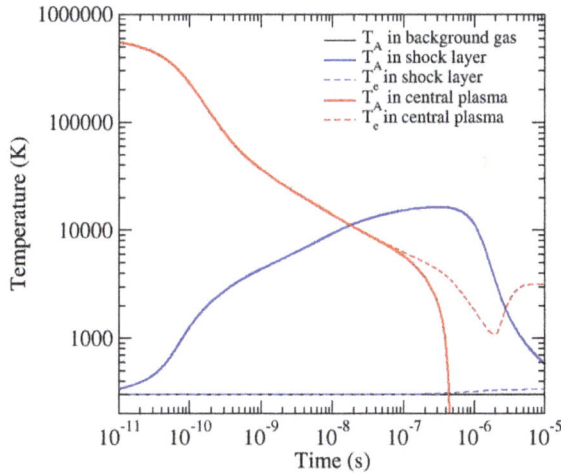

Figure 10. Same as Figure 8, but for the temperatures.

Our state-to-state approach enables the analysis of the departure from excitation equilibrium. Figure 11 displays the Boltzmann plots of the excited states of W and W$^+$ in the central plasma at 200 and 300 ns at atmospheric pressure. Figure 12 displays those related to a 10 Pa argon gas.

On Figure 11, we clearly see that equilibrium is reached. The distribution is perfectly linear. We can also see that temperature is high since W^{2+} ions have a density of the order of 10^{17} m^{-3}, but temperature is too weak to influence the electron density n_e. Indeed, we have $n_e = [W^+] \cong 9 \times 10^{23}$ m^{-3}. At 300 ns, the situation is almost the same. A weak decrease in n_e can be observed. This means that the collisional frequency is high enough to maintain in time the plasma situation.

When the argon background gas pressure is decreased at 10 Pa, the situation is deeply modified. We can see that the main slope of the distribution of neutral or ionic excited states is more negative. The excitation temperature is therefore lower. Electron density is decreased with respect to the atmospheric pressure situation. Indeed, the electron density reaches 5×10^{21} m^{-3} at 200 ns, and 1.4×10^{21} m^{-3} at 300 ns. Moreover, we can see that the distribution departs from a linear behavior. The excited states just below the ionization limit on a 2 eV interval are satisfactorily coupled according to a linear distribution, whereas the lower excited states have a fluctuating behavior increasing with time. We are, in the present case, in a strong recombination situation where the recombination induces a satisfactory coupling of the involved excited states at number density values higher than those expected due to the lower excited states. This very common behavior has been already observed in other situations. In an ionization situation similar to post-shock flows observed in atmospheric entries, the ionization induces the fast depopulating of the excited states of atoms close to the ionization limit, which leads to a depletion of these states in terms of density. Then, this is the exact symmetric case.

Over the whole energy diagram, the distribution cannot be linear. We have also to analyze the influence of radiation. At 200 ns, the order of magnitude of the number density of the excited states close to 5 eV is low. We have $\frac{[W_k]}{g_k} \cong 10^{16}$–$10^{17}$ m^{-3} whereas $\frac{[W_k]}{g_k} \cong 10^{19}$–$10^{20}$ m^{-3} for argon at atmospheric pressure. In these low density conditions, the influence of radiation is much more significant. Radiation causes departures from a linear distribution whose linearity cannot be recovered by the collisional coupling. At 300 ns, the situation is worse since the collisional frequency is collapsing.

These behaviors have important consequences regarding the LIBS diagnostic. In case the LIBS experiments are performed at atmospheric pressure, the equilibrium is rapidly obtained. As a result, the estimate of the excited states population density and of the electron density is sufficient to derive the ground state number density. Using the same number of Boltzmann plots as the number of different species, the composition of the plasma, therefore of the sample, can be identified.

In the case when the experiments are performed at low pressure, the analysis is considerably more difficult. Indeed, it is not directly possible to derive the ground states number density from the analysis of the radiation produced during the deexcitation of the excited states. The distribution of the excited states departs from excitation equilibrium. Then, the development of collisional-radiative models based on state-to-state approaches is therefore mandatory, except if known samples are available whose composition similar to the one to be obtained have been previously determined. In that case, the composition will be directly derived from comparisons with these calibrated samples.

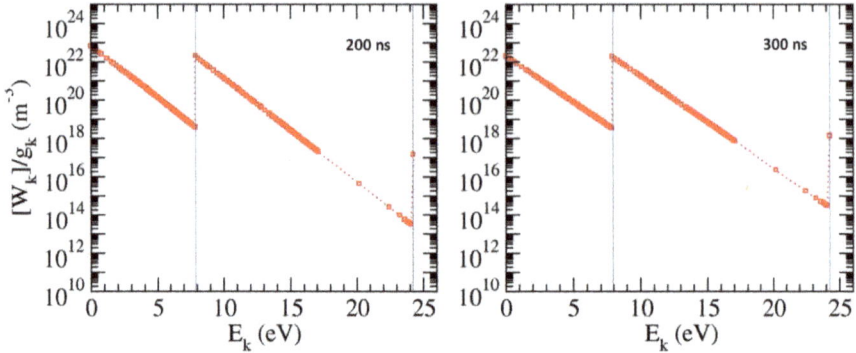

Figure 11. Boltzmann plots at 200 and 300 ns for an argon gas at atmospheric pressure. The first (7.86 eV) and second ionization (24.23 eV) limits are indicated by a vertical blue line. Each state is represented by a square. We see the added states following a hydrogen-like assumption between 20 and 24 eV.

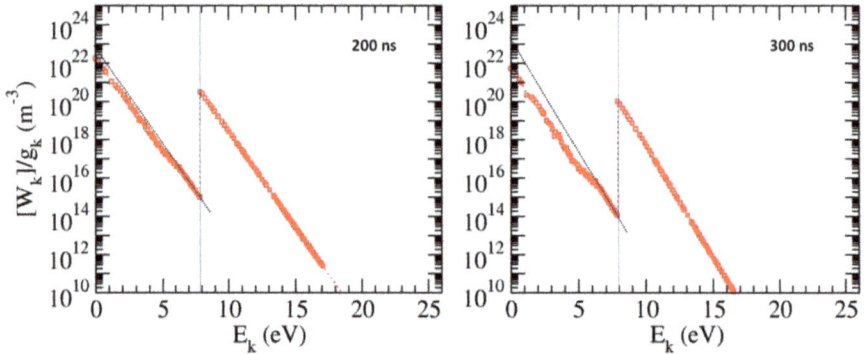

Figure 12. Same as Figure 11, but with an argon gas at 10 Pa. The second ionization limit is not displayed since the corresponding number densities are very weak. The line interpolating the excited states just below the first ionization limit is plotted to easily estimate the departure from excitation equilibrium.

4. Conclusions

Our objective was to give the main information regarding the underlying physics involved by two examples of plasma flows departing from thermochemical equilibrium. The cases of the planetary atmospheric entry plasmas and of the laser-induced plasmas have been discussed.

Modeling strategies have been detailed for two particular situations. For the atmospheric entry plasmas, we have focused our attention on the Martian missions of the EXOMARS type. For the laser-induced plasmas, we have detailed the properties of the plasma flow produced during a LIBS experiment on a divertor tungsten tile. Since the excitation equilibrium condition is not fulfilled, the only relevant way is to develop a state-to-state description of the species, i.e., to consider each excited state of the species as an independent variable and to solve the conservation equations in this framework. This requires the elaboration of a collisional-radiative model taking into account the different elementary processes at the level of each state. This strategy needs enough numerical means since the number of excited species and of elementary processes is often prohibitive.

The results show that the excitation equilibrium can be observed if the collisional frequency is high enough to overcome the perturbative role of radiation. As a result, even in case of thermal non-equilibrium, the observed excitation equilibrium takes place at the translational temperature of the main collision partner.

Author Contributions: The authors have jointly contributed to the work described in this paper.

Funding: This work has been carried out within the framework of the ICOTOM project funded by the French Space Agency CNES (Centre National d'Etudes Spatiales). This work has been also carried out within the framework of the French Federation for Magnetic Fusion Studies (FR-FCM) and of the Eurofusion consortium, and has received funding from the Euratom research and training programme 2014–2018 and 2019–2020 under grant agreement no. 633053. This work has been also carried out within the framework of the TRANSAT project and has received funding from the Euratom Research and Training Programme 2014–2018 under grant agreement no. 754586. The views and opinions expressed herein do not necessarily reflect those of the European Commission. This work has been also funded by the French ANR (Agence Nationale de la Recherche) through the programme "Investissement d'Avenir" (ANR-10-LABX-09-01), LabEx EMC^3, PICOLIBS project.

Conflicts of Interest: The authors declare no conflict of interest.

References

1. Capitelli, M.; Armenise, I.; Bruno, D.; Cacciatore, M.; Celiberto, R.; Colonna, G.; De Pascale, O.; Diomede, P.; Esposito, F.; Gorse, C.; et al. Non-equilibrium plasma kinetics: A state-to-state approach. *Plasma Sources Sci. Technol.* **2007**, *16*, 830. [CrossRef]

2. Capitelli, M.; Celiberto, R.; Colonna, G.; D'Ammando, G.; De Pascale, O.; Diomede, P.; Esposito, F.; Gorse, C.; Laricchiuta, A.; Longo, S.; et al. Plasma kinetics in molecular plasmas and modeling of reentry plasmas. *Plasma Phys. Control. Fusion* **2011**, *53*, 124007. [CrossRef]

3. Anderson, J.D. *Hypersonic and High Temperature Gas Dynamics*; McGraw-Hill: New York, NY, USA, 1988; ISBN 9780070016729.

4. Park, C. *Nonequilibrium Hypersonic Aerothermodynamics*; Wiley: New York, NY, USA, 1990; ISBN 978-0471510932.

5. Cooke, R.D.; Engle, J.H. Entry flight testing the Space Shuttle, 1977 to 1984. In Proceedings of the AIAA Flight Testing Conference, AIAA AVIATION Forum, AIAA 2016-3357, Washington, DC, USA, 13–17 June 2016. [CrossRef]

6. West, T.K.; Brandis, A.M. Updated stagnation point aeroheating correlations for Mars entry. In Proceedings of the 2018 Joint Thermophysics and Heat Transfer Conference, AIAA AVIATION Forum, AIAA 2018-3767, Atlanta, GE, USA, 25–29 June 2018. [CrossRef]

7. Aboudan, A.; Colombatti, G.; Bettanini, C.; Ferri, F.; Lewis, S.; Van Hove, B.; Karatekin, O.; Debei, S. ExoMars 2016 Schiaparelli Module Trajectory and Atmospheric Profiles Reconstruction. *Space Sci. Rev.* **2018**, *214*, 97. [CrossRef]

8. Gülhan, A.; Thiele, T.; Siebe, F.; Kronen, R. Combined Instrumentation Package COMARS+ for the ExoMars Schiaparelli Lander. *Space Sci. Rev.* **2018**, *214*, 12. [CrossRef]

9. Le Brun, A.L.; Omaly, P. Investigation of radiative heat fluxes for Exomars entry in the Martian atmosphere. *ESA Spec. Publ.* **2011**, *689*, 22.

10. Bultel, A.; Annaloro, J. Elaboration of collisional-radiative models for flows related to planetary entries into the Earth and Mars atmospheres. *Plasma Sources Sci. Technol.* **2013**, *22*, 025008. [CrossRef]

11. Annaloro, J.; Bultel, A. Vibrational and electronic collisional-radiative model in air for Earth entry problems. *Phys. Plasmas* **2014**, *21*, 123512. [CrossRef]

12. Bourdon, A.; Bultel, A. Numerical simulation of stagnation line nonequilibrium airflows for reentry applications. *J. Thermophys. Heat Transf.* **2008**, *22*, 168–177. [CrossRef]

13. Miziolek, A.W.; Palleschi, V.; Schechter, I. *Laser-Induced Breakdown Spectroscopy*; Cambridge University Press: Cambridge, UK, 2006; ISBN B000SRGFR2.

14. Morel, V.; Bultel, A.; Schneider, I.F.; Grisolia, C. State-to-state modeling of ultrashort laser-induced plasmas. *Spectrochim. Acta Part B* **2017**, *127*, 7–19. [CrossRef]
15. Morel, V.; Pérès, B.; Bultel, A.; Hideur, A.; Grisolia, C. Picosecond LIBS diagnostics for Tokamak in situ plasma facing materials chemical analysis. *Phys. Scr.* **2016**, *T167*, 014016. [CrossRef]
16. NIST Database. Available online: https://dx.doi.org/10.18434/T4W30F (accessed on 24 September 2018).
17. Morel, V.; Bultel, A.; Annaloro, J.; Chambrelan, C.; Edouard, G.; Grisolia, C. Dynamics of a femtosecond/picosecond laser-induced aluminum plasma out of thermodynamic equilibrium in a nitrogen background gas. *Spectrochim. Acta Part B* **2015**, *103*, 112–123. [CrossRef]

Article

Laser-Induced Plasma Measurements Using Nd:YAG Laser and Streak Camera: Timing Considerations

Maja S. Rabasovic, Mihailo D. Rabasovic, Bratislav P. Marinkovic and Dragutin Sevic *

Institute of Physics, University of Belgrade, 11090 Belgrade, Serbia; majap@ipb.ac.rs (M.S.R.); rabasovic@ipb.ac.rs (M.D.R.); bratislav.marinkovic@ipb.ac.rs (B.P.M.)
* Correspondence: sevic@ipb.ac.rs; Tel.: +381-11-3160-882

Received: 6 November 2018; Accepted: 24 December 2018; Published: 2 January 2019

Abstract: We describe a streak camera system that is capable of both spatial and spectral measurements of laser-induced plasma. The system is based on a Hamamatsu C4334 streak camera and SpectraPro 2300i spectrograph. To improve the analysis of laser-induced plasma development, it is necessary to determine the timing of laser excitation in regard to the time scale on streak images. We present several methods to determine the laser signal timing on streak images—one uses the fast photodiode, and other techniques are based on the inclusion of the laser pulse directly on the streak image. A Nd:YAG laser (λ = 1064 nm, Quantel, Brilliant B) was employed as the excitation source. The problem of synchronization of the streak camera with the Q-switched Nd:YAG laser is also analyzed. A simple modification of the spectrograph enables easy switching between the spectral and spatial measurement modes.

Keywords: laser-induced breakdown; plasma; spectroscopy; streak camera

1. Introduction

Laser-induced breakdown (LIB) is induced by focusing an intense laser beam on a gas, liquid, or solid target. Studying the plasma formation with a high temporal, spectral, and spatial resolution is of a great interest, and the formation of laser-induced breakdown of plasma in air has been studied by many researchers [1–6], including the references presented herein.

After the initial breakdown, plasma plume propagates towards the laser beam focusing lens [1,3]. The bright plasma core of the LIB in the open air is surrounded by a layer of cold, moderately ionized gas called the sheath [1]. The glow of the plasma sheath, although fainter than that of the core, is also visible to the naked eye. An explosive plasma-expansion induces optodynamic phenomena, i.e., the propagation of shock, acoustical, and ultrasonic waves.

Our research of the optical emission of the plasma has been limited thus far to the analysis of time-resolved optical emission spectra acquired by the streak camera [7–9]. To make our study more comprehensive, we saw the need to measure the spatial distribution of plasma optical emissions. The requirement for easy switching between the spectral and spatial measurement modes of our streak camera system soon became apparent to us.

In this paper, we describe an experimental system that is capable of both spatial and spectral measurements of laser-induced plasma with picosecond temporal resolution. Our experimental system is based on the Hamamatsu C4334 streak camera and SpectraPro 2300i spectrograph. A similar streak camera system was used by Hori and Akamatsu [6] in a time-resolved spatial analysis of the optical emissions from laser-induced plasma in air. We performed a simple modification of the spectrograph to enable easy switching between the spectral and spatial measurement modes. Later, we became aware that this modification had already been proposed and successfully used in the study of Siegel et al. [10], where the imaging device used was an intensified charge-coupled device (ICCD) camera.

To improve the analysis of laser-induced plasma development, it is necessary to determine the timing of laser excitation in regard to the time scale on streak images. We present several methods to determine the laser excitation timing on streak images—one uses the fast photodiode, and the other techniques are based on including the laser pulse directly on the streak image. The problem of synchronization of the streak camera with the Q-switched Nd:YAG laser is thoroughly analyzed in this paper.

2. Experimental Setup

2.1. Excitation System

The time-resolved laser-induced breakdown spectroscopy (LIBS) system implemented in our laboratory was based on the Nd:YAG laser, and the Optical Parametric Oscillator (OPO; Vibrant 266). The OPO system, which was pumped by a pulsed Q-switched Nd:YAG laser (Brilliant B) included the second and fourth harmonic generators (SHG and FHG). In this paper, the fundamental output at 1064 nm (pulse energy up to 270 mJ) and the second harmonic output at 532 nm (pulse energy up to 68 mJ) were used to create an optical breakdown in ambient air. The laser pulse width (full width at half maximum, FWHM) was about 5 ns on both wavelengths, as presented in Section 2.2. The plasma plume in air was obtained by focusing the laser beam using a lens with a focal length of 40 mm. A schematic diagram of the experimental apparatus is shown in Figure 1. The OPO system was controlled by OPOTEK software installed on a PC. In our setup, the output energy of the laser was determined by the timing of Q-switch firing, set by OPOTEK software. This characteristic introduced some complications in streak camera triggering, because the detection system was triggered by the Q-switch signal.

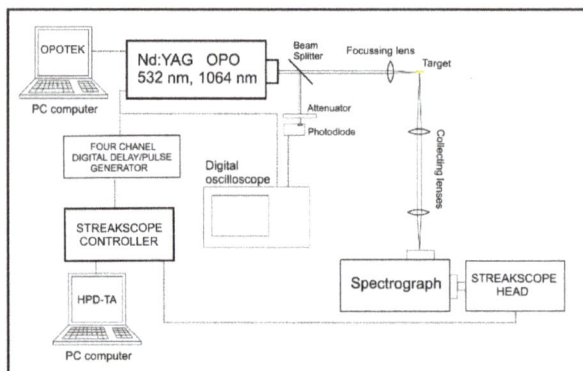

Figure 1. Time-resolved laser-induced breakdown: Experimental setup.

2.2. Detection System

The optical emissions from the plasma plume were collected by a spectrograph (SpectraPro 2300i) and recorded with a Hamamatsu streak camera (model C4334) with an integrated video streak camera (Figure 1). The streak images were time-resolved, thus enabling the monitoring of temporal evolution of the ionic and atomic emission lines [7–9], or spatial development of the plasma. The fundamental advantage of the streak scope was its two-dimensional nature, which was especially important for measuring time-resolved LIBS spectra. The camera had a spectral range from 200 to 850 nm. The CCD chip had a resolution of 640 × 480 pixels. The data were acquired and analyzed using High Performance Digital Temporal Analyzer (HPD-TA) software provided by Hamamatsu.

The spectrograph contained a triple grating turret. Diffraction gratings of 50, 150, and 300 grooves/mm were installed. In the place of the 150 g/mm grating, we mounted the plain mirror (see Figure 2). Thus, when grating of 150 grooves/mm was selected by HPD-TA software, the streak camera, instead of the image of the optical spectrum, took the image of the spatial distribution of the optical emissions of the laser-induced breakdown. To utilize as much of the CCD camera active area as possible, the maximal size of the spectrograph entrance slit was used. The diffraction grating of 50 grooves/mm was used when we required a wide observing wavelength window, and the grating of 300 grooves/mm was used when a better optical resolution was needed. Other optical parts of the acquisition system were chosen to obtain an overall optical magnification of 0.6. In this case, the calibration procedure showed that 1 mm on the target position corresponded to 72 pixels of the CCD camera.

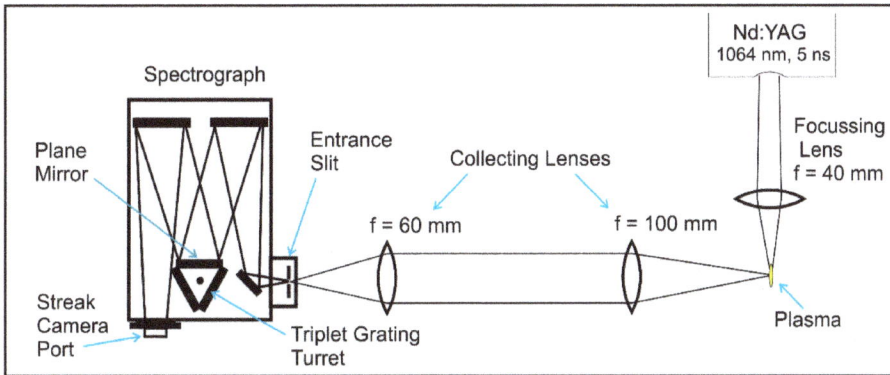

Figure 2. A simple modification of our spectrograph that enables easy switching between the spectral and spatial measurement modes.

To take the streak image in the time frame of interest, the proper delay time must be set on the digital delay generator (*Stanford* DG 535), which triggers the streak camera (see Figure 1). In our setup, for camera time scales up to 200 ns, the laser Q-switch trigger out-signal was used to trigger the streak camera. We used a fast 1-GHz photodiode and digital oscilloscope (Tektronix TDS 5032) to determine the time interval between the Q-switch trigger and the laser pulse (see Figure 3). The laser excitation pulse was partially reflected by the beam splitter, and acquired in an attenuated form by the fast photodiode. The photodiode was chosen to be sensitive both to the fundamental (1064 nm) and second harmonic output (532 nm) of the laser. There was a significant delay between the Q-switch, the top trace (shown in Figure 3), and the laser firing recorded with fast photodiode, as shown in the bottom trace in Figure 3. A similar problem concerning the acquiring of the streak image in the time frame of interest was solved by Mohamed and Kadowaki [11] by using an image light scope.

To determine the time that a streak image begins relative to the Q-switch trigger signal and the laser pulse, we had to tabulate the important time parameters of the detection system, provided in Table 1. The output energy of our laser was varied by a laser controller, by setting the different timings of the Q-switch. Thus, the time interval between the Q-switch signal and the laser pulse was a consequence of the "percent of laser energy" parameter set by the operator on the laser control unit. Moreover, the same "percent of laser energy" corresponded to the same timings and different energy levels of the laser's fundamental output and the laser's second harmonic output. In the measurements presented in this paper, for example, 100% of the laser energy meant 270 mJ on the fundamental harmonic or 68 mJ on the second harmonic. To measure the values presented in Table 1, the second harmonic of the laser was used as an excitation source.

Table 1. Timing parameters of the detection system for different time scales (5, 10, 20, 50, 100, and 200 ns) of the streak camera. "Q-sw" denotes the Q-switch trigger out-signal. "Pulse from top" means the "pulse position from top of the streak camera screen". SH denotes the second harmonic of the laser (532 nm).

Laser Energy SH [mJ]	Time Scale 5 ns Dead Time 189.7 ns			Time Scale 10 ns Dead Time 175.2 ns			Time Scale 20 ns Dead Time 197.0 ns		
	Delay Q-sw Laser [ns]	Delay DG535 Set [ns]	Pulse from Top [ns]	Delay Q-sw Laser [ns]	Delay DG535 Set [ns]	Pulse from Top [ns]	Delay Q-sw Laser [ns]	Delay DG535 Set [ns]	Pulse from Top [ns]
18	581.6	390	1.80	582.4	405	1.89	583.0	383	3.5
33	576.4	384	2.12	576.4	399	2.50	576.5	376	3.8
45	571.2	380	1.73	571.5	394	2.23	571.4	371	3.0
54	567.8	377	1.68	568.1	391	2.00	568.1	368	3.0
62	565.6	374	1.60	565.5	388	2.52	565.4	366	2.7
68	564.5	373	1.80	564.6	386	3.30	564.5	365	2.0

Laser Energy SH [mJ]	Time Scale 50 ns Dead Time 255.1 ns			Time Scale 100 ns Dead Time 337.4 ns			Time Scale 200 ns Dead Time 537.6 ns		
	Delay Q-sw Laser [ns]	Delay DG535 Set [ns]	Pulse from Top [ns]	Delay Q-sw Laser [ns]	Delay DG535 Set [ns]	Pulse from Top [ns]	Delay Q-sw Laser [ns]	Delay DG535 Set [ns]	Pulse from Top [ns]
18	582.4	319	7.96	582.4	238	6.00	582.6	37	7.47
33	576.5	318	3.24	576.4	232	7.06	576.4	33	6.29
45	571.5	313	3.45	571.4	229	5.40	571.7	28	5.89
54	568.0	307	5.68	568.1	223	7.68	568.0	22	7.87
62	565.6	307	3.85	565.5	226	2.28	566.0	19	9.45
68	564.4	305	4.66	564.7	219	8.5	564.3	20	7.47

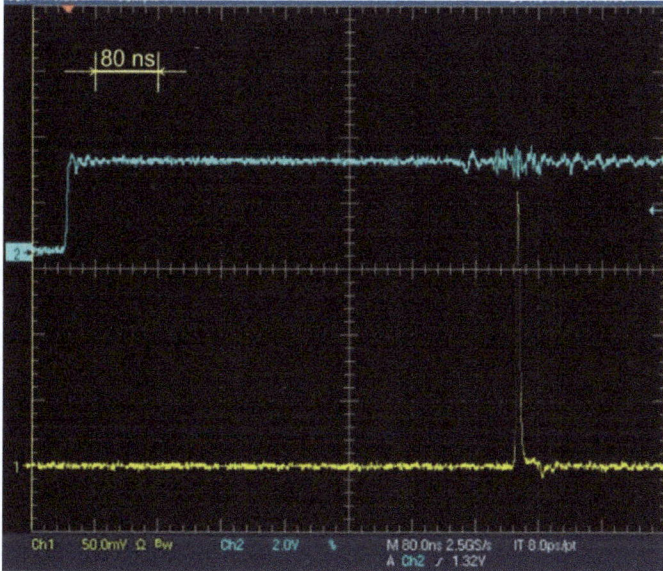

Figure 3. Delay time between the Q-switch trigger and the laser pulse.

Looking at Figure 3 and Table 1, it is easy to see that the time interval between the Q-switch trigger and the laser pulse was more or less longer than the delay set on the delay generator. So, when we say the "camera dead time", we mean the time difference between the camera trigger signal and the moment when the camera is capable of acquiring the streak image. Because of this dead time, the streak camera needed to be triggered in advance of the laser pulse using the delay generator triggered by the Q-switch.

Since the spectral range of our streak camera was in the interval from 200 to 850 nm, the fundamental harmonic of our excitation pulse (at 1064 nm) could not be acquired by the streak camera. We recorded waveforms of the laser's fundamental harmonic (1064 nm) and second harmonic (532 nm) using a fast photodiode and digital oscilloscope. The oscilloscope was triggered by the Q-switch, using an internal trigger delay. The delay between the fundamental and the second harmonic generator (SHG) was determined to be 2 ns, as shown in Figure 4. The delay came from a longer optical path passing through the SHG. Measuring the length of the optical path and calculating the time by using the known value of velocity of light gave the same result as that obtained from Figure 4.

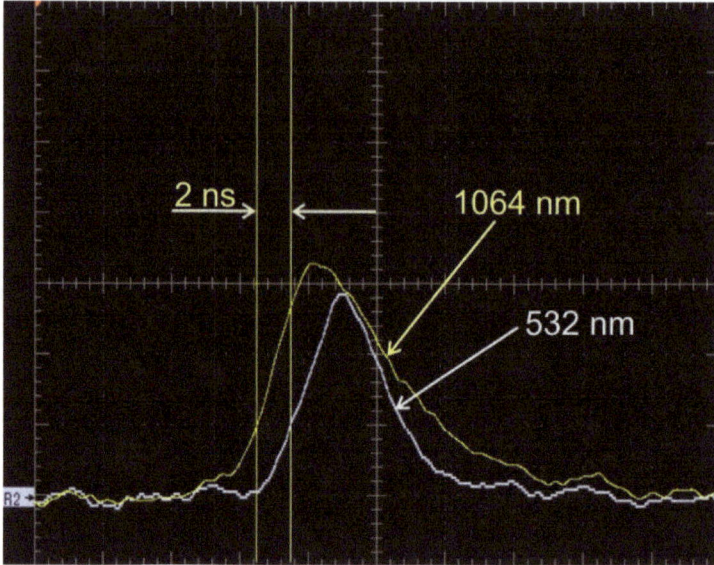

Figure 4. Delay time between laser pulses at 1064 nm and 532 nm obtained by a digital oscilloscope and fast photodiode.

To analyze plasma development, it is necessary to determine the initiation and duration of the excitation-laser pulse on streak images. Excitation-laser pulse is visible in Figure 5 (at 532 nm), where the streak image of the optical spectrum of laser-induced air plasma is presented. The streak images are usually presented in pseudo-color, where different intensities are coded as different colors. However, for laser excitation at 1064 nm, or for spatial streak images, laser excitation is not necessarily visible.

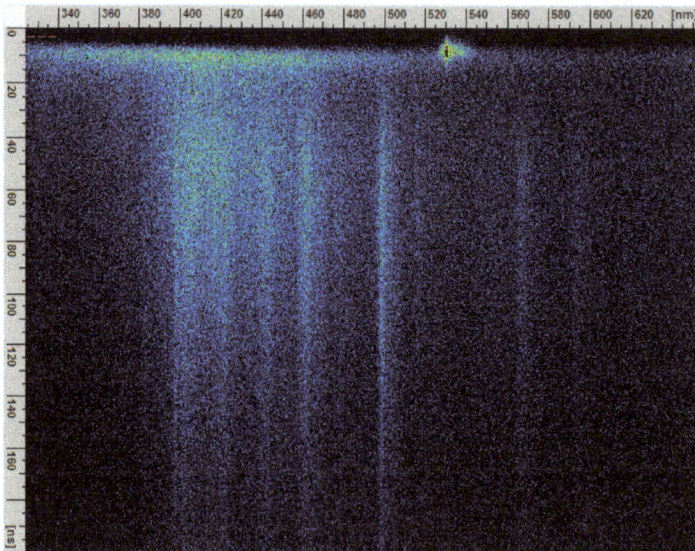

Figure 5. Streak image of the optical spectrum of laser-induced air plasma. Laser excitation at 532 nm is visible.

2.3. Determination of Synchronization Timing Using the Fast Photodiode

Two problems can be solved by acquiring the laser excitation signal using the fast photodiode. First, the proper setting of the delay time generator used for triggering the streak camera can be calculated. If this time is not properly set, the time window of interest will not be acquired by the streak camera; usually, just a blank screen with some noise will be recorded. The proper setting of the delay time generator can be read from Table 1, based on selected experimental parameters, and acquiring the time interval between the Q-switch and laser firing by using the fast photodiode and digital oscilloscope, as depicted in Figure 3.

Moreover, if the start of the laser pulse is not visible on the streak image, the timing of the plasma development (recorded on the streak image) regarding the laser excitation could be calculated using the fast photodiode signal. The calculation was performed with the data provided in Table 1. The laser pulse position from the top of the streak image equaled the difference in the delay between the Q-switch trigger signal (recorded by oscilloscope) and the laser pulse, the sum of the time set on delay generator, and the camera's dead time. If the calculated time is negative, the laser pulse has begun before the time frame visible on the streak camera screen.

2.4. Determination of Synchronization Timing by Recording the Laser Pulse on the Streak Image

There is no doubt that the determination of synchronization timing can be best achieved by recording the laser pulse on the same streak image as the plasma optical emission.

When the second harmonic of the laser at 532 nm was used as an excitation source, the laser signal was made visible and recorded on spatial streak image, as follows. By the appropriate placement of the neutral optical attenuator on the optical axis of the camera detection system, the plasma optical emissions and the elastic scattering of the laser beam from air molecules were recorded simultaneously on the same streak image (see Figure 6). The light attenuator enabled the camera to "see" the plasma breakdown (attenuated by attenuator) on the left-hand side and the laser excitation scattering (not attenuated) on the right-hand side of the focal point. The use of the attenuator was necessary to allow the optical signals of the plasma and scattered laser to have similar values; otherwise, after adjusting the gain of the detection system to match the intensity of the plasma optical emission, the laser scattering signal would not have sufficient intensity to be recorded. We discarded the original idea of transmitting the part of laser beam to the streak camera by a beam splitter as it was too risky.

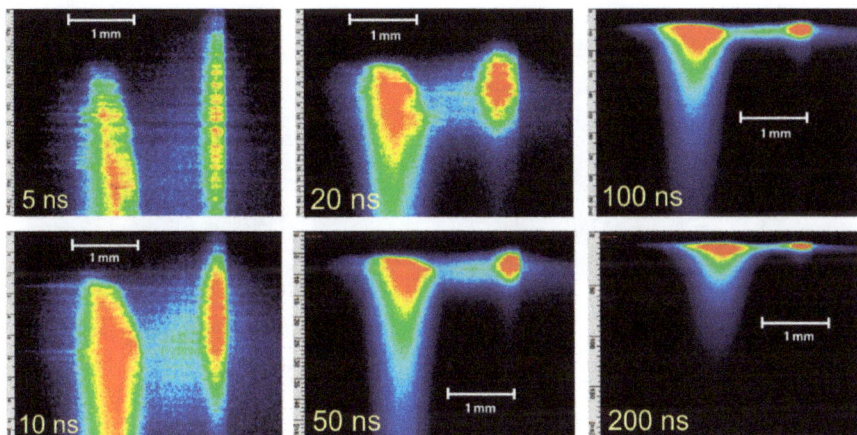

Figure 6. Spatial streak images of a simultaneous recording of the plasma plume and laser pulse. Plasma optical emission is visible on the left-hand side; there is laser excitation at 532 nm on the right-hand side of the image.

Before performing any timing calculations, the streak image was corrected for possible geometric distortion. Looking at Figure 5, the horizontal tilting of the streak image can be easily seen. The tilting of the streak images stems from the fact that the deflection of streak sweep is not completely straight, but rather, elliptic. This results in a geometric distortion of the streak image in the sweep direction. The distortion, which is always present on original streak images, is not obvious when looking at Figure 6. To make time calculations based on streak images, this distortion should be corrected using the curvature correction tool provided by the camera software.

When the more powerful first harmonic of our laser was used as the excitation source, the situation was more complicated. To make laser signal at 1064 nm visible, we used a very low concentration of Rhodamine B dye embedded in PMMA thin film. On the time scales of interest in the present study, the time delay of the Rhodamine B up-conversion fluorescence response was negligible, as proven by our fluorescence measurements. Almost all of the laser beam energy was transmitted through the thin PMMA film, placed about one centimeter from the focal point at an acute (sharp) angle, but not equal to 45 degrees relative to the beam, to avoid even the partial reflection of the laser beam to the camera. The Rhodamine B fluorescence, now visible by the camera, was recorded similarly to the laser scattering shown in Figure 6. It should be noted that, in this case, only the position of the raised edge of the laser excitation was correctly acquired. Again, the use of an optical attenuator for plasma emissions was mandatory.

3. Results and Discussion

3.1. Spectroscopic Streak Images

A set of time-resolved optical emission spectra of the laser-induced breakdown in air is presented in Figure 7. The second harmonic (532 nm) of the Q-switched Nd:YAG laser (nominal energy level of 70% was set on OPOTEK laser controller, measured as ~45 mJ per pulse) was used as an excitation source. The temporal distributions of the laser pulses were also visible.

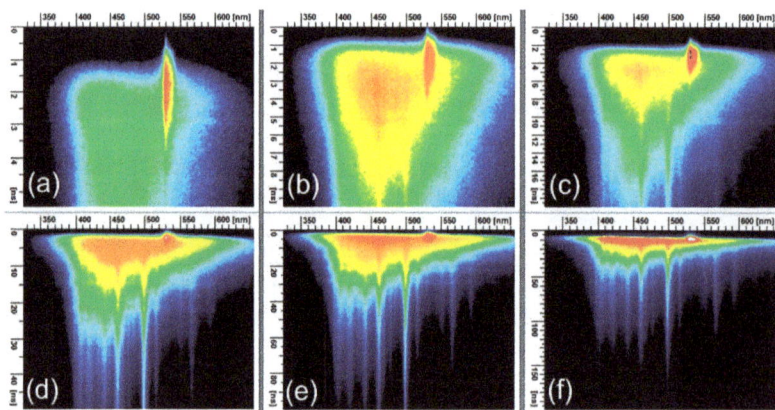

Figure 7. Temporal evolution of air plasma plume with a time range from 5 to 200 ns. (**a**) Time range is 5 ns, (**b**) time range is 10 ns, (**c**) time range is 20 ns, (**d**) time range is 50 ns, (**e**) time range is 100 ns, (**f**) time range is 200 ns.

The images were recorded with the acquire mode operation of the streak camera. The spectrograph grating of 50 grooves/mm was used, which enabled the plasma emission lines to be recorded in the wavelength range of 325 nm (horizontal axis). The vertical axis corresponds to the plasma development over time. The time range of the spectra in Figure 7 was from 5 to 200 ns. These results provide good insight into all stages of the plasma plume development. It is easy to see that the laser pulse was present before the creation of the plasma. The laser pulse at 5 ns and the plasma continuum can be seen in Figure 7a. In Figure 7b, the spectral lines begin to emerge and are quite discernible in Figure 7c. In Figure 7d–f, the spectral lines of the air constituents (N, O, C, etc.) are instantly recognizable.

To make the optical emission spectrum more precise, we used the streak camera photon counting mode of operation with 1000 exposures. The results are presented in Figure 8a. The camera trigger time was delayed, so the streak image did not include the beginning of the plasma to avoid the domination of continuum in the early stage. The profile lines of this image are presented in Figure 8b. The emission lines from the neutral and singly charged nitrogen, oxygen, and carbon atoms are identified in Figure 8b.

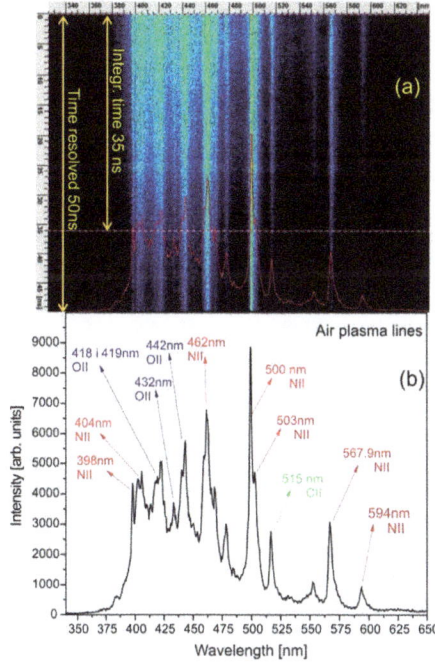

Figure 8. (**a**) The streak image of the emission lines obtained during laser-induced plasma in air. (**b**) One-dimensional profile of streak image (**a**) integrated into the time range of 35 ns.

3.2. Spatial Streak Images

The spatial streak images of the plasma plume in air for different excitation energies are presented in Figure 9. The laser beam was incident from the right-hand side of all spatial streak images of plasma plume. As expected, the plasma plume expanded towards the laser beam, which entered from the right-hand side of the images.

As the input energy gradually increased from 51 to 139 mJ for the streak images shown in Figure 9, more plasma with a longer expansion stage was observed as well as a corresponding increase in the plasma duration.

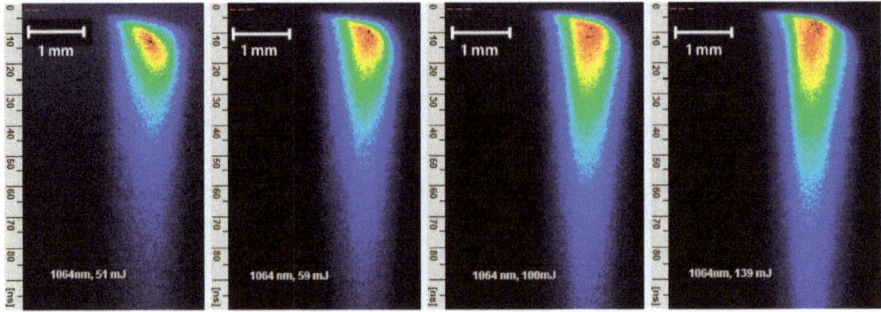

Figure 9. Time evolution of the air plasma obtained at different energy levels of the laser pulse at 1064 nm.

The plasma-expansion process from the laser-induced breakdown can be easily followed over time on the streak image presented in Figure 10. The Nd:YAG laser pulse energy was 140 mJ (λ = 1064 nm). The overall optical magnification of the detection system for this image was chosen to be equal to unity, so the calibration procedure showed that 1 mm on the target area corresponded to 120 pixels of CCD camera. The spatial resolution was enhanced with a smaller streak camera viewing angle as a price to pay. The image was obtained using the maximum entrance slit of 3000 μm. It can be seen from the motion of an emission peak that the laser plasma expanded towards the laser output, shifting gradually in time from the left to the right direction (Figure 11). By tracking the maximum brightness displacement shown in Figure 10, a velocity of plasma expansion of 35 km/s was obtained.

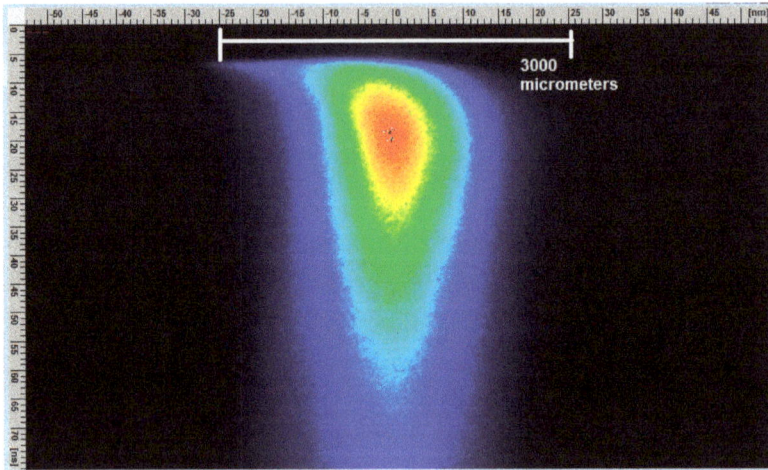

Figure 10. A streak image optimized for plasma-expansion velocity measurements.

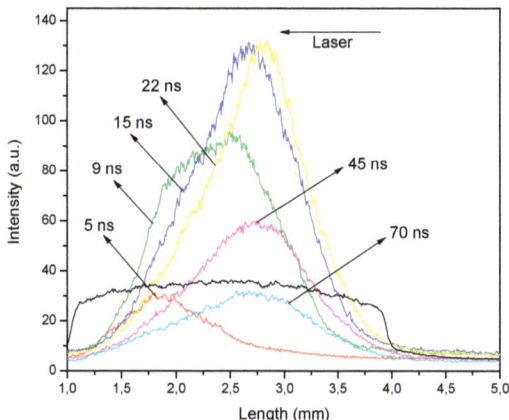

Figure 11. Spatial distributions of the laser-induced plasma for several time points.

4. Conclusions

In this paper, we presented a simple modification of the Hamamatsu 4334 Streak camera and SpectraPro 2300i spectrograph system that enabled easy switching between the spectral and spatial measurement modes, so that spectral and spatial streak images could be taken under equal conditions and with the same acquisition parameters. The problem of synchronization of the streak camera with the Q-switched Nd:YAG laser was also analyzed here.

To improve the analysis of laser-induced plasma development, it is necessary to determine the exact timing of laser excitation relative to the time scale on streak images. We described several methods to determine the laser signal timing on streak images—one using the fast photodiode, and other techniques based on recording the laser pulse directly on streak image, including the excitation at 1064 nm, which is not visible by the streak camera system.

An analysis of streak images acquired using the techniques explained here could be very useful for obtaining important laser plasma parameters such as temperature and electron density, as well as the plasma-expansion velocity and plasma starting times, including their variation over time. As an illustration, the emission lines of elements constituting the air were identified on the spectral streak image. As another example, we estimated the initial air plasma-expansion velocity using the enhanced spatial resolution of the detection system.

Author Contributions: Conceptualization, M.S.R. and D.S.; investigation, M.S.R., M.D.R. and D.S.; writing—original draft, M.S.R. and D.S.; writing—review and editing, M.D.R., B.P.M. and D.S.

Funding: This research was funded by the Ministry of Education, Science and Technology Development of the Republic of Serbia, Project No. OI 171020.

Conflicts of Interest: The authors declare no conflicts of interest.

References

1. Robledo-Martinez, A.; Sobral, H.; Villagrán-Muniz, M.; Bredice, F. Light focusing from large refractive indices in ionized air. *Phys. Plasmas* **2008**, *15*, 093510. [CrossRef]
2. Villagran-Muniz, M.; Sobral, H.; Camps, E. Shadowgraphy and interferometry using a CW laser and a CCD of a laser-induced plasma in atmospheric air. *IEEE Trans. Plasma Sci.* **2001**, *29*, 613–616. [CrossRef]
3. Gregorčič, P.; Možina, J. High-speed two-frame shadowgraphy for velocity measurements of laser-induced plasma and shock-wave evolution. *Opt. Lett.* **2011**, *36*, 2782–2784. [CrossRef] [PubMed]
4. Camacho, J.J.; Díaz, L.; Santos, M.; Juan, L.J.; Poyato, J.M.L. Time-resolved optical emission spectroscopy of laser-produced air plasma. *J. Appl. Phys.* **2010**, *107*, 083306. [CrossRef]

5. Kawahara, N.; Beduneau, J.L.; Nakayama, T.; Tomita, E.; Ikeda, Y. Spatially, temporally, and spectrally resolved measurement of laser-induced plasma in air. *Appl. Phys. B* **2007**, *86*, 605–614. [CrossRef]
6. Hori, T.; Akamatsu, F. Laser-Induced Breakdown Plasma Observed using a Streak Camera. *Jpn. J. Appl. Phys.* **2008**, *47*, 4759–4761. [CrossRef]
7. Sevic, D.; Rabasovic, M.S.; Marinkovic, B.P. Time-Resolved LIBS Streak Spectrum Processing. *IEEE Trans. Plasma Sci.* **2011**, *39*, 2782–2783. [CrossRef]
8. Rabasovic, M.S.; Sevic, D.; Pejčev, V.; Marinkovic, B.P. Detecting indium spectral lines using electron and laser induced breakdown spectroscopy. *Nucl. Instrum. Meth. B* **2012**, *279*, 58–61. [CrossRef]
9. Rabasovic, M.S.; Marinkovic, B.P.; Sevic, D. Time-Resolved Optical Spectra of the Laser-Induced Indium Plasma Detected Using a Streak Camera. *IEEE Trans. Plasma Sci.* **2014**, *42*, 2388–2389. [CrossRef]
10. Siegel, J.; Epurescu, G.; Perea, A.; Gordillo-Vazquez, F.J.; Gonzalo, J.; Afonso, C.N. High spatial resolution in laser-induced breakdown spectroscopy of expanding plasmas. *Spectrochim. Acta Part B* **2005**, *60*, 915–919. [CrossRef]
11. Mohamed, A.I.; Kadowaki, K. Streak Observation System for DC Pre-breakdown Using an Image Guide Scope. *Jpn. J. Appl. Phys.* **2012**, *51*, 028003. [CrossRef]

atoms

MDPI

Article

Measurement of Stark Halfwidths of Spectral Lines of Ionized Oxygen and Silicon Emitted from T-tube Plasma

Lazar Gavanski

Department of Physics, Faculty of Sciences, University of Novi Sad, Trg Dositeja Obradovića 4, 21000 Novi Sad, Serbia; lazar.gavanski@df.uns.ac.rs; Tel.: +381-21-485-2817

Received: 30 November 2018; Accepted: 7 January 2019; Published: 9 January 2019

check for
updates

Abstract: The analysis of experimental Stark halfwidths of spectral lines of singly ionized oxygen and silicon and double ionized silicon is presented in this work. The considered spectral lines were emitted from plasma generated in an electromagnetically driven T-tube, with an electron temperature of 15,000 K and electron density of 1.45×10^{23} m^{-3}. The obtained Stark halfwidths were compared to experimental values given by other authors. In addition, all experimental values were compared to theoretical values. These data are useful for diagnostics of laboratory and astrophysical plasmas as well as verifying theoretical models.

Keywords: plasma; spectral lines; Stark broadening; oxygen; silicon

1. Introduction

Spectral lines of ionized oxygen and silicon are often present in spectra emitted from laboratory [1–26] and astrophysical plasmas [27–33]. Stark parameters of these spectral lines can be used for plasma diagnostics, as it can be seen in the references given above. New and reliable experimental data can also be used for testing existing and new theoretical calculations.

The presence of O II, Si II and Si III spectral lines in our experiments was first observed in the spectra emitted from the T-tube while using pure helium as the working gas [34,35]. Oxygen and silicon occur in our plasma as impurities originating from the glass walls of the discharge vessel. Two examples of these spectra are given in Figure 1.

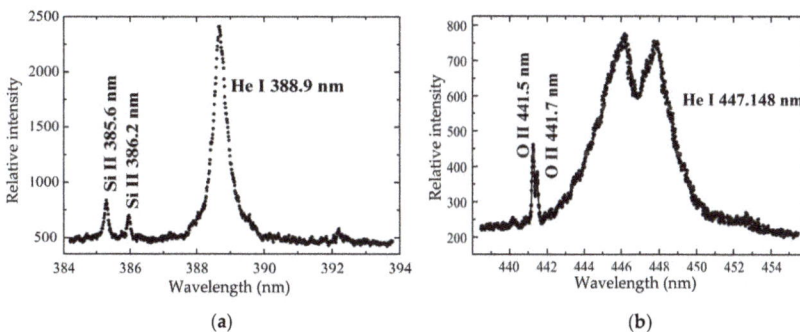

Figure 1. Examples of recorded He spectra with (**a**) Si II and (**b**) O II lines present as impurities in plasma.

Since the plasma is produced in a T-tube, which operates on the principle of shockwave propagation in a glass tube, oxygen and silicon appear in the plasma due to an ablation of the glass walls of the tube [35].

The intention of this paper is to give a comprehensive analysis of all experimental results of spectral line halfwidths presented in [35] and give more details about the experiment used in [35].

2. Details of the Experiment

2.1. Plasma Source

The plasma was produced in an electromagnetically driven T-tube [35,36]. The T-tube consists of a glass vessel with an internal radius of 27 mm with electrodes placed in the vertical part of the tube. A reflector made of quartz glass is positioned in the horizontal part of the tube, 14 mm from the electrode axis. The cross section of the T-tube is given in Figure 2.

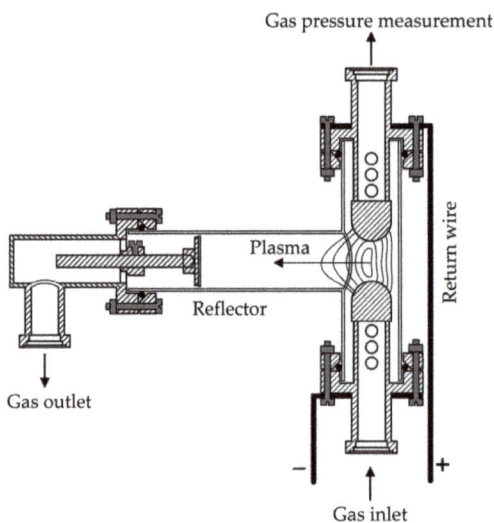

Figure 2. The cross section of the T-tube.

The plasma is produced by electrical discharge between the electrodes. The discharge current runs between the electrodes and through the return wire. Due to the opposing current flow directions, a repelling force exists between those two current flows.

Since a large amount of energy is released in a short amount of time, a shockwave is formed when the discharge occurs. This shockwave is then driven towards the reflector and produces plasma in the horizontal part of the T-tube. The front of the shockwave is flat, which results in a radially homogenous plasma [37]. Once the shockwave reaches the reflector it reflects and propagates in the opposite direction, further heating, exciting and ionizing the gas. More details on shockwave propagation in the used T-tube can be found in [36].

Most of the glass wall ablation occurs in the corners where the horizontal and vertical parts of the tube are joined [35], due to the impact of the incident shock wave front. The ablated silicon and oxygen atoms and ions are carried by the moving plasma toward the reflector.

The accompanying gas system is schematically presented in Figure 3. The T-tube has openings on the electrodes and behind the reflector for gas input, output and pressure measurement. The tube is first evacuated to a base pressure of 0.5 Pa. The working gas, in this case pure helium, is then released into the system and it flows through the T-tube at an operating pressure of 300 Pa. The pressure in the tube is monitored by a U-tube manometer.

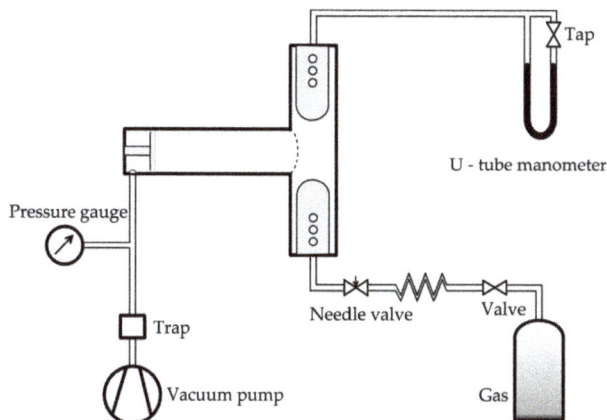

Figure 3. The T-tube, the system for supplying gas and the vacuum system.

A schematic representation of the electrical system is given in Figure 4. Four capacitors, 1 µF each, are charged up to a voltage of 20 kV. The voltage on the capacitors is measured by a high voltage probe, and led to a voltage comparator. Once the voltage reaches 20 kV, a 400 V pulse is generated and sent to the pulse transformer. There, a 10 kV pulse is formed and sent to the spark plug of the spark gap. This pulse is sufficient to initiate a discharge of the capacitors through the spark gap and the T-tube.

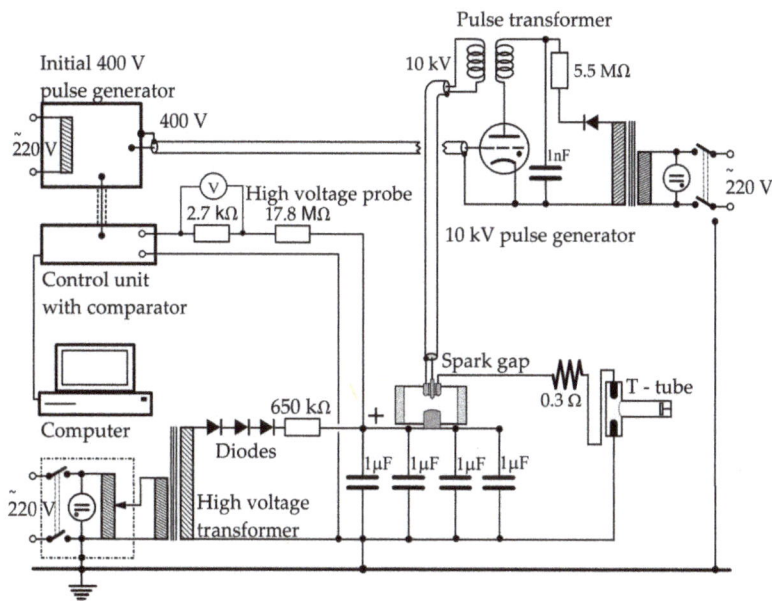

Figure 4. Schematic representation of the electrical system used for performing the discharge through the T-tube.

The discharge current has to be critically dampened, which was achieved by serially connecting a 0.3 Ω resistor in the discharge circuit. In this way, a single, short lasting, discharge current pulse is provided at the desired moment, resulting in the formation of a single, well defined, shock wave.

2.2. The Experimental Setup with the Spectral Recording System

A block scheme of the experimental system is given in Figure 5. The radiation emitted from the T-tube was collected radially, by means of an optical fiber which was placed at 2 cm from the reflector. At that position, optimum intensities of ionized oxygen and silicon spectral lines were observed. All measurements were carried out for the incident shock wave. The collected radiation was led to the entrance slit of the spectrometer. The used spectrometer was a Czerny-Turner type device, with 1 m focal length and a dispersion grating of 1200 g mm^{-1}. The inverse linear dispersion was 0.83 nm/mm.

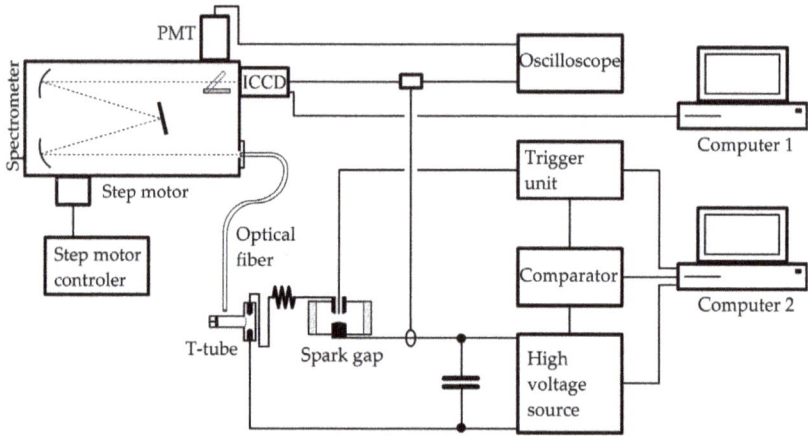

Figure 5. Block scheme of the T-tube experimental setup with the spectral recording system.

A photomultiplier tube and an ICCD camera were placed at the exit of the spectrometer, and a selection mirror allowed for alternating between those two detectors. The photomultiplier tube was used to observe the time evolution of the optical signal emitted from the plasma, as shown in Figure 6.

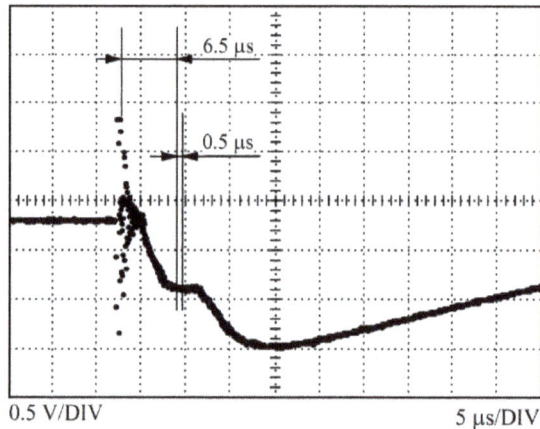

Figure 6. Time evolution of the optical signal from the plasma, recorded by the photomultiplier. The exposure time of 0. 5 μs in the incident shock wave is denoted by the arrows.

According to the observed signal, a delay time of 6.5 μs and an exposure time of 0.5 μs were chosen, in order to record radiation originating from stable plasma [35]. An ICCD camera was used for the recording of the spectra, see example in Figure 7.

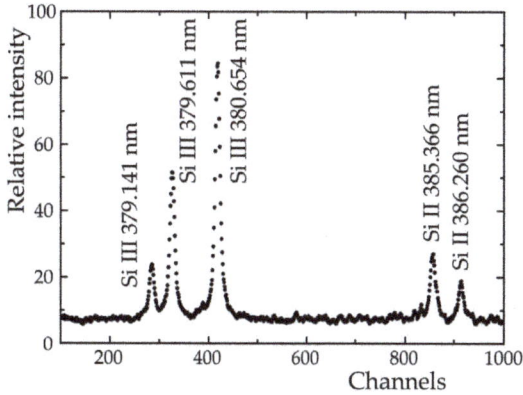

Figure 7. Example of the recorded spectra containing some Si II and Si III spectral lines.

2.3. Plasma Diagnostics

The electron density was determined by measuring the peak separation of the He 447.1 nm line and calculated by using the formula given in [38]. The obtained value was 1.45×10^{23} m^{-3}. The experimental uncertainty was estimated as 15% and was caused by the error in peak separation measurement [35].

The electron temperature was determined by using the Boltzmann plot method. This required the plasma to be in thermodynamic equilibrium. This was checked according to the criteria given in [39]. It was found that the plasma observed in this work did not meet the conditions for local thermodynamic equilibrium, but did meet the conditions for partial local thermodynamic equilibrium. In order to determine the electron temperature, 6 Si II and 12 O II spectral lines were used. The necessary atomic data were taken from the NIST atomic spectra database [40]. The obtained Boltzmann plots are presented in Figure 8.

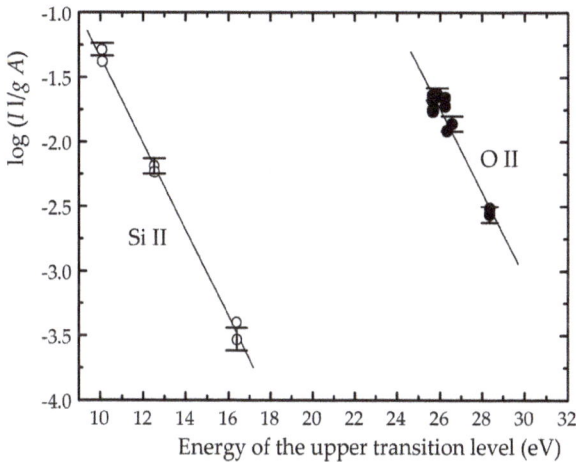

Figure 8. Boltzmann plots of Si II and O II spectral lines [35].

The electron temperature was determined from the slope of these plots by using the formula given in [41,42]. The obtained values were 14,900 K for Si II lines, and 15,300 K for O II lines. The estimated errors were within 10.8 and 17.2 percent [35]. The necessary recorded spectral line intensities are an average of ten shots. The variation of line intensities in successive shots is less than 5%. The spectral sensitivity uncertainty is below 1%. These uncertainties are included in the estimated temperature errors.

3. Spectral Line Broadening

The shape of O II, Si II and Si III spectral lines emitted from the plasma is best described by a theoretical Voigt profile. This profile is a convolution of the Gaussian and Lorentzian profiles. The Gaussian profile is due to Doppler and instrumental broadening, while the Lorentzian profile originates from Stark, Van der Waals and resonant broadening mechanisms. The broadening contributions of these individual mechanisms werecalculated for the observed lines in [35]. The Doppler broadening was between 0.005 nm and 0.01 nm for all lines. This was calculated by using the formula given in [43]. Van der Waals halfwidths were obtained by using the formula in [41] and [44], and ranged from 2×10^{-5} nm to 12×10^{-5} nm. Resonance halfwidths, where this effect was possible, were in the range between 6×10^{-6} nm and 2.3×10^{-3} nm and were calculated by using the formula given in [45] and [46]. Since the instrumental halfwidth for our experimental setup was around 0.45 nm, it can be claimed that the Gaussian part of the line profile is due to instrumental broadening. Similarly, the Lorentzian part of the line profile is a consequence of Stark broadening. Thus, by carrying out the deconvolution procedure [47], Stark halfwidths are obtained.

Although oxygen and silicon are present in the observed plasma only as impurities, the self-absorption effect was checked. Details of the used procedure are given in [48,49]. It was found that self-absorption was negligible in this experiment [35].

4. Results of Stark Halfwidth Measurements for O II, Si II and Si III Spectral Lines

Experimental Stark halfwidth data for 37 O II, 10 Si II and 12 Si III spectral lines were determined and are published in [35]. Available Stark halfwidth data, which have been published by other authors, can be found in the Critical Review papers [49–55]. According to these Critical Review papers, data on Stark halfwidths of O II spectral lines were found in 7 papers [1–7]. Data on Stark halfwidths of Si II and Si III spectral lines were found in 15 papers [8–22] and 6 papers [18,20,23–26], respectively. Each of those papers contains only a small number of spectral lines. Different experiments were performed under different conditions and by using different plasma sources. In contrast, Stark halfwidths of a large number of spectral lines from a wide spectral region and from one plasma source were obtained in [35].

The results were analyzed in several different ways. First, according to the regularities and similarities in plasma broadened spectral line widths [56]. Second, Stark halfwidths obtained in [35] were compared to values published by other authors [1,4–10,12–15,17,18,20,23,25,26]. Third, all experimental halfwidth values, both from [2] and from [1,4–10,12–15,17,18,20,23,25,26], were compared to theoretical values [57,58] by calculating the ratio of experimental to theoretical halfwidths. The analysis in this paper is performed only for those experimental results of other authors which contain data for more than one line in the observed multiplets.

4.1. Analysis of Stark Halfwidths of O II Spectral Lines

The analysis of O II Stark halfwidths inside a multiplet, according to [56], is given in Table 1. The line halfwidths in a multiplet usually agree within a few percent [56]. Taking into account experimental errors, we propose that if the variation of line halfwidths inside a multiplet is within 10%, the multiplet criterion is satisfied.

Table 1. Overview of the analysis of Stark halfwidth results for O II spectral lines.

Reference	Transition	Multiplet	No. of Measured Lines in Multiplet	Proposed Multiplet Criterion
[35]	$(^3P)3s-(^3P)3p$	$^4P-^4D^o$	7	Satisfied
	$(^3P)3s-(^3P)3p$	$^4P-^4P^o$	3	Satisfied
	$(^3P)3s-(^3P)3p$	$^4P-^4S^o$	3	Satisfied
	$(^3P)3s-(^3P)3p$	$^2P-^2D^o$	2	Satisfied
	$(^3P)3s-(^3P)3p$	$^2P-^2P^o$	4	Satisfied
	$(^3P)3p-(^3P)3d$	$^4D^o-^4F$	2	Satisfied
	$(^1D)3s-(^1D)3p$	$^2D-^2F^o$	2	Satisfied
	$(^3P)3p-(^3P)3d$	$^4P^o-^4P$	2	Satisfied
	$(^3P)3p-(^3P)3d$	$^4S^o-^4P$	2	Satisfied
	$(^1D)3p-(^1D)3d$	$^2F^o-^2G$	2	Satisfied
[1]	$(^3P)3s-(^3P)3p$	$^4P-^4D^o$	2	Satisfied
	$(^3P)3s-(^3P)3p$	$^4P-^4S^o$	3	Satisfied
	$(^1D)3s-(^1D)3p$	$^2D-^2F^o$	2	Satisfied
[4]	$(^3P)3s-(^3P)3p$	$^4P-^4D^o$	2	Satisfied
[5]	$(^3P)3s-(^3P)3p$	$^2P-^2D^o$	2	Satisfied
[6]	$(^3P)3s-(^3P)3p$	$^4P-^4D^o$	4	Satisfied
	$(^3P)3s-(^3P)3p$	$^4P-^4P^o$	3	Satisfied
	$(^3P)3p-(^3P)3d$	$^4D^o-^4F$	2	Satisfied
	$(^1D)3s-(^1D)3p$	$^2D-^2F^o$	2	Satisfied
[7]	$(^3P)3s-(^3P)3p$	$^2P-^2P^o$	2	Not satisfied, 32.1%
	$(^3P)3p-(^3P)3d$	$^4D^o-^4F$	2	Not satisfied, 26.7%
	$(^1D)3p-(^1D)3d$	$^2F^o-^2G$	2	Not satisfied, 19.9%

All results regarding OII lines, obtained in [35], meet the proposed multiplet criterion. Lines within the same multiplet deviate within a maximum of 5% from the average value. Most of the results published by other authors also meet this criterion. The results in [7], however, do not meet the proposed criterion for multiplets.

When comparing Stark halfwidths obtained in [35] to those published by other authors, the general halfwidth ratio range was found to be between 0.6 and 1.2 [35]. The results of [35] agree best with the results in [5], for which the halfwidth ratio is between 0.92 and 1.04. However, halfwidths published in [3,6] are significantly larger than those obtained in [35]. A strong disagreement was found with the results of [7] as well. For this analysis, the results of other authors were recalculated for the same electron temperature and density.

In order to compare experimental halfwidth results [35] with theoretical ones, halfwidth values were calculated according to the modified semi empirical formula (MSE) [57], the simplified modified semi empirical formula (SMSE) [58] and, where it was possible, were taken from the semi classical calculations [59].

The ratio of all experimental halfwidths and theoretical halfwidths [56] is generally between 0.7 and 1.18 [35]. Only the results given in [7] deviate from this range, with halfwidth ratios up to 1.6. Apart from that, it can be concluded that the experimental results, in general, agree best with theoretical values in [59].

In comparison to experimental Stark halfwidth values, the MSE theory provides systematically larger halfwidth values. The general halfwidth ratio is from 0.6 to 1.0 [35]. In contrast, the SMSE theory predicts values which are smaller than the experimental ones, with a general halfwidth ratio from 1.3 to 2.5 [35]. The results of [3,6] differ from SMSE results even more, resulting in halfwidth ratios from 4.0 to 5.0.

An example of the experimental and theoretical data distribution for one spectral line is given in Figure 9.

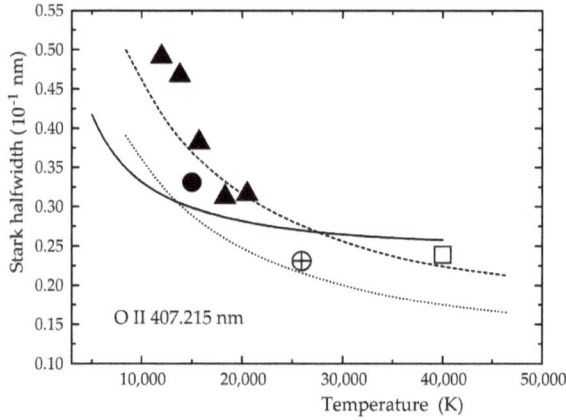

Figure 9. Experimental and theoretical Stark halfwidth dependence on temperature for the O II 407.215 nm spectral line for the electron density of 10^{23} m^{-3}. The presented data are: $\oplus$ [1], $\square$ [6], $\blacktriangle$ [7], $\bullet$ [35], ---- [57], ······ [58] and —— [59].

4.2. Analysis of Stark Halfwidths of Si II Spectral Lines

Stark halfwidths of SI II spectral lines, which were measured in [35], satisfy the proposed multiplet criterion very well, with deviations from the average halfwidth value being less than 3.5%. The results published in [9,10,13,17] also meet this criterion, while some results in [8,12,14,15,18] do not. An overview is given in Table 2.

Table 2. Overview of the analysis of Stark halfwidth results for Si II spectral lines.

Reference	Transition	Multiplet	No. of Measured Lines in Multiplet	Proposed Multiplet Criterion
[35]	$3s3p^2$-$3s^24p$	^2D-$^2P^o$	3	Satisfied
	$3s^24s$-$3s^24p$	^2S-$^2P^o$	2	Satisfied
	$3s^23d$-$3s^24f$	^2D-$^2F^o$	2	Satisfied
[8]	$3s3p^2$-$3s^24p$	^2D-$^2P^o$	3	Not satisfied, 14%
[9]	$3s3p^2$-$3s^24p$	^2D-$^2P^o$	2	Satisfied
	$3s^24s$-$3s^24p$	^2S-$^2P^o$	2	Satisfied
[10]	$3s^24s$-$3s^24p$	^2S-$^2P^o$	2	Satisfied
[12]	$3s^24s$-$3s^24p$	^2S-$^2P^o$	2	Satisfied
	$3s^23d$-$3s^24f$	^2D-$^2F^o$	2	Not satisfied, 25%
[13]	$3s3p^2$-$3s^24p$	^2D-$^2P^o$	2	Satisfied
	$3s^24s$-$3s^24p$	^2S-$^2P^o$	2	Satisfied
[14]	$3s3p^2$-$3s^24p$	^2D-$^2P^o$	2	Not satisfied, 28%
[15]	$3s3p^2$-$3s^24p$	^2D-$^2P^o$	2	Not satisfied, 11.7%
	$3s^24s$-$3s^24p$	^2S-$^2P^o$	2	Not satisfied, 18%
	$3s^23d$-$3s^24f$	^2D-$^2F^o$	2	Not satisfied, 11.5%
[17]	$3s3p^2$-$3s^24p$	^2D-$^2P^o$	3	Satisfied
	$3s^24s$-$3s^24p$	^2S-$^2P^o$	2	Satisfied
	$3s^23d$-$3s^24f$	^2D-$^2F^o$	2	Satisfied
[18]	$3s3p^2$-$3s^24p$	^2D-$^2P^o$	3	Not satisfied, 11.3%
	$3s^24s$-$3s^24p$	^2S-$^2P^o$	2	Satisfied
	$3s^23d$-$3s^24f$	^2D-$^2F^o$	2	Satisfied
[19]	$3s3p^2$-$3s^24p$	^2D-$^2P^o$	2	Satisfied
	$3s^24s$-$3s^24p$	^2S-$^2P^o$	2	Satisfied
	$3s^23d$-$3s^24f$	^2D-$^2F^o$	2	Satisfied

By comparing the experimental results of [35] to those of other authors, a common halfwidth ratio from 0.4 to 1.5 was found. Results given in [20] deviate significantly from this, with a ratio of 5.0.

When compared to the theoretical results of [59], most experimental values provide a halfwidth ratio from 0.5 to 1.2. Some results in [12] are outside of this range (1.95). Comparing experimental results to values obtained from the MSE theory, a common halfwidth ratio from 0.5 to 1.7 is obtained. Again, the results given in [12] deviate with a ratio up to 2.58. The SMSE theory gives systematically smaller values than experimental ones, which is especially noticeable for transitions with a higher orbital quantum number. The experimental and theoretical Stark halfwidth data for one Si II spectral line are presented in Figure 10.

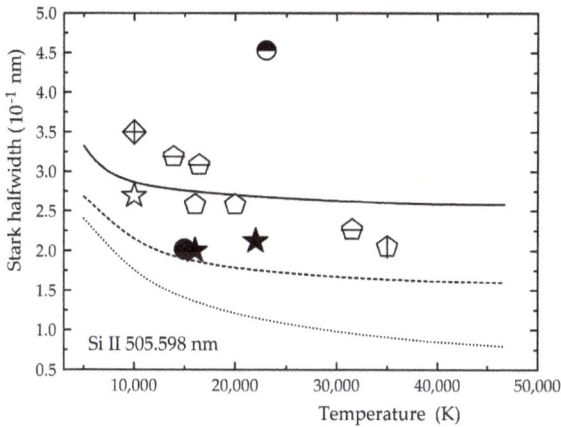

Figure 10. Experimental and theoretical Stark halfwidth dependence on temperature, for the Si II 505.598 nm spectral line for the electron density of 10^{23} m^{-3}. The presented data are: ⊖ [12], ★ [15], ◇ [16], ⌂ [17], ☆ [19], ⬤ [20], ● [35], ---- [57], ⋯⋯ [58] and ── [59].

4.3. Analysis of Stark Halfwidths of Si III Spectral Lines

The experimental results for Si III from [35] satisfy the proposed multiplet criterion very well, with line Stark halfwidths differing 4% or less from the multiplet average value. Some of the other authors satisfy this criterion as well [18,23,25] while some do not [20,26], as it can be seen in Table 3.

Table 3. Overview of the analysis of Stark halfwidth results for Si III spectral lines.

Reference	Transition	Multiplet	No. of Measured Lines in Multiplet	Proposed Multiplet Criterion
[35]	3s3d-3s4p	^{3}D-^{3}P^o	3	Satisfied
	3s4s-3s4p	^{3}S-^{3}P^o	3	Satisfied
	3s4p-3s4d	^{3}P^o-^{3}D	3	Satisfied
[20]	3s4p-3s4d	^{3}P^o-^{3}D	3	Not satisfied, 16%
[18]	3s3d-3s4p	^{3}D-^{3}P^o	3	Satisfied
	3s4s-3s4p	^{3}S-^{3}P^o	3	Satisfied
	3s4p-3s4d	^{3}P^o-^{3}D	3	Satisfied
[23]	3s4s-3s4p	^{3}S-^{3}P^o	3	Satisfied
	3s4p-3s4d	^{3}P^o-^{3}D	3	Satisfied
[25]	3s4s-3s4p	^{3}S-^{3}P^o	3	Satisfied
[26]	3s4s-3s4p	^{3}S-^{3}P^o	2	Not satisfied, 11.5%

The experimental halfwidths obtained in [35], when compared to those of other authors, are mostly in a ratio from 0.6 to 1.55. For the results published in [20], however, this ratio is 0.14.

For the observed Si III spectral lines, no data could be found in [59], so that no comparison could be made. When comparing all experimental values to MSE theory values, a common halfwidth ratio was found to be from 0.6 to 1.6. Again, some results given in [20] deviate significantly from most other results with a ratio of 6.08. The SMSE theory mostly gives smaller values, especially for higher orbital quantum numbers, as in the case of Si II. This is not always the case, as it can be seen in Figure 11. For the spectral line Si III 456.782, the SMSE theoretical values are higher than both the MSE theory and experimental data.

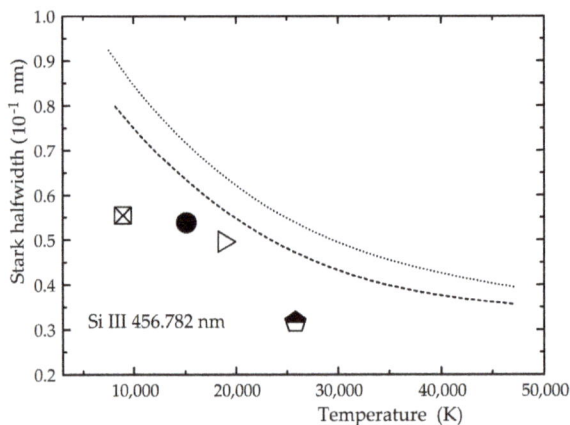

Figure 11. Experimental and theoretical Stark halfwidth dependence on temperature, for the Si III 456.782 nm spectral line for the electron density of 10^{23} m^{-3}. The presented data are: ▷ [23], ⬗ [25], ⊠ [26], ● [35], ---- [57] and ······· [58].

5. Conclusions

In this work, a comprehensive analysis of the experimental Stark halfwidth data of ionized oxygen and silicon is presented. The analysis is performed on experimental results measured in [2] and the available results of other authors [1,4–10,12–15,17,18,20,23,25,26]. A comparison with theoretical results [57–59] is also given.

In measurements performed in [35], special attention was paid to the experiment, the spectral recording and the data analysis. Stark halfwidths of a large number of spectral lines were obtained, covering a wide spectral region. All of these lines were emitted from the same plasma source, under the same conditions, and recorded with the same equipment. Some additional details about the experiment are given in this paper.

The analysis of the results published by other authors, revealed a few papers that contain results that strongly deviate from the results measured in [35], from available experimental data by other authors, and from theoretical values (see Sections 4.1–4.3). These halfwidth values should be used with caution for e.g., plasma diagnostics purposes.

The data obtained in [35] could be used for plasma diagnostics of both laboratory and astrophysical plasmas. Existing, as well as new theories could also be tested using these data. More independent experimental work is still needed to further improve the database of experimental Stark halfwidth results.

Funding: This research was funded by the Ministry of Education, Science and Technological development of Republic Serbia, grant number 171014.

Conflicts of Interest: The author declares no conflict of interest.

References

1. Platiša, M.; Popović, M.; Konjević, N. Stark broadening of O II and O III Lines. *Astron. Astrophys.* **1975**, *45*, 325–327.
2. Purić, J.; Djeniže, S.; Srećković, A.; Platiša, M.; Labat, J. Stark-broadening regularities of prominent multiply-ionized-oxygen spectral lines in plasma. *Phys. Rev. A* **1998**, *37*, 498–503. [CrossRef]
3. Djeniže, S.; Srećković, A.; Labat, J.; Platiša, M. Stark broadening and shift of OII spectral lines in higher multiplets. *Z. Phys. D* **1991**, *21*, 295–297. [CrossRef]
4. Djeniže, S.; Milosavljević, V.; Srećković, A. Measured Stark widths and shifts of singly ionized oxygen spectral lines in lower multiplets. *J. Quant. Spectrosc. Radiat. Transf.* **1998**, *59*, 71–75. [CrossRef]
5. Blagojević, B.; Popović, M.V.; Konjević, N. Stark Broadening of Spectral Lines of Singly Ionized C, N, O, F and Ne. *Phys. Scr.* **1999**, *59*, 374–378. [CrossRef]
6. Del Val, J.A.; Aparicio, J.A.; González, V.; Mar, S. Measurement of Stark widths in single ionized oxygen. *Astron. Astrophys. Suppl. Ser.* **1999**, *140*, 171–176. [CrossRef]
7. Srećković, A.; Drinčić, V.; Bukvić, S.; Djeniže, S. Stark Broadening Parameters and Transition Probabilities in the O II Spectrum. *Phys. Scr.* **2001**, *63*, 306–312. [CrossRef]
8. Wollschläger, F.; Mitsching, J.; Meiners, D.; Depiesse, M.; Richou, J.; Lesage, A. Measurements of the ionised silicon multiplet (1) Stark parameters by two different methods. *J. Quant. Spectrosc. Radiat. Transf.* **1997**, *58*, 135–140. [CrossRef]
9. Chiang, W.T.; Griem, H.R. Measurements of the Stark broadening of ionized silicon lines from a plasma. *Phys. Rev. A* **1978**, *18*, 1169–1175. [CrossRef]
10. Pérez, C.; de la Rosa, I.; Frutos, A.M.; González, V.R.; Mar, S. Stark broadening of several SiII lines. *Ann. Phys.* **1990**, *15*, 115–116.
11. Lesage, A.; Redon, R. Stark widths of faint Si II lines. *Astron. Astrophys.* **2004**, *418*, 765–769. [CrossRef]
12. Pérez, C.; de la Rosa, I.; Frutos, A.M.; Mar, S. Temperature dependence of Stark broadening for several Si II lines. *Phys. Rev. E* **1993**, *47*, 756–759. [CrossRef]
13. Konjević, N.; Purić, J.; Ćirković, L.; Labat, J. Measurements of the Stark broadening parameters of several Si II lines. *J. Phys. B* **1970**, *3*, 999–1003. [CrossRef]
14. Purić, J.; Djeniže, S.; Labat, J.; Ćirković, L. Stark shift of neutral and ionized silicon spectral lines. *Phys. Lett. A* **1973**, *45*, 97–98. [CrossRef]
15. Lesage, A.; Rathore, B.A.; Lakićević, I.S.; Purić, J. Stark widths and shifts of singly ionized silicon spectral lines. *Phys. Rev. A* **1983**, *28*, 2264–2268. [CrossRef]
16. Lesage, A.; Miller, M.H. Détermination expérimentale de l'élargissment du déplacement Stark de raies d'atomes de silicium ionizes. *C. R. Acad. Sci. Paris Ser. B* **1975**, *280*, 645–647.
17. González, V.R.; Aparicio, J.A.; del Val, J.A.; Mar, S. Stark broadening and shift measurements of visible Si II lines. *J. Phys. B* **2002**, *35*, 3557–3573.
18. Bukvić, S.; Djeniže, S.; Srećković, A. Line broadening in the Si I, Si II, Si III, and Si IV spectra in the helium plasma. *Astron. Astrophys.* **2009**, *508*, 491–500. [CrossRef]
19. Lesage, A.; Sahal-Brechot, S.; Miller, M.H. Stark broadening of singly ionized-silicon. *Phys. Rev. A* **1977**, *16*, 1617–1624. [CrossRef]
20. Kusch, H.J.; Schröder, K. Experimental Stark Broadening Data of Si II and Si III Lines. *Astron. Astrophys.* **1982**, *116*, 255–259.
21. Miller, M.H. *Technical Note BN-550*; University of Maryland: College Park, MD, USA, 1968.
22. Chapelle, J.; Czernichowski, A. Stark Broadening of the Spectral Lines 6371.4 and 6347.1 A for Si II. *Acta Phys. Pol. A* **1972**, *41*, 753.
23. González, V.R.; Aparicio, J.A.; del Val, J.A.; Mar, S. Stark width and shift measurements of visible SI III lines. *Astron. Astrophys.* **2000**, *363*, 1177–1185.
24. Djeniže, S.; Srećković, A.; Labat, J.; Purić, J.; Platiša, M. Measured Stark widths of doubly and triply ionized silicon spectral lines. *J. Phys. B* **1992**, *25*, 785–790. [CrossRef]
25. Platiša, M.; Dimitrijević, M.S.; Popović, M.; Konjević, N. Stark broadening of Si III and Si IV lines. *J. Phys. B* **1977**, *10*, 2997–3004. [CrossRef]
26. Purić, J.; Djeniže, S.; Labat, J.; Ćirković, L. Stark broadening parameters of Si I, Si II and Si III lines. *Z. Phys.* **1974**, *267*, 71–75. [CrossRef]

27. Sollerman, J.; Fynbo, J.P.U.; Gorosabel, J.; Halpern, J.P.; Hjorth, J.; Jakobsson, P.; Mirabal, N.; Watson, D.; Xu, D.; Castro-Tirado, A.J.; et al. The nature of the X-ray flash of 24 August 2005—Photometric evidence for an on-axis z = 0.83 burst with continuous energy injection and an associated supernova. *Astron. Astrophys.* **2007**, *466*, 839–846. [CrossRef]

28. Muzzin, A.; Wilson, G.; Yee, H.K.C.; Gilbank, D.; Hoekstra, H.; Demarco, R.; Balogh, M.; van Dokkum, P.; Franx, M.; Ellingson, E.; et al. The Gemini Cluster Astrophysics Spectroscopic Survey (GCLASS): The Role of Environment and Self-regulation in Galaxy Evolution at z ~ 1. *Astrophys. J.* **2012**, *746*, 188. [CrossRef]

29. Lorenzo, J.; Negueruela, I.; Castro, N.; Norton, A.J.; Vilardell, F.; Herrero, A. Astrophysical parameters of the peculiar X-ray transient IGR J11215−5952. *Astron. Astrophys.* **2014**, *562*, A18. [CrossRef]

30. Berdyugina, S.V.; Ilyin, I.; Tuominen, I. The long-period RS Canum Venaticorum binary IM Pegasi I. Orbital and stellar parameters. *Astron. Astrophys.* **1999**, *347*, 932–936.

31. Asplund, M. Line formation in solar granulation III. The photospheric Si and meteoritic Fe abundances. *Astron. Astrophys.* **2000**, *359*, 755–758.

32. Briquet, M.; Aerts, C.; Lüftinger, T.; De Cat, P.; Piskunov, N.E.; Scuflaire, R. He and Si surface inhomogeneities of four Bp variable stars. *Astron. Astrophys.* **2004**, *413*, 273–283. [CrossRef]

33. Khalack, V.; Landstreet, J.D. Partial Paschen–Back splitting of Si II and Si III lines in magnetic chemically peculiar stars. *Mon. Not. R. Astron. Soc.* **2012**, *427*, 569–580. [CrossRef]

34. Gigosos, M.A.; Djurović, S.; Savić, I.; González-Herrero, D.; Mijatović, Z.; Kobilarov, R. Stark broadening of lines from transition between states n = 3 to n = 2 in neutral helium. *Astron. Astrophys.* **2014**, *561*, A135–A147. [CrossRef]

35. Gavanski, L.; Belmonte, M.T.; Savić, I.; Djurović, S. Experimental Stark halfwidths of the ionized oxygen and silicon spectral lines. *Mon. Not. R. Astron. Soc.* **2016**, *457*, 4038–4050. [CrossRef]

36. Djurović, S.; Mijatović, Z.; Vujičić, B.; Kobilarov, R.; Savić, I.; Gavanski, L. Measurement of the shock front velocity produced in a T-tube. *Phys. Plasmas* **2015**, *22*, 013505. [CrossRef]

37. Kolb, A.C. Production of High-Energy Plasmas by Magnetically Driven Shock Waves. *Phys. Rev.* **1957**, *107*, 345–350. [CrossRef]

38. Ivković, M.; González, M.Á.; Jovićević, S.; Gigosos, M.A.; Konjević, N. A simple line shape technique for electron number density diagnostics of helium and helium-seeded plasmas. *Spectrochim. Acta B* **2010**, *65*, 234–240. [CrossRef]

39. Griem, H.R. Validity of Local Thermal Equilibrium in Plasma Spectroscopy. *Phys. Rev.* **1963**, *131*, 1170–1176. [CrossRef]

40. NIST Atomic Spectra Database. Available online: https://www.nist.gov/pml/atomic-spectra-database (accessed on 13 November 2018).

41. Griem, H.R. *Plasma Spectroscopy*; McGraw-Hill Book Company: New York, NY, USA, 1964.

42. Mitchner, M.; Kruger, C.H. *Partially Ionized Gases*; John Wiley & Sons: New York, NY, USA, 1973.

43. Konjević, N.; Roberts, J.R. A critical review of the Stark widths and shifts of spectral lines from non-hydrogenic atoms. *J. Phys. Chem. Ref. Data* **1976**, *5*, 209–257. [CrossRef]

44. Kelleher, D.E. Stark broadening of visible neutral helium lines in a plasma. *J. Quant. Spectrosc. Radiat. Transf.* **1981**, *25*, 191–220. [CrossRef]

45. Ali, A.W.; Griem, H.R. Theory of Resonance Broadening of Spectral Lines by Atom-Atom Impacts. *Phys. Rev.* **1965**, *140*, A1044–A1049. [CrossRef]

46. Ali, A.W.; Griem, H.R. Errata: Theory of Resonance Broadening of Spectral Lines by Atom-Atom Impacts. *Phys. Rev.* **1966**, *144*, 366. [CrossRef]

47. Davies, J.T.; Vaughan, J.M. A New Tabulation of the Voigt Profile. *Astrophys. J.* **1963**, *137*, 1302–1305. [CrossRef]

48. Djurović, S.; Kobilarov, R.; Vujičić, B. Experimental difficulties in determination of the spectral line shapes emitted from plasma. *Bull. Astron. Belgrade* **1996**, *153*, 41.

49. Konjević, N. Plasma broadening and shifting of non-hydrogenic spectral lines: Present status and applications. *Phys. Rep.* **1999**, *316*, 339–401. [CrossRef]

50. Konjević, N.; Wiese, W.L. Experimental Stark widths and shifts for non-hydrogenic spectral lines of ionized atoms. *J. Phys. Chem. Ref. Data* **1976**, *5*, 259–308. [CrossRef]

51. Konjević, N.; Dimitrijević, M.S.; Wiese, W.L. Experimental Stark Widths and Shifts for Spectral Lines of Neutral Atoms (A Critical Review of Selected Data for the Period 1976 to 1982). *J. Phys. Chem. Ref. Data* **1984**, *13*, 619–647. [CrossRef]
52. Konjević, N.; Dimitrijević, M.S.; Wiese, W.L. Experimental Stark Widths and Shifts for Spectral Lines of Positive Ions (A Critical Review and Tabulation of Selected Data for the Period 1976 to 1982). *J. Phys. Chem. Ref. Data* **1984**, *13*, 649–686. [CrossRef]
53. Konjević, N.; Wiese, W.L. Experimental Stark Widths and Shifts for Spectral Lines of Neutral and Ionized Atoms. *J. Phys. Chem. Ref. Data* **1990**, *19*, 1307–1385. [CrossRef]
54. Konjević, N.; Lesage, A.; Fuhr, J.R.; Wiese, W.L. Experimental Stark Widths and Shifts for Spectral Lines of Neutral and Ionized Atoms (A Critical Review of Selected Data for the Period 1989 Through 2000). *J. Phys. Chem. Ref. Data* **2002**, *31*, 819–927. [CrossRef]
55. Lesage, A. Experimental Stark widths and shifts for spectral lines of neutral and ionized atoms A critical review of selected data for the period 2001–2007. *New Astron. Rev.* **2009**, *52*, 471–535. [CrossRef]
56. Wiese, W.L.; Konjević, N. Regularities and similarities in plasma broadened spectral line widths. *J. Quant. Spectrosc. Radiat. Transf.* **1982**, *28*, 185–198. [CrossRef]
57. Dimitrijević, M.S.; Konjević, N. Stark widths of doubly- and triply-ionized atom lines. *J. Quant. Spectrosc. Radiat. Transf.* **1980**, *24*, 451–459. [CrossRef]
58. Dimitrijević, M.S.; Konjević, N. Simple estimates for Stark broadening of ion lines in stellar plasmas. *Astron. Astrophys.* **1987**, *172*, 345–349.
59. Griem, H.R. *Spectral Line Broadening by Plasmas*; Academic Press: New York, NY, USA, 1974.

![atoms logo] *atoms*

MDPI

Article

The Study of Ar I and Ne I Spectral Line Shapes in the Cathode Sheath Region of an Abnormal Glow Discharge

Nikola V. Ivanović

Faculty of Agriculture, University of Belgrade, 11080 Belgrade, Serbia; nikolai@ff.bg.ac.rs; Tel.: +381-64-229-2435

Received: 30 November 2018; Accepted: 3 January 2019; Published: 9 January 2019

check for updates

Abstract: The cathode sheath (CS) region is the most important part of abnormal glow discharge (GD), where various processes relevant for the operation and application occur. The most important parameter of the CS is the distribution of electric field strength E which is of crucial importance for charged particles acceleration, their trajectories, kinetic energies, and collisions with other particles and cathode sputtering. All these processes are relevant for the operation of GD as well as for numerous applications in the field of spectroscopic analysis, plasma etching, thin film deposition, and depth profiling of cathode material. Thus, the importance of non-perturbing technique for E distribution measurement in the CS region was recognized long time ago. Within this article, a simple technique based on standard optical emission spectroscopy (OES) and typical laboratory equipment has been used for E mapping in the CS region of an abnormal glow discharge.

Keywords: spectroscopy; gas discharges; plasma applications

1. Introduction

Over several past decades, glow discharges (GD) have been successfully used as the excitation sources in analytical spectroscopy of metal and alloy samples, as well as in their depth profiling [1–3]. For successful operation and maintenance of GD, their cathode sheath (CS) region turns out to be of crucial importance. The condition necessary for successful use of the CS region for investigation of electric field was first given by Lo Surdo see e.g., [4]. By reduction of the distance between cathode and anode, a great potential drop located in a few millimeters is attained. In such a way, an appreciable density of the electric current arises, and this is of vital importance for the intensity of light emitted from the discharge.

In the cathode sheath region, the emitted spectral lines exhibit a direct current (dc) Stark effect under the influence of electric field E, causing splitting of the radiation into Stark components and their Stark shifting. A well-established method for the measurement of electric field is based on the spectroscopy of the first two hydrogen Balmer lines (H_α, H_β), which exhibit a linear dc Stark effect, see for instance [4].

Recent studies of He I lines, see e.g., [5–12], indicate that some of the He I lines may be used for low electric field E measurement with an accuracy of several percent. Other experiments were performed with argon and neon as the working gas, where the spectral lines of argon and neon atoms, Ar I and Ne I, exhibit a quadratic Stark effect, see e.g., [13–16]. Most of these experiments were performed at relatively high E values exceeding 100 kVcm^{-1}. A theoretical study of neutral neon lines was given by Ziegelbecker and Schnizer [17].

Because glow discharge sources were intensively used for applications in optical emission spectroscopy (OES), it has become evident that reliable wavelength tables of glow discharge spectra are needed. A group of authors decided to overcome this lack of data and began measuring and comparing GD wavelengths and line intensity data with other available sources of information [18]. For the measurement of line intensity and wavelength several instruments were used, see [18] for details. In order to achieve a high spectral resolution a Fourier transform spectrometer is used (FTS). As a light source, a Grimm-type GD with 4 mm anode hole operated at the pressure of about 15 mbar in argon or neon is used. This analytical GD source is built on the basis of the Grimm original design [19]. All discharge observations are carried out end-on through the anode opening, perpendicular to the cathode surface, with argon and neon as working gas. These two gases were used most frequently to operate an analytical GD. This investigation showed that some Ar I and Ne I lines have large widths, exceeding 4–5 times the instrumental width, and in addition to line broadening, red shifting of lines has also been detected [20].

Theoretically speaking, there are two possible reasons for the broadening and shifting of non-hydrogenic spectral lines in a Grimm GD source observed perpendicular to the cathode surface. These are: (i) the dc Stark effect within the cathode sheath region; and (ii) plasma broadening in the negative glow (NG) region. The aim of this work is to study and explain the origin of large widths and shifts detected in [18], as well as to supply experimental Stark shift data for several Ne I lines not studied before. The explanation is of importance for the application of new wavelength tables and may have a considerable importance to preventing possible misinterpretation of the profile wavy features as a weak radiation emitted by some elements that are contained in traces in the samples under study. Also, an appropriated method for measurement of electric field in CS region was developed.

2. Description of Experimental Setup

The discharge source, a modified Grimm GD, was laboratory made after a Ferreira et al. design [21]. A hollow anode 30 mm long with inner and outer diameter 8 mm and 13 mm, respectively, has a longitudinal slot (16 mm long and 1.5 mm wide) for side-on observations along the discharge axis, see Figure 1. The water-cooled cathode holder has an exchangeable electrode (Fe, W), 18 mm long and 7.60 mm in diameter, which screws tightly into its holder to ensure good cooling.

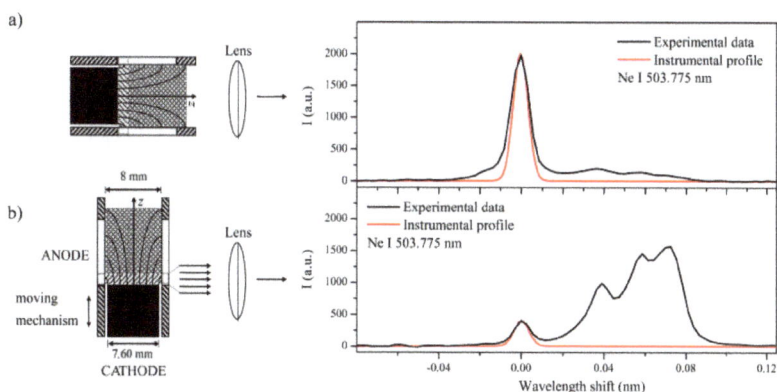

Figure 1. (a) Schematic description of the end-on recordings setup (**left**), and the end-on profile of the Ne I 503.775 nm spectral line (**right**). (b) Schematic description of the side-on recordings setup (**left**), and the side-on profile of the same line recorded in the same experiment in the vicinity of the cathode surface.

Experiments were carried out with argon (purity 99.999%) and mixture of neon and hydrogen (vol. 99.2% Ne + 0.8% H_2). The continuous flow of about 300 cm^3/min of working gases was sustained in the pressure range 5–10 mbar by means of needle valve and two two-stage mechanical vacuum pumps. The reported results for gas pressure represent an average between gas inlet and outlet pressure measurements.

To run the discharge a current stabilized dc power supply (0–2 kV, 0–100 mA) is used. A ballast resistor of 5.3 kΩ is placed in series with the discharge and the power supply. Spectroscopic observations of Grimm GD were performed end-on, while for axial intensity distribution of side-on radiation observed through the anode slot, the discharge tube was moved in ≈0.125 mm steps. The light from the discharge was focused with an achromatic lens (focal length 75.8 mm) with 1:1 magnification onto the 20 μm entrance slit (height restriction 2 mm) of 2 m focal length Ebert type spectrometer with 651 g/mm reflection grating blazed at 1050 nm. For the line shape measurements, the reciprocal dispersion of 0.37 nm/mm in second diffraction order is used. All spectral measurements were performed with an instrumental profile very close to Gaussian form with measured full width at half maximum (FWHM) of 0.0082 nm. Signals from CCD detector (29.1 mm length, 3648 pixels, 1 pixel ≈ 0.00278 nm) were A/D converted, collected, and processed by PC.

The spectral lines were observed along the axis of a cylindrical glow discharge perpendicular to the cathode surface, see end-on profile in Figure 1a, and parallel to the cathode surface, see side-on profile in Figure 1b. The overall profile of the Ne I 503.775 nm spectral line recorded end-on comprises of the pronounced central peak and a wavy red wing, see Figure 1a.

3. Historical Background

3.1. The dc Stark Effect on Ar I Spectral Lines

The line shifting of ninety-one Ar I lines with electric field were studied in [13]. Only two Ar I lines from this study are detected in our experimental conditions, see Table 1 containing their configuration, term values and total angular momentum. The shifting of the non-degenerate upper level of Ar I 522.127 nm as a function of electric field strength E can be evaluated by means of the approximate formulas [13]:

$$\Delta \bar{v}_i = \frac{a_1 E^2}{a_2 - \Delta \bar{v}_i} \text{ or } E = \sqrt{\frac{\Delta \bar{v}_i (a_2 - \Delta \bar{v}_i)}{a_1}} \tag{1}$$

and for the upper level of Ar I 518.775 nm line by the following Formula [13]:

$$\Delta \bar{v}_i = A_1 E + A_2 E^2 + A_3 E^3 + \dots \tag{2}$$

where: $\Delta \bar{v}_i$ in both formulas represents the change of wave number of level i; E is the strength of electric field, while a_1, a_2, A_1, and A_2 are the algebraic coefficients for the studied energy levels taken from Tables III and IV in [13]. The list of studied Ar I lines with relevant spectroscopic data is given in Table 1, and the list of algebraic coefficients for these lines is given in Table 2.

Table 1. List of studied Ar I lines [13].

Emitter	Wavelength (nm)	Lower Level Conf., Term, J	Upper Level Conf., Term, J
Ar I	518.775	$3s^2 3p^5 (^2P°_{3/2}) 4p^2[^1/_2]$ 1	$3s^2 3p^5 (^2P°_{1/2}) 5d'^2[^3/_2]°$ 2
	522.127	$3s^2 3p^5 (^2P°_{3/2}) 4p^2[^5/_2]$ 3	$3s^2 3p^5 (^2P°_{3/2}) 7d^2[^7/_2]°$ 4

Table 2. Wavelengths and dc Stark effect coefficients for studied Ar I lines [13].

Wavelength (nm)	a_1 (kV^{-2})	a_2 (cm^{-1})	A_1 (kV^{-1})	A_2 (kV^{-2}cm^{-2})	A_3 (kV^{-3}cm^{-2})
518.775	-	-	1.25×10^{-1}	1.51×10^{-3}	-4.19×10^{-6}
522.127	4.23×10^{-1}	-169.34	-	-	-

Examples of Stark shifts for the Ar I 522.127 nm and the Ar I 518.775 nm lines are given in Figure 2a,b, respectively, as electric field strength E against the change of wave number $\Delta \bar{v}_i$.

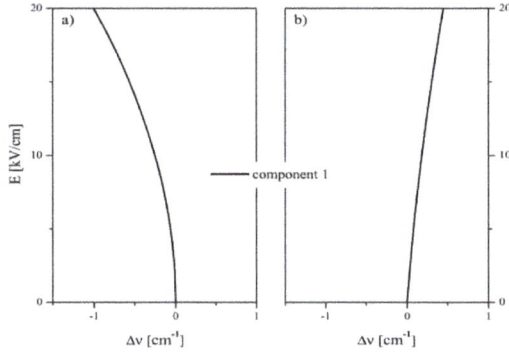

Figure 2. Behavior of terms of group: (**a**) 7*d* (Ar I 522.127 nm) and (**b**) 5*d'* (Ar I 518.775 nm) in a dc Stark field [13].

3.2. The dc Stark Effect on Ne I Spectral Lines

In Table 3, the list of studied Ne I spectral lines are presented. For the two Ne I spectral lines denoted with asterisk, the relevant dc Stark shift data are taken from [14].

Table 3. The list of studied Ne I lines. The lines denoted with an asterisk were studied in [14].

Emitter	Wavelength (nm)	Lower Level Conf., Term, J	Upper Level Conf., Term, J
	503.135	$2s^2 2p^5 (^2P^\circ_{3/2}) 3p^2 [^5/_2] 3$	$2s^2 2p^5 (^2P^\circ_{3/2}) 5d^2 [^5/_2]^\circ 3$
	503.775	$2s^2 2p^5 (^2P^\circ_{3/2}) 3p^2 [^5/_2] 3$	$2s^2 2p^5 (^2P^\circ_{3/2}) 5d^2 [^7/_2]^\circ 4$
	507.420	$2s^2 2p^5 (^2P^\circ_{3/2}) 3p^2 [^5/_2] 2$	$2s^2 2p^5 (^2P^\circ_{3/2}) 5d^2 [^5/_2]^\circ 2$
	508.038	$2s^2 2p^5 (^2P^\circ_{3/2}) 3p^2 [^5/_2] 2$	$2s^2 2p^5 (^2P^\circ_{3/2}) 5d^2 [^7/_2]^\circ 3$
Ne I	511.367*	$2s^2 2p^5 (^2P^\circ_{3/2}) 3p^2 [^1/_2] 1$	$2s^2 2p^5 (^2P^\circ_{1/2}) 4d'^2 [^3/_2]^\circ 1$
	515.196	$2s^2 2p^5 (^2P^\circ_{3/2}) 3p\ ^2 [^3/_2] 1$	$2s^2 2p^5 (^2P^\circ_{3/2}) 5d\ ^2 [^5/_2]^\circ 2$
	515.443	$2s^2 2p^5 (^2P^\circ_{3/2}) 3p\ ^2 [^3/_2] 1$	$2s^2 2p^5 (^2P^\circ_{3/2}) 5d\ ^2 [^3/_2]^\circ 1$
	520.390	$2s^2 2p^5 (^2P^\circ_{3/2}) 3p\ ^2 [^3/_2] 2$	$2s^2 2p^5 (^2P^\circ_{3/2}) 5d^2 [^5/_2]^\circ 3$
	520.886	$2s^2 2p^5 (^2P^\circ_{3/2}) 3p\ ^2 [^3/_2] 2$	$2s^2 2p^5 (^2P^\circ_{3/2}) 5d\ ^2 [^3/_2]^\circ 2$
	534.109*	$2s^2 2p^5 (^2P^\circ_{3/2}) 3p^2 [^1/_2] 1$	$2s^2 2p^5 (^2P^\circ_{3/2}) 4d^2 [^1/_2]^\circ 1$

In reference [14], Jäger and Windholz presented the results of experimental study of dc Stark shift for 141 Ne I lines corresponding to 31 levels of neon atom. This study showed that the relation between electric field strength E and measured wave number shifts $\Delta\sigma$ can be approximated in a rather wide range of E by the analytical expressions:

$$\Delta\sigma = \Delta\sigma_1 + \Delta\sigma_2,$$
$$\Delta\sigma_1 = (A_1 E^2)/(A_2 - \Delta\sigma_1); \ \Delta\sigma_2 = A_3 E^2. \tag{3}$$

resulting in explicit form:

$$\Delta\sigma = \frac{A_2 - \sqrt{A_2^2 - 4A_1 E^2}}{2} + A_3 E^2. \tag{4}$$

In Equations (3) and (4), the parameters A_1, A_2, and A_3 are the line-specific coefficients chosen in such way to obtain the best agreement with experimental data. The values of these three coefficients for each Stark component of Ne I 511.367 nm (3 components) and Ne I 534.109 nm line (2 components), are presented in Table 4 in accordance with [14].

Table 4. The A_1, A_2, and A_3 parameters for Ne I 511.367 nm and Ne I 534.109 nm lines [14].

Wavelength (nm)	Comp.	A_1 (kV^{-2})	A_2 (cm^{-1})	A_3 (cm kV^{-2})
511.367	1	-6.111×10^{-2}	2.783×10^1	-4.700×10^{-5}
	2	-8.889×10^{-2}	5.165×10^1	-1.370×10^{-4}
	3	-1.416×10^{-1}	4.707×10^1	-6.700×10^{-15}
534.109	1	-7.935×10^{-2}	7.138×10^1	-1.470×10^{-4}
	2	-2.330×10^{-1}	1.883×10^2	-5.200×10^{-5}

The Stark shift dependence versus electric field strength E for the Ne I 511.367 nm line and the Ne I 534.109 nm line are given in Figure 3a,b, respectively.

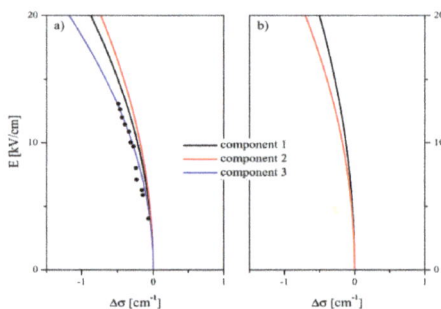

Figure 3. Behavior of terms of group: (**a**) $4d'$ (Ne I 511.367 nm) and (**b**) $4d$ (Ne I 534.109 nm) in a dc Stark field [14].

4. Results

Taking into consideration the results of analytical GD investigation presented in [22], the first end-on recordings were performed on Ne I 503.775 nm line. One may notice that the overall profile of Ne I 503.775 nm line comprises the central peak and a wavy red wing, see Figure 1a. The central peak dominantly consists of the unshifted radiation from the NG region, while the wavy wing consists of the red-shifted radiation from the CS region due to quadratic dc Stark effect. When observed side-on, the profile of Ne I 503.775 nm line emitted in the vicinity of cathode surface shows a complex structure, see Figure 1b. From this Ne I profile, one can easily recognize a small central peak of the unshifted radiation, followed by a pronounced red wing consisting of three components with different red shifts. The unshifted radiation is emitted from the discharge protruding outside of electric field region, and this phenomenon was noticed earlier in the study of He I lines in the same type of discharge [6]. The presence of unshifted radiation enabled us to measure dc Stark shift of studied lines. Increasing the distance to the cathode, the position of shifted profiles maxima move towards the central peak due to decrease of E, which falls to zero at the boundary between CS and NG region. For that reason, the three wavy peaks of the end-on profile, representing the integral radiation both from CS and NG, are shifted somewhat towards the central peak but remain correlated with the maxima of the presented Ne I side-on profile. On the basis of this research, one may conclude that the wavy forms observed during end-on recordings, are the result of superposition of line intensities emitted from the CS region (the influence of the dc Stark effect) and the NG region. It is important to mention that similar behavior was observed as well for other investigated spectral lines (Ar I and Ne I), but the Ne I 503.775 nm line represent the best example. The primary goals of this work are the study of Ar I and Ne I spectral lines emitted from CS region, the measurement of dc Stark shifts and determination of the electric field distribution along CS region of Grimm type abnormal glow discharge.

4.1. Results of the dc Stark Shift Measurements for Ar I Lines

Although, a numerous Ar I lines in the presence of dc electric field were already investigated ($E > 60$ kV/cm) [13], under our experimental conditions ($E<15$ kV/cm), only two Ar I spectral lines with noticeable dc Stark effect are detected. From Figure 2a one may notice that Ar I 522.127 nm line shifts toward higher wavelengths (smaller wave numbers). For the Ar I 518.775 nm line, the dc Stark shift is smaller and directed toward the blue spectral region (shorter wavelengths), see Figure 2b. The dc Stark shift of Ar I 518.775 nm line as well as other Ar I lines originating from versatile upper levels shows a complex dependence on E [13]. For these levels, a maximum shift takes place at certain field strength, and above this value the shift decreases. A thorough investigation of Ar I 518.775 nm line was made in [23]. Here, for the sake of electric field distribution measurement, Ar I 522.127 nm line was used.

In order to confirm the influence of the dc Stark effect on the profiles of Ar I 522.127 nm line, a simple experiment was carried out. The profile of Ar I spectral line was recorded side-on near the cathode surface, where the maximum strength of the electric field is expected. The change in observed Ar I 522.127 nm spectral line profiles with the input electrical power (primarily by increasing the discharge current) is shown in Figure 4. For different values of electrical power, the profile of Ar I 522.127 nm spectral line shows different values of dc Stark shifts and the width, which indicates the presence of dc Stark effect.

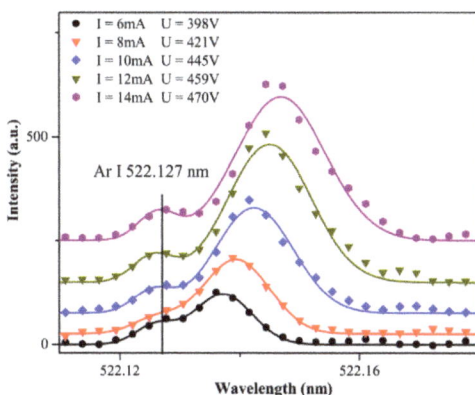

Figure 4. The side-on recording of Ar I 522.127 nm line shapes in the vicinity of cathode for different currents and the best fits. For better representation, the profiles are moved upward deliberately. The discharge pressure was $p = 6.5$ mbar.

A systematic study of Ar I spectral line profiles in the CS region is performed in the scope of this study. Spectroscopic observations are performed side-on i.e., along the direction normal to direction of the electric field strength, see Figure 1. Ar I spectral lines was observed end-on but for more details see [24]. The change in Ar I 522.127 nm line profiles along CS region is shown in Figure 5. The spectra shown in Figure 5 a–c depict the spectral line shape at three GD axial positions starting from the vicinity of the cathode surface. The unshifted component in Figure 5 appears in side-on spectra as well and is successfully used to measure dc Stark shift.

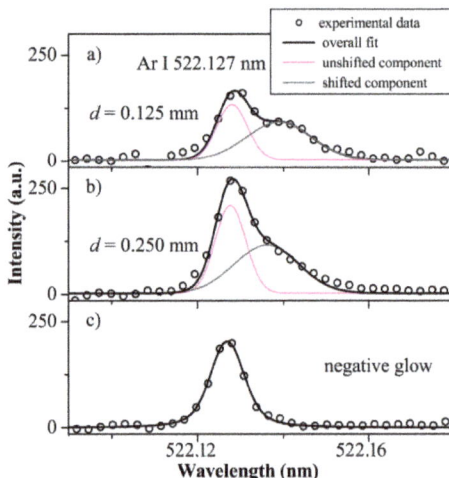

Figure 5. Spectral line shapes and best fits of Ar I 522.127 nm line at different axial positions from cathode: (**a**) 0.125 mm, (**b**) 0.25 mm and (**c**) NG region. Discharge conditions: iron cathode, $p = 6.5$ mbar, $I = 12$ mA, $U = 472$ V.

In order to accurately determine the position of the shifted Stark component, the profiles of Ar I 522.127 nm line, shown in Figure 5a,b, are fitted with Gaussian profiles. The unshifted component has a Gaussian instrumental profile with constant width for all recorded spectra, while the width of shifted Stark component is considered to be twice as large as the unshifted component, in accordance with results given in [25]. The profile recorded in the NG region is fitted with a symmetric Voigt function. Also, one should mention that the sensitivity of line detection system and geometrical factors of the optical system are identical for each spectra presented in Figure 5a–c.

Finally, using the results of Ar I 522.127 nm line dc Stark shifts in conjunction with Equation (1), it is possible to determine the electric field strength distribution in the CS region, see Figure 6. The voltage drop across the CS shown in Figure 6, calculated by integrating the measured field strength distribution, agrees well within 15% with the applied discharge voltage for iron as cathode material.

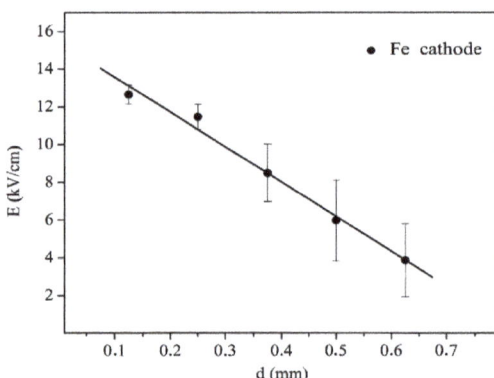

Figure 6. Distribution of electric field strength in the GD CS region determined using measured shifts for the Ar I 522.127 nm line and Equation (1). Discharge conditions: iron cathode, $p = 6.5$ mbar, $I = 12$ mA and $U = 472$V.

4.2. Results of the dc Stark Shift Measurements for Ne I Lines

During investigation of Ne I spectral lines, see [26], it was shown that Ne I lines are good candidates for electric field strength diagnostics in the CS region of an abnormal Grimm type glow discharge. In relation to this, the H_α line of hydrogen is analyzed first in order to measure E, while dc Stark shifts are determined using Ne I spectral lines, recorded from the CS region.

In the study of Grimm type discharge CS region in neon, a small amount (vol. 0.8%) of hydrogen was added as the admixture to perform the spectroscopic diagnostics of electric field using π-polarized side-on profiles of the H_α line. In Figure 7, an example of such profile recorded from the cathode surface region is presented. The unperturbed central peak corresponds to radiation emitted from the protruding discharge through the anode slot, as already mentioned previously. This peak is superimposed with the π-polarized profile of the H_α shifted radiation, originating from a non-zero electric field CS region visible through the slot. The presence of the central peak enabled us to present data in Figure 7 in the function of wavelength shift $\Delta\lambda$ instead of wavelength λ. The solid (red) line represent the best fit of experimental data obtained by advanced model function, see Equation (8), given in [27]. The corresponding best fit value of E together with the distance d from cathode surface is shown in Figure 7. For more details about the measurement of electric field strength distribution in CS region analyzing the H_α line profiles, see [27].

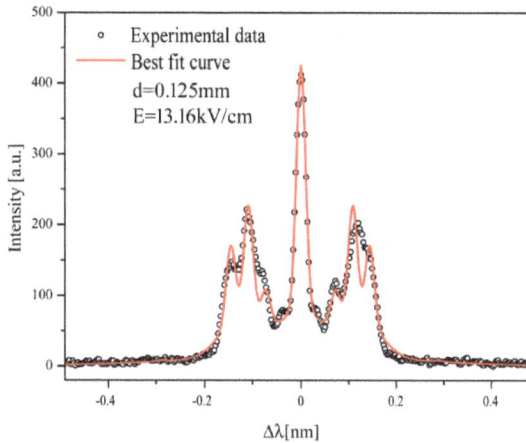

Figure 7. The π polarized profile of the H_α spectral line, recorded from the CS region. Discharge conditions: tungsten cathode, $p = 6$ mbar, $I = 12.11$ mA, and $U = 914$ V.

The analysis of Ne I 511.367 nm line dc Stark shift, previously reported by Jäger and Windholz [14], revealed that the line splits into three Stark components (Figure 3a) but under experimental conditions in [14], only one component is detected. In Figure 8, the Ne I 511.367 nm line profiles recorded side-on at four different positions within the CS, as well as the profile recorded in the NG region are presented. In order to measure dc Stark shifts of the studied Ne I line the following model function is applied:

$$I_{Ne}(\Delta\lambda; H, c, b) = A\Im(\Delta\lambda) + \Im \times G(\Delta\lambda; H_{Ne}, c_{Ne}, w_{Ne}) + b_{Ne} \tag{5}$$

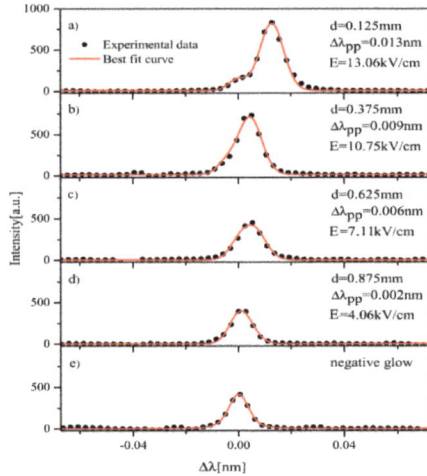

Figure 8. The Ne I 511.367 nm spectral line (dots) recorded from the CS at four different distances *d* from the cathode surface (panels (**a–d**)), and from the NG region (panel (**e**)). Solid (red) line represents the best fit obtained by model function (5) for the profiles recorded in CS, and by symmetric Voigt profile in NG region. Discharge conditions: same as in Figure 7.

As shown in Equation (5), the model function comprises the three fitting parameters: the height H_{Ne} and the center c_{Ne} of Gaussian profile, see Equation (6), as well as the base line level b_{Ne}

$$G(\Delta\lambda; H_{Ne}, c_{Ne}, w_{Ne}) = H_{Ne} \exp\left[-\left(2\sqrt{2\ln 2}\frac{\Delta\lambda - c_{Ne}}{w_{Ne}}\right)^2\right] \tag{6}$$

The function shown in Equation (6) represents the Doppler broadening profile of the Stark shifted Ne I 511.367 nm line. The two remaining parameters appearing in Equation (5), the area of protruded radiation *A* and the FWHM of Stark shifted neon line w_{Ne}, maintain the constant values for all profiles recorded from the CS region. The profile observed from the NG region is well described with symmetric Voigt function.

The dc Stark shifts determined using mentioned numerical procedure show the reasonable agreement with prediction of Jäger and Windholz's for the Ne I 511.367 nm line third Stark component, see black dots in Figure 3a.

In addition to experimentally obtained Ne I lines, an additional six Ne I lines were detected with the primary task to supply dc Stark shift data for these lines. To measure dc Stark shift, the model function (5) is employed once again, while the distribution of electric field is calculated by means of the H_α line, see the text above. As an example, the side-on profiles of the Ne I 503.134 nm line recorded at the same set of distances from the cathode as in Figure 8 are given in Figure 9. Due to presence and crossing of more than one Stark component, experimental data designated with hollow dots in Figure 9 are excluded from the fitting curve. This was accomplished with the aid of the advanced numerical algorithm of outlier detection [28]. The Ne I 503.134 nm line (as well as five other Ne I lines) has significantly higher wavelength shifts then the studied Ne I 511.367 nm line, showing the full separation between the Stark shifted component and the unshifted peak of protruded radiation. In the position where the clear separation between unshifted and shifted component occurs, the simple "peak to peak" dc Stark shift measurement technique is used, see panels in Figure 9a–c. The same procedure is applied on five other spectral lines. It is a well-known fact that spectral lines of neon atom exhibit a quadratic Stark effect. This was also supported by the Jäger and Windholz Equation (4) which, in the small field range reduces to:

$$\Delta\sigma \approx \left(\frac{A_1}{A_2} + A_3\right) E^2 = CE^2, \tag{7}$$

or equally

$$\Delta\lambda \approx -\lambda_0^2 CE^2. \tag{8}$$

The measured Stark shifts are fitted with the simple quadratic function given by Equation (7), and the best fit values of the constant *C* are listed in Table 5.

Table 5. The Stark shift coefficients *C* of quadratic function (7) for the six Ne I lines listed in Table 3. E_{max} is the maximum value of measured electric field strength.

Wavelength (nm)	E_{max} [kV/cm]	Upper Level	C [cm/kV^2]	ΔC [cm/kV^2]
503.135		$2s^2 2p^5(^2P°_{3/2})\, 5d\,^2[^5/_2]°\ 3$	−0.0238	
520.390				
507.420	13.4	$2s^2 2p^5(^2P°_{3/2})\, 5d\,^2[^5/_2]°\ 2$	−0.0229	0.005
515.196				
515.442		$2s^2 2p^5(^2P°_{3/2})\, 5d^2[^3/_2]°\ 1$	−0.0189	
520.886		$2s^2 2p^5(^2P°_{3/2})\, 5d^2[^3/_2]°\ 2$	−0.0138	

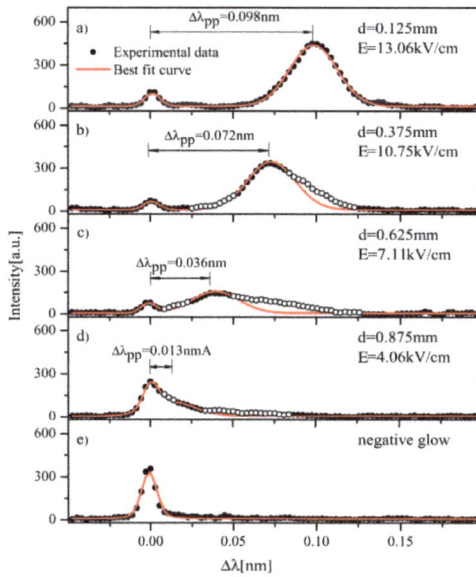

Figure 9. Experimental profiles (points) of the Ne I 503.134 nm line recorded at different distances *d* from the cathode surface (panels (**a–d**)), and from the NG region (panel (**e**)), and their best fits (red lines); hollow points are discarded in fitting. In legends, I give the corresponding values of electric field strength *E* and Stark wavelength shift $\Delta\lambda_{pp}$ measured between the unshifted and shifted components on the recorded line shapes. Discharge conditions: same as in Figure 7.

5. Conclusions

In this work, the diagnostics of the CS region of Grimm type abnormal glow discharge in argon and neon–hydrogen mixture are shown. The broadening and shifting of argon and neon spectral lines in the Grimm type observed perpendicular to cathode surface (end-on) is explained by (i) dc Stark effect within the cathode sheath and (ii) plasma broadening in the negative glow region.

By comparison of the end-on with the side-on profiles, it has been demonstrated that the wavy form of the far red wing of end-on profile originates from the dc Stark effect, causing splitting of the radiation into Stark components and their Stark shifting. Taking into account that the commercial GD sources used for the elemental trace analysis permits only end-on observations, this finding could be valuable because it may prevent possible misinterpretation of the wavy features of the end-on profile as a weak radiation from some elements that are contained in traces in the sample under study.

The investigation of dc Stark effect in the CS region was of particular interest in this study. The distribution of electric field in the CS region is measured using Ar I 522.127 nm spectral line. In another experimental investigation, the Ne I lines were employed in conjunction with the dedicated model function for the dc Stark shift measurement under the conditions of low electric field. The electric field strength in the CS region is determined from the H_α line π-polarized profile using the Stark polarization spectroscopy technique.

On the other hand, the results of the Ne I 511.367 nm line analysis are used for testing the best fit formula derived from measured Stark shifts at high electric fields (about 400 kV/cm). Concerning the other six Ne I lines, the coefficients in the best fit Formula (7) are determined and may be applied for low electric field measurement (up to 13.4 kV/cm) with uncertainty below 5%.

Acknowledgments: This work is supported by the Ministry of Education, Science and Technological Development of the Republic of Serbia under Projects 171014 and 171027. I would like to thank Gordana Majstorović, Đorđe Spasojević, and Nikola Konjević, my coworker Jovica Jovović, and in particular Nikola Šišović who was my supervisor on Ph.D. studies.

Conflicts of Interest: The author declares no conflict of interest. The funders had no role in the design of the study; in the collection, analyses, or interpretation of data; in the writing of the manuscript, or in the decision to publish the results.

References

1. Jakubowski, N.; Bogaerts, A.; Hoffmann, V. *Glow Discharges in Emission and Mass Spectrometry Atomic Spectroscopy in Elemental Analysis*; Cullen, M., Ed.; Blackwell: Sheffield, UK, 2003.
2. Chapman, B. *Glow Discharge Processes Sputtering and Plasma Etching*; John Wiley & Sons: New York, NY, USA, 1980.
3. Marcus, R.K.; Broeckaert, J.A.C. *Glow Discharge Plasmas in Analytical Spectroscopy*; Wiley: Chichester, UK, 2003.
4. Ryde, N. *Atoms and Molecules in Electric Fields*; Almqvist & Wiksell International: Stockholm, Sweden, 1976.
5. Kuraica, M.M.; Konjević, N.; Videnović, I.R. Spectroscopic study of the cathode fall region of Grimm-type glow discharge in helium. *Spectrochim. Acta B* **1997**, *52*, 745–753. [CrossRef]
6. Kuraica, M.M.; Konjević, N. Electric field measurement in the cathode fall region of a glow discharge in helium. *Appl. Phys. Lett.* **1997**, *70*, 1521–1523. [CrossRef]
7. Windholz, L.; Winklhofer, E.; Drozdowski, R.; Kwela, J.; Wąsowicz, T.J.; Heldt, J. Stark effect of atomic helium second triplet series in electric fields up to 1600 kV cm^{-1}. *Phys. Scr.* **2008**, *78*, 065303. [CrossRef]
8. Windholz, L.; Wąsowicz, T.J.; Drozdowski, R.; Kwela, J. Stark effect of atomic Helium singlet lines. *J. Opt. Soc. Am. B* **2012**, *29*, 934–943. [CrossRef]
9. Cvetanović, N.; Martinović, M.M.; Obradović, B.M.; Kuraica, M.M.J. Electric field measurement in gas discharges using stark shifts of He I lines and their forbidden counterparts. *Phys. D Appl. Phys.* **2015**, *48*, 205201. [CrossRef]
10. Yatom, S.; Stambulchik, E.; Vekselman, V.; Krasik, Y.E. Spectroscopic study of plasma evolution in runaway nanosecond atmospheric-pressure He discharges. *Phys. Rev. E* **2013**, *88*, 013107. [CrossRef] [PubMed]
11. Yatom, S.; Tskhai, S.; Krasik, Y.E. Electric field in a plasma channel in a high-pressure nanosecond discharge in hydrogen: A coherent anti-stokes Raman scattering study. *Phys. Rev. Lett.* **2013**, *111*, 255001. [CrossRef] [PubMed]
12. Levko, D.; Yatom, S.; Krasik, Y.E. Particle-in-cell modeling of the nanosecond field emission driven discharge in pressurized hydrogen. *J. Appl. Phys.* **2018**, *123*, 083303. [CrossRef]
13. Windholz, L. Stark effect of Ar I'-Lines. *Phys. Scr.* **1980**, *21*, 67–74. [CrossRef]
14. Jäger, H.; Windholz, L. Stark Effect of NeI-Lines (I). *Phys. Scr.* **1984**, *29*, 344. [CrossRef]

15. Windholz, L.; Neureiter, C. Laser spectroscopic measurements of the Stark shift of the Ne i lines at 5852 and 5882 A. *Phys. Rev. A* **1988**, *37*, 1978–1982. [CrossRef]

16. Jäger, H.; Windholz, L.; Ziegelbecker, R.C. Stark effect of neon I–lines (II): designation of Stark levels according to theoretical results. *Phys. Scr.* **1989**, *40*, 740–744. [CrossRef]

17. Ziegelbecker, R.C.; Schnizer, B.Z. Calculation of the Stark effect of neon I using *jl* coupled wave functions. *Z. Phys. D Atoms Mol. Cluster* **1987**, *6*, 327–335. [CrossRef]

18. Weiss, Z.; Steers, E.B.M.; Šmíd, P.; Hoffmann, V. Towards a catalogue of glow discharge emission spectra. *J. Anal. At. Spectrom.* **2009**, *24*, 27. [CrossRef]

19. Grimm, W. Eine neue glimmentladungslampe für die optische emissionsspektralanalyse. *Spectrochim. Acta B* **1968**, *23*, 443. [CrossRef]

20. Weiss, Z.; (LECO Instrumente Plzeň, spol. s r.o., Plzeň, Czech Republic). Private communication with Konjević, N., (University of Belgrade, Belgrade, Serbia). 2011.

21. Ferreira, N.P.; Human, H.G.C.; Butler, L.R.P. Kinetic temperatures and electron densities in the plasma of a side view Grimm-type glow discharge. *Spectrochim. Acta B* **1980**, *35*, 285–295. [CrossRef]

22. Weiss, Z.; Steers, E.B.M.; Šmid, P.; Pickering, J.C. The GLADNET Catalogue of Glow Discharge Spectra: Part 1 finished! (almost). In Proceedings of the EW-GDS Meeting, Kingston, UK, 3–4 September 2012.

23. Vasiljević, M.M.; Spasojević, Dj.; Šišović, N.M.; Konjević, N. Stark effect of Ar I lines for electric field strength diagnostics in the cathode sheath of glow discharge. *Europhys. Lett.* **2017**, *119*, 55001. [CrossRef]

24. Majstorović, G.L.; Ivanović, N.V.; Šišović, N.M.; Djurović, S.; Konjević, N. Ar I and Ne I spectral line shapes for an abnormal glow discharge diagnostics. *Plasma Sources Sci. Technol.* **2013**, *22*, 045015. [CrossRef]

25. Nakajima, T.; Uchitomi, N.; Adachi, Y.; Maeda, S.; Hirose, C. Stark Shift and Broadening of Atomic Lines as Observed on Optogalvanic Spectra of Noble Gases. *J. Phys. Colloque.* **1983**, *44*, C7-497–C7-504. [CrossRef]

26. Šišović, N.M.; Ivanović, N.V.; Majstorović, G.L.; Konjević, N. Ne I spectral line shapes in Grimm-type glow discharge. *J. Anal. At. Spectrom.* **2014**, *29*, 2058–2063. [CrossRef]

27. Ivanović, N.V.; Šišović, N.M.; Spasojević, D.; Konjević, N. Measurement of the DC Stark shift for visible NeI lines and electric field distribution in the cathode sheath of an abnormal glow discharge. *J. Phys. D Appl. Phys.* **2017**, *50*, 125201. [CrossRef]

28. Bukvić, S.; Spasojević, D.; Žigman, V. Advanced fit technique for astrophysical spectra-Approach insensitive to a large fraction of outliers. *Astron. Astrophys.* **2008**, *477*, 967–977. [CrossRef]

atoms

MDPI

Article

BEAMDB and MOLD—Databases at the Serbian Virtual Observatory for Collisional and Radiative Processes

Bratislav P. Marinković [1,*], Vladimir A. Srećković [1], Veljko Vujčić [2], Stefan Ivanović [1,3], Nebojša Uskoković [1,3], Milutin Nešić [3], Ljubinko M. Ignjatović [1], Darko Jevremović [2], Milan S. Dimitrijević [2,4] and Nigel J. Mason [5,6]

[1] Institute of Physics Belgrade, University of Belgrade, Pregrevica 118, 11080 Belgrade, Serbia; vlada@ipb.ac.rs (V.A.S.); stefan.ivanovic992@gmail.com (S.I.); nesauskokovic@gmail.com (N.U.); ljubinko.ignjatovic@ipb.ac.rs (L.M.I.)

[2] Astronomical Observatory Belgrade, Volgina 7, 11000 Belgrade, Serbia; veljko@aob.rs (V.V.); darko@aob.rs (D.J.); mdimitrijevic@aob.rs (M.S.D.)

[3] The School of Electrical Engineering and Computer Science of Applied Studies, Vojvode Stepe 283, 11000 Belgrade, Serbia; nesic@viser.edu.rs

[4] Sorbonne Université, Observatoire de Paris, Université PSL, CNRS, LERMA, F-92190 Meudon, France

[5] Department of Physical Sciences, The Open University, Milton Keynes MK7 6AA, UK; N.J.Mason@open.ac.uk

[6] School of Physical Sciences, University of Kent, Canterbury, Kent CT2 7NZ, UK

* Correspondence: bratislav.marinkovic@ipb.ac.rs; Tel.: +381-11-316-0882

Received: 1 December 2018; Accepted: 9 January 2019; Published: 14 January 2019

check for updates

Abstract: In this contribution we present a progress report on two atomic and molecular databases, BEAMDB and MolD, which are web services at the Serbian virtual observatory (SerVO) and nodes within the Virtual Atomic and Molecular Data Center (VAMDC). The Belgrade Electron/Atom (Molecule) DataBase (BEAMDB) provides collisional data for electron interactions with atoms and molecules. The Photodissociation (MolD) database contains photo-dissociation cross sections for individual rovibrational states of diatomic molecular ions and rate coefficients for the chemi-ionisation/recombination processes. We also present a progress report on the major upgrade of these databases and plans for the future. As an example of how the data from the BEAMDB may be used, a review of electron scattering from methane is described.

Keywords: databases; virtual observatory; cross sections; rate coefficients

1. Introduction

Databases in atomic and molecular physics have become essential for developing models and simulations of complex physical and chemical processes and for the interpretation of data provided by observations and measurements, e.g., in laboratory plasma [1], and studying plasma chemistries and reactions in planetary atmospheres [2]. In the last decade large amounts of data have been collected for medical applications including stopping powers in different media and tissues as well as cross sections for atomic particles (photons, electrons, positrons, ions) interacting with biomolecules and their constituents in order to achieve an insight, at the molecular level, of the radiation damage and radiotherapy [3]. In order to solve the problem of analysis and mining of such large amounts of data, the creation of Virtual Observatories and Virtual Data Centres have been crucial ([4] and refs. therein). In this contribution we present a progress report of two atomic and molecular databases, the Belgrade Electron/Atom (Molecule) DataBase (BEAMDB) and Photodissociation (MolD), which are web services at the Serbian virtual observatory (SerVO) [5] and nodes within the Virtual Atomic and Molecular Data Centre (VAMDC) [6].

This branch of science often entitled 'Data management' or 'Data mining' is undergoing rapid expansion and development such that nowadays it is not enough for these databases to satisfy the standards of Virtual centres, etc., but they have to deal with new challenges such as the input of large amounts of data, i.e., Big Data. Thus, we can expect major investment and activity in this field in the next decade. Indeed, in September 2018, the European Strategy Forum on Research Infrastructures (ESFRI) presented its Strategy Report and Roadmap 2018. As a strategic instrument that identifies Research Infrastructures (RI) of strategic interest for Europe and the wider research community, the document presented Big data and e-infrastructure needs and highlighted the Virtual Observatory (VO) and The International VO Alliance (IVOA) as very important initiatives with VAMDC relevant data standards as well as state-of-the-art data analysis tools being highlighted as evidence of good practice.

2. BEAMDB and MolD Database Nodes

The Belgrade nodes of VAMDC are hosted by SerVO (see Figure 1a) and currently consists of two databases BEAMDB (servo.aob.rs/emol) and MolD (servo.aob.rs/mold). These databases have been developed using the standards developed and operated by the VAMDC project [6] (see Figure 1b). VAMDC and SerVO have been through several different stages of development. SerVO (http://servo.aob.rs/) is a project formally created in 2008 but it originated in 2000, when the first attempts to organize data and to create a kind of web service were made in the BELDATA project, the precursor of SerVO. VAMDC started on 1 July 2009 as a FP7-funded project and originally was to be developed with about 20 databases, but the portal now has more than 33 operational databases [7].

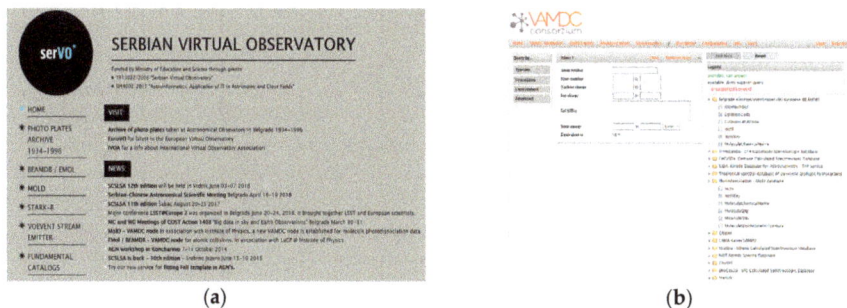

(a) (b)

Figure 1. The home pages of databases: (**a**) The SerVO [5]; (**b**) the VAMDC [6] portal query snapshot (http://www.portal.vamdc.eu).

We are currently in a transition phase updating the software "platform" (Python update, Django, XSAMS evolution, new Query Store on VAMDC, etc.) as the consequence of the rapid development and expansion of our two databases. Some current technical characteristics and aspects of these databases will be briefly introduced here (for details see [4]). Access to the BEAMDB and MolD data is possible via Table Access Protocol (TAP), a Virtual Observatory standard for a web service or via AJAX (Asynchronous JavaScript and XML)-enabled web interface (http://servo.aob.rs/). Both queries return data in XSAMS (XML Schema for Atoms, Molecules and Solids) format. The XSAMS schema provides a framework for a structured presentation of atomic, molecular, and particle-solid interaction data in an XML file. The underlying application architecture is written in Django, a Python web framework, and represents a customization and extension of VAMDC's NodeSoftware [8,9].

2.1. BEAMDB—Belgrade Electron/Atom(Molecule) DataBase

The origins of this database date from the early ideas of developing an Information System in Atomic Collision Physics [10] and at first it provided only cross sections for electron interactions with neutral atoms and molecules [11]. However the database has now been extended to cover electron spectra (energy loss and threshold) and ionic species [4].

Maintaining databases on cross sections and other collisional data, such as different types of spectra, is important for several reasons. One is to provide a comprehensive set of data to both researchers and applied scientists or engineers who need such data to design and make better devices and products. On the other hand, we need basic data to be able to include them in sophisticated models and to understand more complex processes, one of recent example is the use of electron cross section data for oxygen and water molecules in order to reveal for the role of electron induced processing in the coma of Comet 67P/Churyumov-Gerasimenko during the Rosetta mission [12]. Such an analysis clearly demonstrated the need for comprehensive datasets of electron-molecule collisions in format that is readily accessible and understandable to the space community. Electron collision cross sections are also a subject of databases with the particular interest in plasma processes data. An overview of such databases is given by Huo and Kim [13] and White et al. [14], with the emphasis on the role of electron collisional data in gases and surfaces in plasma processes. Another database specialized for modelling in low-temperature plasma is the LXCat database [15]. The compilation of electron scattering data from atoms and molecules has a rich history. After the discovery of the electron in 1897 by J. J. Thompson (see more about his route and how he conducted the experiments in [16]), a series of experiments on how electron beams behave by passing through a gas started to develop. The discovery of electron was followed by intense research of its interactions with matter, such that several Nobel Prizes were awarded for such studies: in 1904 to Philipp E. A. von Lenard for his "work on cathode rays"; in 1906 to Joseph J. Thomson "in recognition of the great merits of his theoretical and experimental investigations on the conduction of electricity by gases"; and in 1925 to James Franck and Gustav Ludwig Hertz "for their discovery of the laws governing the impact of an electron upon an atom." Carl W. Ramsauer at Danzig Technische Hochschule and Sir John S. Townsend at Oxford University independently studied the scattering of electrons of low energy by atoms and discovered an effect of occurrence of minima in total cross section that was named after them. The Ramsauer–Townsend effect was also observed in electron scattering from the methane molecule at the end of the 1920s and beginning of the 1930s, when electron collision studies were pioneered at St. John's College and Trinity College, Cambridge. Dymond and Watson [17] made the first direct determination of the scattering curve for slow electrons by helium atom. Arnot [18] performed scattering experiments in mercury vapour, while Bullard and Massey [19] performed experiments over a wider range of scattering angles in order to observe maxima and minima in scattering curves by argon atoms demonstrating diffraction phenomena. In parallel, the theoretical description of elastic scattering emerged on the basis of quantum wave mechanics by Mott [20] and later was fully developed by Mott and Massey [21]. Theoretical calculations producing cross sections have advanced over the years as the use of quantum mechanics allowed new methods to be developed, such as the time-independent close-coupling approach and R-matrix approach used to study low-energy collisions [22], relativistic convergent close-coupling method [23], distorted-wave Born approach [24] absorption potential [25] and optical potential method [26] to study high-energy collisions [27]. A special advantage of the time-independent close-coupling approach is its possibility of providing a guide to uncertainty estimates of the calculated values.

As an example of how BEAMDB may be used, we will discuss a review of electron scattering from methane. The methane molecule as well as other hydrocarbons have been identified as sources of infrared absorption in the atmospheres of giant planets. Novel measurements of IR spectra allow more precise determination of the methane content in these atmospheres [28]. It is considered that it is constituent of the atmosphere of Uranus, with an abundance of 2.3%, and of Neptune of 1.5%. It plays a very important role in the photochemistry processes on Neptune [29]. It is also considered as one of the major greenhouse gases in the Earth's atmosphere [30].

2.1.1. Elastic Electron Scattering by Methane Molecule—Early Experiments

The methane molecule is one of the molecules for which we have a relatively complete dataset for electron interactions. The electronic structure of methane is representative of most bio-molecules with its valence molecular orbitals being delocalized over the entire nuclear frame. As discussed by Herzberg [31], at first glance it is not obvious which conformation of C and H atoms would be the most stable and to which point group it should be attributed: a regular tetrahedron (T_d), a non-regular tetrahedron (C_{3v}), or square planar form (D_{4h}). In the T_d point group representation the highest occupied molecular orbitals (HOMO) of methane are of a_1 and t_1 symmetry, so the ground state of methane has a configuration $1(a_1)^2 2(a_1)^2 1(t_{2x})^2 1(t_{2y})^2 1(t_{2z})^2$ [32]. The pairing of all electrons in the HOMO makes methane a closed-shell compound. The binding energy of the lowest MO $2(a_1)$ is -18.8 eV, the energy of the three $1t_2$ orbitals is -10.6 eV, while the first LUMO $3(a_1)$ is at $+1.99$ eV and the second $2t_2$ is $+3.90$ eV [33].

The first measurements of electron differential elastic cross sections (DCS) for methane was performed in 1931 by Arnot [34] and by Bullard and Massey [35]. Arnot measured DCSs at higher impact energies of 30, 84, 205, 410 and 820 eV, while Bullard and Massey used lower incident electron energies of 4, 6, 10, 20 and 30 eV and covered an angular range from 20° to 120° in steps of 10°, except at an incident energy of 10 eV, where an additional point was measured at 125°. A close resemblance of between the DCS of methane and that of an argon atom was recognized. From this observation, the authors opened up the possibility of considering scattering by heavy atoms in terms of successive electron shells [35]. In 1932 Mohr and Nicoll [36] also measured DCS for methane at incident energies of 30, 52 and 84 eV, covering the full accessible angular range up to 150°. Hughes and McMillen [37] investigated the interference effects between the electron waves scattered by individual atoms as indicated by the presence of maxima in the curves for the ratios of the DCSs for different hydrocarbon molecules. For methane they measured DCS at 11 incident energies in the range from 10 eV to 800 eV and in the angular range from 10° to 150°.

2.1.2. Elastic Electron Scattering by Methane Molecule—Modern Experiments and Calculations

Gianturco and Thompson [32] used a model of scattering by a rigid molecule with inclusion of exchange and polarisation (with an ad hoc short range cutoff parameter) effects in an approximate way in order to calculate differential cross sections at low electron energies. For CH_4 they used the scattering states of symmetry A_1, T_1, T_2 and E, except A_2, was found to be of little importance for low-energy scattering, with A_1 and T_2 being found to be the most important. The exchange effects were included in each of these states. The results were presented graphically for 9.5 eV incident electron energy and for three cutoffs, $r_0 = 0.92$; 0.88 and 0.84. The authors concluded that it was not possible to give a single r_0 that gives complete agreement with the wide variety of experimental data, which reflects the crudeness of their model, but a good feature of the model is the correct incorporation of the static potential, calculated from good molecular wavefunctions. For intermediate and higher electron energies, from 205 to 820 eV, Dhal et al. [38] calculated DCS using the first Born, the Eikonal and the two-potential approximations including the polarisation and exchange effects. In these calculations no account was taken of the absorption effects that would certainly improve the accuracy of the calculated data.

New experiments exploring vibrationally elastic cross sections were performed by Rohr [39] at energies below 10 eV, i.e., at 1, 2 and 5 eV and from 10° to 120° using an spectrometer that consists of an electron monochromator to produce a high-resolution electron beam and a rotatable electron analyser capable of resolving the vibrational modes in energy loss mode, both systems had 127° electrostatic selectors. Absolute DCS were obtained by a normalization to the integral cross sections obtained in transmission experiments by other authors. A comprehensive study of differential, integral and momentum transfer cross sections was reported by Tanaka et al. [40] for elastic e/CH_4 scattering. They performed measurements using a crossed electron beam, molecular beam apparatus with the relative flow technique allowing the elastic DCS of CH_4 to be derived by comparison with those of

He. DCSs were measured at electron impact energies of 3, 5, 6, 7.5, 9, 10, 15, and 20 eV for scattering angles from 30° to 140°. The authors concluded that the angular distribution in the energy region of 3 to 7.5 eV is dominated by a *d*-wave scattering as was theoretically predicted and also established experimentally at 5 eV. Later the same authors remeasured the elastic DCS with a new spectrometer for impact energies from 1.5–100 eV and scattering angles from (10°–130°) and it was found that the previous values were systematically lower by about 30–35% [41].

Using the same set of data for normalization, Vušković and Trajmar [42] obtained DCSs for elastic scattering as well as inelastic cross sections by recording energy-loss spectra at incident energies of 20, 30 and 200 eV. The set of data obtained at 200 eV was normalized to the data by Dhal et al. [38]. All measured relative angular dependences were corrected for effective path length variation with scattering angle.

A group from University College London used an electron spectrometer, incorporating hemispherical electrostatic energy analysers, and a crossed beam of target molecules in order to measure elastic and vibrational excitation cross sections at low incident electron energies from 7.5 to 20 eV and scattering angles from 32° to 142° [43].

After the measurements by Rohr [39] at the University of Kaiserslautern, Sohn et al. [44] investigated threshold structures in the cross sections of low-energy electron scattering of methane. They also presented the measurements of angular dependences (DCS) at 0.6 and 1.0 eV from 35° to 105° scattering angles. Müller et al. [45] investigated the rotational excitation in vibrationally elastic e/CH$_4$ collisions and presented vibrationally elastic (rotationally summed) differential cross sections at the primary energies 5, 7.5 and 10 eV. They normalized their results to the previous measurements of Tanaka et al. [40]. Sohn et al. [46], with an improved (in comparison with [45]) crossed-beam spectrometer, measured DCSs at low energies in the range from 0.2 to 5.0 eV in the angular range between 15° and 138°. With the aid of a phase-shift analysis, integrated cross sections were calculated as well, but the absolute DCS scale was obtained by the relative flow technique, using He as a reference gas.

Further DCS measurements at intermediate and high electron incident energies were made by Sakae et al. [47]. The angular range was 5°–135° and the electron energies were 75, 100, 150, 200, 300, 500 and 700 eV. Absolute DCS were determined by using He as the known reference DCS. Data were presented as rotationally and vibrationally summed elastic DCS, with the overall uncertainty being estimated at approximately 10%. Shyn and Cravens [48] measured differential vibrationally elastic scattering cross sections from 5 to 50 eV and from 12° to 156°. The beam of methane molecules was modulated at a frequency of 150 Hz so that the pure beam signal could be separated from the background using a phase-sensitive detector. The overall uncertainty of data was about 14%, including the uncertainty of the He cross sections (filled into the chamber for normalization), from which relative curves were placed on an absolute scale.

Jain and Thompson [49] calculated cross sections for low electron-molecule scattering using a local exchange potential and polarisation potential introducing a first-order wavefunction. DCS were calculated at 3 and 5 eV using three different types of potentials, one parameter-free polarisation potential and two phenomenological potentials. Abusalbi et al. [50] calculated ab initio interaction potentials for e/CH$_4$ scattering at 10 eV impact energy. Jain [51] exploited a spherical optical complex potential model to investigate, over a wide energy range (0.1–500 eV), electron interactions with methane. The whole energy range was divided into three regions; (i) from 0.1 to 1.0 eV in which a Ramsauer-Townsend minimum is observed in the total cross section; (ii) between 2 and 20 eV where the scattering is dominated by a *d*-wave broad structure around 7–8 eV; and (iii) from 20 to 500 eV, where ionization and dissociation dominate over the elastic process. It was found that an absorption potential using the distorted charge density is more successful than one with polarized density and that the elastic cross sections are reduced significantly by including the imaginary part in the optical potential.

Gianturco et al. [52] used a parameter-free treatment of the interaction e/CH_4 and calculated cross sections for low energies. Functional forms of exchange and polarisation interactions were examined to find their importance over the whole range of collision energies. McNaughten et al. [53] reported rotationally elastic DCSs in the 0.1–20 eV energy region using the parameter-free model polarisation potential with electron exchange treated exactly and distortion effects included. Lengsfield, [54] used the complex Kohn method with polarized trial functions at incident energies from 0.2 to 10 eV. The latter was the first ab initio study to accurately characterize low-energy electron-methane scattering.

Later Gianturco et al. [55] calculated vibrational elastic, rotationally summed cross sections with ab initio static-exchange interactions and using a symmetry-adapted, single-centre expansion (SCE) representation for the close-coupling (CC) equations. Elastic DCS were obtained at energies 10, 15, 20, 30 and 50 eV. Nishimura and Itikawa [56] calculated vibrationally elastic DCS at 10 to 50 eV impact electron energies using an ab initio electrostatic potential and treating exchange and polarization in approximate way. Nestmann et al. [57] employed the variational R-matrix theory based on the fixed-nuclei approximation in order to calculate DCSs at low energies, i.e., 0.2, 0.5, 0.7, 1.5, 2.5, 3.5 and 5.0 eV. The structures in the calculated DCSs are shifted to smaller angles compared with the experimental results due to the omission of nonadiabatic effects.

Mapstone and Newell [58] reported measurements of hydrocarbon molecules at incident energies from 3.2 to 15.4 eV using electron spectrometer with hemispherical analysers both in monochromator and analyser. They first determined volume correction factors by using a phase shift analysis for helium DCS as a reference gas and then normalized relative values to the data of Tanaka et al. [32]. Bundschu et al. [59] performed a combined experimental and theoretical study for low-energy electron interactions with methane molecule. They determined absolute DCS at energies from 0.6 to 5.4 eV and within the angular range from 12° to 132.5°. Elastic differential cross sections were calculated using a body-fixed, SCE for the CC equations.

A group at Wayne State University, although primarily interested in positron scattering by methane [60], also measured electron elastic cross sections at 15, 20 and 200 eV. Their electron beam was produced as secondary electrons from the moderator with the energy spread of several electronvolts. Maji et al. [61] measured elastic DCSs for a number of carbon-containing molecules at the high-energy region from 300 to 1300 eV by the crossed-beam technique. They wanted to test the validity of the independent atom model for polyatomic molecules. The measurements of DCS were carried out with an energy resolution of about 1 eV and by using the relative flow method at 30° where the overall uncertainty was 15%. Basavaraju et al. [62] gave tabulated values for measured DCSs in [61] and obtained a scaled DCSs regarded as a universal function of a scaled momentum transfer for a number molecular targets.

Iga et al. [63] performed a joint theoretical (for 1–500 eV) and experimental (100–500 eV) investigation on e/CH_4 elastic scattering. Within the complex optical potential method they used the Schwinger variational iterative procedure combined with the distorted-wave approximation to calculate the scattering amplitudes. Experimentally, they used the relative flow technique and neon as the reference gas. The overall experimental uncertainty in the obtained absolute DCSs was about 10.3%. Lee et al. [64], on the basis of previous calculations, tested an improved version of the quasifree scattering model (QFSM) potential proposed by Blanco and García [25].

2.1.3. Elastic Electron Scattering by Methane Molecule—The Twenty-First Century Results

Bettega et al. [65] reported elastic DCSs for a class of molecules, (XH_4) among them methane, at incident electron energies between 3 and 10 eV using the Schwinger multichannel method with pseudopotentials. They demonstrated the importance of polarization effects in elastic collisions. Absolute differential elastic and vibrational excitation cross sections were measured by Allan [66] who exploited the improved resolution of the electron spectrometer in Fribourg to achieve the separation for all four vibrational modes within 0.4 eV from elastic peak at the impact energies from 0.1 to 1.5 eV. Varambhia et al. [67] and Tennyson [27] presented a sophisticated R-matrix approach to calculations of low-energy electron alkane collisions. DCSs were obtained for rotationally summed elastic scattering and the graphs were presented at 3.0 and 5.0 eV incident energies. Brigg et al. [68] performed R-matrix calculations at energies between 0.02 and 15 eV using a series of different ab initio models for both the target and the full scattering system.

Fedus and Karwasz [69] investigated the depth and position of the Ramsauer-Townsend minimum in methane by applying the MERT theory (Modified Effective Range Theory). They presented the results at incident energies from 0.2 to 1.5 eV and compared them with other results. They were able to put forward the recommended set of data for integral and momentum transfer data for methane at energies from 10^{-3} to 2.0 eV. Sun et al. [70] used a difference converging method (DCM) to predict accurate values of experimentally unknown DCSs. They presented vibrationally elastic cross section at 5.0 eV.

2.1.4. Elastic Electron Scattering by Methane Molecule—Coverage in the BEAMDB

Elastic cross sections for electron scattering by molecular targets comprise the majority of data items within the BEAMDB. Molecular targets covered by the present database are: alanine, formamide, tetrahydrofuran, hydrogen sulfide, pyrimidine, N-methylformamide, water, furan, nitrous oxide, and newly added datasets for methane. Currently there are 17 datasets for elastic DCSs for methane, spanning from 1931 (Arnot [34] and Bullard and Massey [35]) to the most recent work by Iga et al. [63]. For example, in Figure 2 we present in 3D graphical form one of the rather complete sets of data by Boesten and Tanaka [41] that have been used by many researchers for comparison and/or normalization.

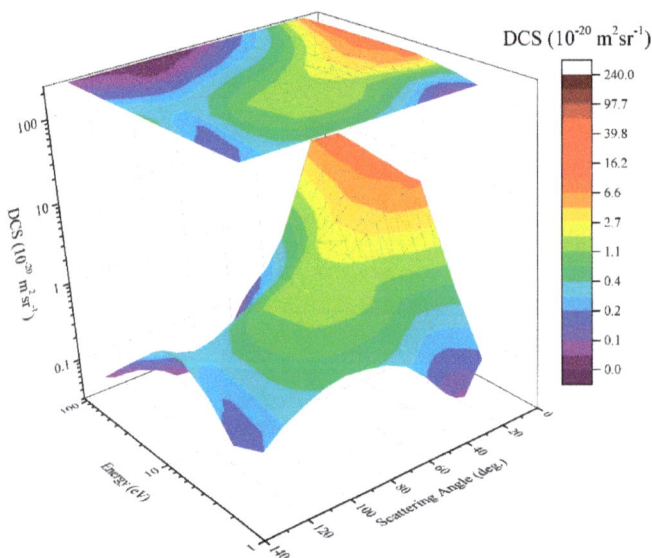

Figure 2. The DCS surface for elastic electron scattering by methane molecule. Data are obtained by Boesten and Tanaka [41] in the range from 1.5 to 100 eV.

2.2. Photodissociation—The MolD Database

MolD as a part of SerVO and VAMDC is intensively used by astrophysicists for model atmosphere calculations of solar and near solar-type stars, atmospheric parameter determinations, etc. as well as for theoretical and laboratory plasma research [71–76]. Such data are also important for astrochemistry and especially for studies of early Universe chemistry (see e.g., Heathcote et al. [74]). MolD consists of several components such as data collection, and user interface tools (e.g., on-site AJAX enabled queries and visualizations).

The database contains photodissociation cross sections for the individual rovibrational states of the diatomic molecular ions as well as corresponding data on molecular species and molecular state characterizations (rovibrational energy states, etc.). These cross sections can be summed and averaged (Figure 3) for further applications, e.g., obtaining rate coefficients (see Figure 4) for non-local thermal equilibrium models of early universe chemistry (Coppola et al. [77]), models of the solar atmosphere, or models of the atmospheres of white dwarfs (Wen & Han [78]), etc.

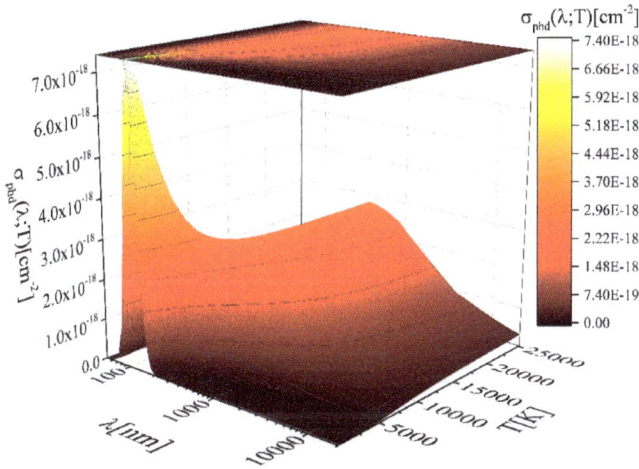

Figure 3. The surface plot of the averaged cross section σ_{phd} for photodissociation of the hydrogen molecular ion H_2^+ as a function of λ and T.

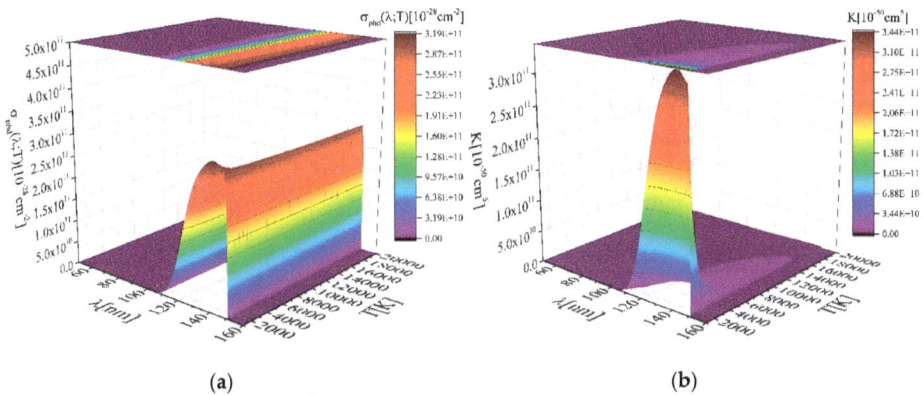

Figure 4. The surface plot of the: (a) averaged cross section σ_{phd} for photodissociation of the HLi$^+$; (b) rate coefficients for photodissociation of the HLi$^+$.

The cross sections are obtained using a quantum mechanical method where the photodissociation process is treated as result of radiative transitions between the ground and the first excited adiabatic electronic state of the molecular ion (see e.g., Ignjatović et al., 2014b [79]). The transitions are the outcome of the interaction of the electronic component of ion-atom systems with the electromagnetic field in the dipole approximation.

MolD offers on-site services that include calculation of the average thermal cross sections based on temperature for a specific molecule and wavelength. Besides acting as a VAMDC compatible web service, accessible through VAMDC portal and other tools implemented using VAMDC standards, MolD offers additional on-site utilities enable the plotting of average thermal cross sections along available wavelengths for a given temperature.

The MolD database was developed in three stages [80,81]. The first, completed at the end of 2014, was characterized by the construction of the service for all photodissociation data for hydrogen H_2^+ and helium He_2^+ molecular ions, together with the development of web interface and some utility programs. In stage 2, completed at the end of 2016, to MolD have been added averaged thermal photodissociation cross sections for H_2^+ and helium He_2^+ molecular ions as well as new cross sections for processes that involve species like diatomic molecular ions HX^+, where X = Mg, Li, Na. During 2017, in stage 3, MolD implemented cross section data for processes involving MgH^+, HeH^+, LiH^+, NaH^+, H_2^+, He_2^+. In this third stage, the design of the web interface was also improved and utility programs that allow online data visualization of a wide range of data were developed. The third stage of the MolD development was completed at the beginning of 2018 and has been followed by work on a major upgrade of the MolD database, including inserting new photodissociation data (for Na_2^+, Li_2^+ and $LiNa^+$). All of these data have possible applications in spectroscopy, low-temperature laboratory plasma created in gas discharges, e.g., in microwave discharges at atmospheric pressure [82]. Processes that involve alkali metals are also important for the optical properties and modelling of weakly ionized layers of different stellar atmospheres and rate coefficients are also needed as input parameter for models of the Io atmosphere [83]. The data may also be important in the investigation of metal-polluted white dwarfs and dusty white dwarfs, interstellar gas chemistry, etc. [84].

2.3. Node Maintenance

As VAMDC recently introduced *Query Store*, a new paradigm for dataset citation (see Zwölf et al. [85]), NodeSoftware upgrade was necessary at the Belgrade server. In order for latest pull of NodeSoftware repository (v12.7) from GitHub to work, Django was upgraded from version 1.4 to 1.11.2. The code is still running on Python 2.7, due to requirements of some other services running at the same server, but will be transferred to Python 3.x in the near future [86].

VAMDC has accepted the suggestion of Research Data Alliance, a research community organization [87], to implement the concept of Query Store. Now that Query Store is enabled, each query is persisted as a unique resource (with an identifier) with its pertinent citations and can be recreated even if data or schema at the host node change, as explained by Moreau et al. [88]. In this way, the connection between the citation and the dataset is straightforward and it can be interconnected with existing scientific infrastructures via Zenodo DOI request. This will increase the impact of data producers and will give more reliable citation of datasets.

3. Conclusions

This review presents the continuation of the work performed on database development at Serbian Virtual Observatory. The SVO is now addressing the challenge of upgrading software and continuous improvements of data processing. The changes since the last publication in 2017 include data for new targets like methane, hydrogen sulfide and rare gas atoms. The upgrades and new standards for databases include:

- Developing of the VAMDC Portal as a Major Enabler of Atomic and Molecular Data Citation;
- Python, Django updates;
- Installing the Query Store on VAMDC node that could have a plan store for holding the execution plan information, and a runtime stats store for carrying on the execution statistics information.

- XSAMS evolution to deal with Big Data (resources to be accessed by diverse client platforms across the network; generating and transferring data over a network without requiring human-to-human or human-to-computer; provide security and data quality; etc.).

In this paper, as examples of exploitation of the datasets in the database, we have reviewed the available results on elastic scattering of electron by methane molecule for both experiments and theoretical treatments, covering the ranges of incident energies and scattering angles. By comparing the collected datasets for methane we have been able to show that, although it has been extensively studied in the past, there is a need for new measurements in the intermediate electron impact energy range. Hydrogen sulfide data should be contrasted with data for water molecule and that will be one of our future goals. Data for rare gas atoms, especially helium and argon, may serve as reference gases with the well-established cross sections in relative flow method for upbringing unknown cross sections on absolute scale.

We have also shown that MolD may be used to reveal the surface plot of the averaged cross section and rates for photodissociation of the HLi$^+$ are given, as well as averaged cross sections for photodissociation of the hydrogen molecular ion H$_2{}^+$ as a function of λ and T.

Author Contributions: Conceptualization, B.P.M. and V.A.S.; software, V.V. and D.J.; data curation, S.I., N.U. and M.N.; writing—original draft preparation, B.P.M., V.A.S. and V.V.; writing—review and editing, All Authors; visualization, V.A.S.; supervision, L.M.I. and M.S.D.; project administration, D.J., M.S.D. and N.J.M.; funding acquisition, N.J.M.

Funding: This research was funded by MESTD of the Republic of Serbia, grant numbers (OI171020) and (III44002). NJM recognizes support from Europlanet 2020 RI, which has received funding from the European Union's Horizon 2020 research and innovation programme under grant agreement number 654208 and ELEvaTE grant agreement number 692335, as well as the support of the UK STFC and the Leverhulme trust.

Acknowledgments: Part of this work was supported by the VAMDC and the SUP@VAMDC projects funded under the 'Combination of Collaborative Projects and Coordination and Support Actions' Funding Scheme of The Seventh Framework Program.

Conflicts of Interest: The authors declare no conflict of interest. The funders had no role in the design of the study; in the collection, analyses, or interpretation of data; in the writing of the manuscript, or in the decision to publish the results.

References

1. Mason, N.J. The status of the database for plasma processing. *J. Phys. D* **2009**, *42*, 194003. [CrossRef]
2. Tennyson, J.; Rahimi, S.; Hill, C.; Tse, L.; Vibhakar, A.; Akello-Egwel, D.; Brown, D.B.; Dzarasova, A.; Hamilton, J.R.; Jaksch, D.; et al. QDB: A new database of plasma chemistries and reactions. *Plasma Sources Sci. Technol.* **2017**, *26*, 055014. [CrossRef]
3. Sanche, L. Interaction of low energy electrons with DNA: Applications to cancer radiation therapy. *Radiat. Phys. Chem.* **2016**, *128*, 36–43. [CrossRef]
4. Marinković, B.P.; Jevremović, D.; Srećković, V.A.; Vujčić, V.; Ignjatović, L.M.; Dimitrijević, M.S.; Mason, N.J. BEAMDB and MolD—Databases for atomic and molecular collisional and radiative processes: Belgrade nodes of VAMDC. *Eur. Phys. J. D* **2017**, *71*, 158. [CrossRef]
5. Jevremović, D.; Dimitrijević, M.S.; Popović, L.Č.; Dačić, M.; Benišek, V.P.; Bon, E.; Gavrilović, N.; Kovačević, J.; Benišek, V.; Kovačević, A.; et al. The project of Serbian Virtual Observatory and data for stellar atmosphere modeling. *New Astron. Rev.* **2009**, *53*, 222–226. [CrossRef]
6. Dubernet, M.L.; Antony, B.K.; Ba, Y.A.; Babikov, Y.L.; Bartschat, K.; Boudon, V.; Braams, B.J.; Chung, H.-K.; Daniel, F.; Delahaye, F.; et al. The virtual atomic and molecular data centre (VAMDC) consortium. *J. Phys. B* **2016**, *49*, 074003. [CrossRef]
7. VAMDC Consortium—Active Databases. Available online: https://portal.vamdc.eu/vamdc_portal/home.seam (accessed on 26 October 2018).
8. GitHub—VAMDC/NodeSoftware: Python/Django-Based Software for Running VAMDC Data Nodes. Available online: https://github.com/VAMDC/NodeSoftware (accessed on 26 October 2018).
9. Django: The Web Framework for Perfectionists with Deadlines. Available online: https://www.djangoproject.com/ (accessed on 26 October 2018).

10. Cvjetković, V.; Marinković, B.; Šević, D. Information System in Atomic Collision Physics. In *Book Advances and Innovations in Systems, Computing Sciences and Software Engineering*; Elleithy, K., Ed.; Springer: Dordrecht, The Netherlands, 2007; pp. 485–490. ISBN 978-1-4020-6263-6. [CrossRef]

11. Marinković, B.P.; Vujčić, V.; Sushko, G.; Vudragović, D.; Marinković, D.B.; Đorđević, S.; Ivanović, S.; Nešić, M.; Jevremović, D.; Solov'yov, A.V. Development of Collisional Data Base for Elementary Processes of Electron Scattering by Atoms and Molecules. *Nucl. Instrum. Methods Phys. Res. B* **2015**, *354*, 90–95. [CrossRef]

12. Marinković, B.P.; Bredehöft, J.H.; Vujčić, V.; Jevremović, D.; Mason, N.J. Rosetta Mission: Electron Scattering Cross Sections—Data Needs and Coverage in BEAMDB Database. *Atoms* **2017**, *5*, 46. [CrossRef]

13. Huo, W.M.; Kim, Y.K. Electron collision cross-section data for plasma modelling. *IEEE Trans. Plasma Sci.* **1999**, *27*, 1225–1240. [CrossRef]

14. White, R.D.; Cocks, D.; Boyle, G.; Casey, M.; Garland, N.; Konovalov, D.; Philippa, B.; Stokes, P.; de Urquijo, J.; González-Magaña, O.; et al. Electron transport in biomolecular gaseous and liquid systems: Theory, experiment and self-consistent cross-sections. *Plasma Sources Sci. Technol.* **2018**, *27*, 053001. [CrossRef]

15. Pitchford, L.C.; Alves, L.L.; Bartschat, K.; Biagi, S.F.; Bordage, M.-C.; Bray, I.; Brion, C.E.; Brunger, M.J.; Campbell, L.; Chachereau, A.; et al. LXCat: An Open-Access, Web-Based Platform for Data Needed for Modeling Low Temperature Plasmas. *Plasma Process. Polym.* **2017**, *14*, 1600098. [CrossRef]

16. Falconer, I. JJ Thomson and the discovery of the electron. *Phys. Educ.* **1997**, *32*, 226–231. [CrossRef]

17. Dymond, E.G.; Watson, E.E. Electron scattering in helium. *Proc. R. Soc. Lond. A* **1929**, *122*, 571–582. [CrossRef]

18. Arnot, F.L. Electron Scattering in Mercury Vapour. *Proc. R. Soc. Lond. A* **1929**, *125*, 660–669. [CrossRef]

19. Bullard, E.C.; Massey, H.S.W. The elastic scattering of slow electrons in argon. *Proc. R. Soc. Lond. A* **1931**, *130*, 579–590. [CrossRef]

20. Mott, N.F. Elastic Collisions of Electrons with Helium. *Nature* **1929**, *123*, 717. [CrossRef]

21. Mott, N.F.; Massey, H.S.W. *The Theory of Atomic Collisions*, 3rd ed.; Oxford University Press: Oxford, UK, 1965.

22. Bartschat, K.; Tennyson, J.; Zatsarinny, O. Quantum-Mechanical Calculations of Cross Sections for Electron Collisions with Atoms and Molecules. *Plasma Process. Polym.* **2017**, *14*, 1600093. [CrossRef]

23. Fursa, D.V.; Bray, I. Fully Relativistic Convergent Close-Coupling Method for Excitation and Ionization Processes in Electron Collisions with Atoms and Ions. *Phys. Rev. Lett.* **2008**, *100*, 113201. [CrossRef]

24. Madison, D.H.; Al-Hagan, O. The Distorted-Wave Born Approach for Calculating Electron-Impact Ionization of Molecules. *Hindawi J. At. Mol. Opt. Phys.* **2010**, *2010*, 367180. [CrossRef]

25. Blanco, F.; García, G. Improved non-empirical absorption potential for electron scattering at intermediate and high energies: 30–10000 eV. *Phys. Lett. A* **1999**, *255*, 147–153. [CrossRef]

26. Das, T.; Stauffer, A.D.; Srivastava, R. A method to obtain static potentials for electron-molecule scattering. *Eur. Phys. J. D* **2014**, *68*, 102. [CrossRef]

27. Tennyson, J. Electron–molecule collision calculations using the R-matrix method. *Phys. Rep.* **2010**, *491*, 29–76. [CrossRef]

28. Boudon, V.; Pirali, O.; Roy, P.; Brubach, J.-B.; Manceron, L.; Vander Auwera, J. The high-resolution far-infrared spectrum of methane at the SOLEIL synchrotron. *J. Quant. Spectrosc. Radiat. Trans.* **2010**, *111*, 1117–1129. [CrossRef]

29. Romani, P.N.; Atreya, S.K. Methane photochemistry and haze production on Neptune. *Icarus* **1988**, *74*, 424–445. [CrossRef]

30. Kirschke, S.; Bousquet, P.; Ciais, P.; Saunois, M.; Canadell, J.G.; Dlugokencky, E.J.; Bergamaschi, P.; Bergmann, D.; Blake, D.R.; Bruhwiler, L.; et al. Three decades of global methane sources and sinks. *Nat. Geosci.* **2013**, *6*, 813–823. [CrossRef]

31. Herzberg, G. *Molecular Spectra and Molecular Structure. III. Electronic Spectra of Polyatomic Molecules*; Van Nostrand Reinhold Company: New York, NY, USA, 1966; p. 392. ISBN 0-442-03387-7.

32. Gianturco, F.A.; Thompson, D.G. The scattering of slow electrons by polyatomic molecules. A model study for CH_4, H_2O and H_2S. *J. Phys. B* **1980**, *13*, 613–625. [CrossRef]

33. Hollister, C.; Sinanoglu, O. Molecular Binding Energies. *J. Am. Chem. Soc.* **1966**, *88*, 13–21. [CrossRef]

34. Arnot, F.L. The Diffraction of Electrons in Gases. *Proc. R. Soc. Lond. A* **1931**, *133*, 615–636. [CrossRef]

35. Bullard, E.C.; Massey, H.S.W. The Elastic Scattering of Slow Electrons in Gases—II. *Proc. R. Soc. Lond. A* **1931**, *133*, 637–651. [CrossRef]

36. Mohr, C.B.O.; Nicoll, F.H. The large angle scattering of electrons in gases—II. *Proc. R. Soc. Lond. A* **1932**, *138*, 469–478. [CrossRef]

37. Hughes, A.L.; McMillen, J.H. Electron Scattering in Methane, Acetylene and Ethylene. *Phys. Rev.* **1933**, *44*, 876–882. [CrossRef]

38. Dahl, S.S.; Srivastava, B.B.; Shingal, R. Elastic scattering of electrons by methane molecules at intermediate energies. *J. Phys. B* **1979**, *12*, 2727–2734. [CrossRef]

39. Rohr, K. Cross beam experiment for the scattering of low-energy electrons from methane. *J. Phys. B* **1980**, *13*, 4897–4905. [CrossRef]

40. Tanaka, H.; Okada, T.; Boesten, L.; Suzuki, T.; Yamamoto, T.; Kubo, M. Differential cross sections for elastic scattering of electrons by CH_4 in the energy range of 3 to 20 eV. *J. Phys. B* **1982**, *15*, 3305–3319. [CrossRef]

41. Boesten, L.; Tanaka, H. Elastic DCS for e+CH_4 collisions, 1.5–100 eV. *J. Phys. B* **1991**, *24*, 821–832. [CrossRef]

42. Vušković, L.; Trajmar, S. Electron impact excitation of methane. *J. Chem. Phys.* **1983**, *78*, 4947–4951. [CrossRef]

43. Curry, P.J.; Newell, W.R.; Smith, A.C.H. Elastic and inelastic scattering of electrons by methane and ethane. *J. Phys. B* **1985**, *18*, 2303–2318. [CrossRef]

44. Sohn, W.; Jung, K.; Ehrhardt, H. Threshold structures in the cross sections of low-energy electron scattering of methane. *J. Phys. B* **1983**, *16*, 891–901. [CrossRef]

45. Müller, R.; Jung, K.; Kochem, K.-H.; Sohn, W.; Ehrhardt, H. Rotational excitation of CH_4 by low-energy-electron collisions. *J. Phys. B* **1985**, *18*, 3971–3985. [CrossRef]

46. Sohn, W.; Kochem, K.-H.; Scheuerlein, K.-M.; Jung, K.; Ehrhardt, H. Elastic electron scattering from CH_4 for collision energies between 0.2 and 5 eV. *J. Phys. B* **1986**, *19*, 3625–3632. [CrossRef]

47. Sakae, T.; Sumiyoshi, S.; Murakami, E.; Matsumoto, Y.; Ishibashi, K.; Katase, A. Scattering of electrons by CH_4, CF_4 and SF_6 in the 75–700 eV range. *J. Phys. B* **1989**, *22*, 1385–1394. [CrossRef]

48. Shyn, T.W.; Cravens, T.E. Angular distribution of electrons elastically scattered from CH_4. *J. Phys. B* **1990**, *23*, 293–300. [CrossRef]

49. Jain, A.; Thopmson, D.G. Elastic scattering of slow electrons by CH_4 and H_2O using a local exchange potential and new polarisation potential. *J. Phys. B* **1982**, *15*, L631–L637. [CrossRef]

50. Abusalbi, N.; Eades, R.A.; Nam, T.; Thirumalai, D.; Dixon, D.A.; Truhlar, D.G. Electron scattering by methane: Elastic scattering and rotational excitation cross sections calculated with ab initio interaction potentials. *J. Chem. Phys.* **1983**, *78*, 1213–1227. [CrossRef]

51. Jain, A. Total (elastic+absorption) cross sections for e-CH_4 collisions in a spherical model at 0.10–500 eV. *Phys. Rev. A* **1986**, *34*, 3707–3722. [CrossRef]

52. Gianturco, F.A.; Jain, A.; Pantano, L.C. Electron-methane scattering via a parameter-free model interaction. *J. Phys. B* **1987**, *20*, 571–586. [CrossRef]

53. McNaughten, P.; Thompson, D.G.; Jain, A. Low-energy electron-CH_4 collisions using exact exchange plus parameter-free polarisation potential. *J. Phys. B* **1990**, *23*, 2405S. [CrossRef]

54. Lengsfield, B.H., III; Rescigno, T.N.; McCurdy, C.W. Ab initio study of low-energy electron-methane scattering. *Phys. Rev. A* **1991**, *44*, 4296–4308. [CrossRef]

55. Gianturco, F.A.; Rodriguez-Ruiz, J.A.; Sanna, N. Elastic scattering of electrons by methane molecules. *Phys. Rev. A* **1995**, *52*, 1257–1265. [CrossRef]

56. Nishimura, T.; Itikawa, Y. Elastic scattering of electrons by methane molecules. *J. Phys. B* **1994**, *27*, 2309–2316. [CrossRef]

57. Nestmann, B.M.; Pfingst, K.; Peyerimhoff, S.D. R-matrix calculation for electron-methane scattering cross sections. *J. Phys. B* **1994**, *27*, 2297–2308. [CrossRef]

58. Mapstone, B.; Newell, W.R. Elastic differential electron scattering from CH_4, C_2H_4 and C_2H_6. *J. Phys. B* **1992**, *25*, 491–506. [CrossRef]

59. Bundschu, C.T.; Gibson, J.C.; Gulley, R.J.; Brunger, M.J.; Buckman, S.J.; Sanna, N.; Gianturco, F.A. Low-energy electron scattering from methane. *J. Phys. B* **1997**, *30*, 2239–2259. [CrossRef]

60. Przybyla, D.A.; Kauppila, W.E.; Kwan, C.K.; Smith, S.J.; Stein, T.S. Measurements of positron-methane differential scattering cross sections. *Phys. Rev. A* **1997**, *55*, 4244–4247. [CrossRef]

61. Maji, S.; Basavaraju, G.; Bharathi, S.M.; Bhushan, K.G.; Khare, S.P. Elastic scattering of electrons by polyatomic molecules in the energy range 300–1300 eV: CO, CO_2, CH_4, C_2H_4 and C_2H_6. *J. Phys. B* **1998**, *31*, 4975–4990. [CrossRef]

62. Basavaraju, G.; Bharathi, S.M.; Bhushan, K.G.; Maji, S.; Patil, S.H. A Unified Description of Elastic, High Energy Electron–Molecule Scattering. *Phys. Scr.* **1999**, *60*, 28–31. [CrossRef]

63. Iga, I.; Lee, M.-T.; Homem, M.G.P.; Machado, L.E.; Brescansin, L.M. Elastic cross sections for CH$_4$ collisions at intermediate energies. *Phys. Rev. A* **2000**, *61*, 022708. [CrossRef]

64. Lee, M.-T.; Iga, I.; Machado, L.E.; Brescansin, L.M. Model absorption potential for electron-molecule scattering in the intermediate-energy range. *Phys. Rev. A* **2000**, *62*, 062710. [CrossRef]

65. Bettega, M.H.F.; Varella, M.T. do, N.; Lima, M.A.P. Polarization effects in the elastic scattering of low-energy electrons by XH$_4$ (X=C, Si, Ge, Sn, Pb). *Phys. Rev. A* **2003**, *68*, 012706. [CrossRef]

66. Allan, M. Excitation of the four fundamental vibrations of CH$_4$ by electron impact near threshold. *J. Phys. B At. Mol. Opt. Phys.* **2005**, *38*, 1679–1685. [CrossRef]

67. Varambhia, H.N.; Munro, J.J.; Tennyson, J. R-matrix calculations of low-energy electron alkane collisions. *Int. J. Mass Spectrom.* **2008**, *271*, 1–7. [CrossRef]

68. Brigg, W.J.; Tennyson, J.; Plummer, M. R-matrix calculations of low-energy electron collisions with methane. *J. Phys. B* **2014**, *47*, 185203. [CrossRef]

69. Fedus, K.; Karwasz, G.P. Ramsauer-Townsend minimum in methane—Modified effective range analysis. *Eur. Phys. J. D* **2014**, *68*, 93. [CrossRef]

70. Sun, W.; Wang, Q.; Zhang, Y.; Li, H.; Feng, H.; Fan, Q. Predicting differential cross sections of electron scattering from polyatomic molecules. *J. Phys. B* **2015**, *48*, 125201. [CrossRef]

71. Zammit, M.C.; Savage, J.S.; Colgan, J.; Fursa, D.V.; Kilcrease, D.P.; Bray, I.; Fontes, C.J.; Hakel, P.; Timmermans, E. State-resolved Photodissociation and Radiative Association Data for the Molecular Hydrogen Ion. *Astrophys. J.* **2017**, *851*, 64. [CrossRef]

72. Mihajlov, A.A.; Sakan, N.M.; Srećković, V.A.; Vitel, Y. Modeling of continuous absorption of electromagnetic radiation in dense partially ionized plasmas. *J. Phys. A* **2011**, *44*, 095502. [CrossRef]

73. Ignjatović, L.M.; Mihajlov, A.A.; Srećković, V.A.; Dimitrijević, M.S. Absorption non-symmetric ion–atom processes in helium-rich white dwarf atmospheres. *Mon. Not. R. Astron. Soc.* **2014**, *439*, 2342–2350. [CrossRef]

74. Heathcote, D.; Vallance, C. Total electron ionization cross-sections for neutral molecules relevant to astrochemistry. *J. Phys. B* **2018**, *51*, 195203. [CrossRef]

75. Mihajlov, A.A.; Sakan, N.M.; Srećković, V.A.; Vitel, Y. Modeling of the continuous absorption of electromagnetic radiation in dense Hydrogen plasma. *Baltic Astron.* **2011**, *20*, 604–608. [CrossRef]

76. Babb, J.F. State resolved data for radiative association of H and H+ and for Photodissociation of H$_2$+. *Astrophys. J. Suppl. Ser.* **2015**, *216*, 21. [CrossRef]

77. Coppola, C.M.; Galli, D.; Palla, F.; Longo, S.; Chluba, J. Non-thermal photons and H$_2$ formation in the early Universe. *Mon. Not. R. Astron. Soc.* **2013**, *434*, 114–122. [CrossRef]

78. Wen, Z.L.; Han, J.L. A sample of 1959 massive galaxy clusters at high redshifts. *Mon. Not. R. Astron. Soc.* **2018**, *481*, 4158–4168. [CrossRef]

79. Ignjatović, L.M.; Mihajlov, A.A.; Srećković, V.A.; Dimitrijević, M.S. The ion–atom absorption processes as one of the factors of the influence on the sunspot opacity. *Mon. Not. R. Astron. Soc.* **2014**, *441*, 1504–1512. [CrossRef]

80. Vujčič, V.; Jevremović, D.; Mihajlov, A.A.; Ignjatović, L.M.; Srećković, V.A.; Dimitrijević, M.S.; Malović, M. MOL-D: A Collisional Database and Web Service within the Virtual Atomic and Molecular Data Center. *J. Astrophys. Astron.* **2015**, *36*, 693–703. [CrossRef]

81. Srećković, V.A.; Ignjatović, L.M.; Jevremović, D.; Vujčić, V.; Dimitrijević, M.S. Radiative and Collisional Molecular Data and Virtual Laboratory Astrophysics. *Atoms* **2017**, *5*, 31. [CrossRef]

82. Pichler, G.; Makdisi, Y.; Kokaj, J.; Mathew, J.; Rakić, M.; Beuc, R. Superheating effects in line broadening of dense alkali vapors. *J. Phys. Conf. Ser.* **2017**, *810*, 012013. [CrossRef]

83. Strobel, D.F.; Zhu, X.; Summers, M.E. On the vertical thermal structure of Io's atmosphere. *Icarus* **1994**, *111*, 18–30. [CrossRef]

84. Dalgarno, A.; Black, J.H. Molecule formation in the interstellar gas. *Rep. Prog. Phys.* **1976**, *39*, 573. [CrossRef]

85. Zwölf, C.-M.; Moreau, N.; Dubernet, M.-L. New model for datasets citation and extraction reproducibility in VAMDC. *J. Mol. Spectrosc.* **2016**, *327*, 122–137. [CrossRef]

86. Regandell, S.; Marquart, T.; Piskunov, N. Inside a VAMDC data node—Putting standards into practical software. *Phys. Scr.* **2018**, *93*, 035001. [CrossRef]

87. Asmi, A.; Rauber, A.; Pröll, S.; van Uytvanck, D. Citing Dynamic Data-Research Data Alliance working group recommendations. In Proceedings of the EGU General Assembly Conference Abstracts, Vienna, Austria, 17–22 April 2016; Volume 18.

88. Moreau, N.; Zwolf, C.-M.; Ba, Y.-A.; Richard, C.; Boudon, V.; Dubernet, M.-L. The VAMDC Portal as a Major Enabler of Atomic and Molecular Data Citation. *Galaxies* **2018**, *6*, 105. [CrossRef]

atoms

MDPI

Article

Experimental Runaway Electron Current Estimation in COMPASS Tokamak

Milos Vlainic [1,*], Ondrej Ficker [2], Jan Mlynar [2], Eva Macusova [2] and the COMPASS Tokamak Team

[1] Institute of Physics, University of Belgrade, Pregrevica 118, P.O. Box 68, 11080 Belgrade, Serbia
[2] Institute of Plasma Physics, Czech Academy of Sciences—Institute of Plasma Physics, Za Slovankou 1782/3, 182 00 Prague 8, Czech Republic; ficker@ipp.cas.cz (O.F.); mlynar@ipp.cas.cz (J.M.); macusova@ipp.cas.cz (E.M.)
* Correspondence: milos.vlainic@ipb.ac.rs

Received: 30 November 2018; Accepted: 13 January 2019; Published: 16 January 2019

check for
updates

Abstract: Runaway electrons present a potential threat to the safe operation of future nuclear fusion large facilities based on the tokamak principle (e.g., ITER). The article presents an implementation of runaway electron current estimations at COMPASS tokamak. The method uses a theoretical method developed by Fujita et al., with the difference in using experimental measurements from EFIT and Thomson scattering. The procedure was explained on the COMPASS discharge number 7298, which has a significant runaway electron population. Here, it was found that at least 4 kA of the plasma current is driven by the runaway electrons. Next, the method aws used on the set of plasma discharges with the variable electron plasma density. The difference in the plasma current was explained by runaway electrons, and their current was estimated using the aforementioned method. The experimental results are compared with the theory and simulation. The comparison presented some disagreements, showing the possible direction for the code development. Additional application on runaway electron energy limit is also addressed.

Keywords: runway electron; plasma current; fusion plasma; tokamak

1. Introduction

In the last decade, nuclear fusion started to shift from science to industry, where the tokamak-based ITER is playing the leading role. Runaway electrons (REs) present a potential threat to the safe operation of the future nuclear fusion power plants based on the tokamak principle. Namely, in the latest issue of ITER Physics Basis [1], REs are considered as the second-highest priority for the ITER disruption mitigation.

COMPASS tokamak [2] is a suitable and low-risk test bed for studying RE disruption physics and RE model benchmarking. However, one of the common missing RE parameters for comparison of experiment and simulation is RE current or RE density. This article addresses this issue for certain cases.

The basic idea behind the used model was reported on ORMAK tokamak by Fujita [3]. In contrast to Fujita's full-theoretical approach, we implemented experimental data from EFIT [4] and Thomson Scattering [5] to the calculation. The obtained results are in the acceptable range and their application on density scans and RE localization is presented.

The article starts with a description of an experimental observation of the particular discharge. Theoretical models and detailed description of the settings used in the models for comparison with the experiments are then reported. An estimation of the RE current I_{RE} is the first experimental analysis. Subsequently, the influence of the REs on the current ramp-up phase of the discharge is evaluated. Following that, a principle on how to localize RE beam is done in the next section, using the knowledge of the RE beam current and its maximum energy. Finally, discussions and conclusions are given in the last section with an outlook towards expected future works.

2. Experimental Setup: COMPASS Tokamak

The COMPASS tokamak is a compact experimental fusion facility with a major radius $R_0 = 0.56$ m and a minor radius $a_p = 0.23$ m. The toroidal magnetic field B_{tor} is in the 0.9–1.6 T range (typically set to 1.15 T), coming from 16 toroidal field coils. The plasma current I_p can reach up to 400 kA using an air-cored transformer. The range for the electron densities is flexible and is typically in the 10^{19}–10^{20} m^{-3} domain. Plasma shaping varies from circular and elliptical to single-null D-shaped ITER-like plasmas. When circular, the plasma is limited by a carbon HFS wall. The regular pulse length is ~ 0.4 s, although low current circular discharges with RE can last almost 1 s [6,7].

The Thomson Scattering diagnostic system in COMPASS uses two Nd:YAG lasers of wavelength 1064 nm with an energy of 1.5 J for each laser. The whole system is oriented vertically and the scattered light is recorded in the radial direction. Each laser has 30 Hz repetition rate, which offers a ~ 16.7 ms time resolution if the two lasers are operated equidistantly.

The EFIT reconstruction provides relevant quantities on a 2D mesh cross-section of the poloidal plasma plane solving the Grad–Shafranov equation for plasma MHD equilibrium, constrained by the magnetic diagnostics at the COMPASS tokamak (current measured in plasma and field coils, pick-up coils). Additionally, in the analysis addressed here, a toroidal loop voltage in plasma V_{plasma} for computing a maximum RE energy W_{max} is calculated using METIS [8], a Multi-Element Tokamak-oriented Integrated Simulator.

3. Experiment: High RE Current Discharge

The very first magnetic observations of high RE current in the COMPASS tokamak are reported in Ref. [9] for discharge #7298. The discharge had circular-shaped limited plasma, with toroidal magnetic field $B_{tor} = 1.15$ T and plasma current $I_p = 120$ kA. The plasma density n_e was lower than 1.5×10^{19} m^{-3} in the plasma core as measured by the Thomson Scattering system.

The COMPASS tokamak is well-supplied with the general plasma measurements, which provided us with the following list of observations in the discharge:

- Loop voltage measured plasma voltage V_{plasma} bellow 1 V, while the typical range is 1.5–2 V;
- Electron temperature from Thomson Scattering was 1 order of magnitude colder than ordinary tokamak plasma;
- Hard X-ray scintillators observed a significant amount of RE losses;
- Poloidal internal pick-up coils detected inner structure that is moving outwards;
- Vertical field for feedback of horizontal plasma position was increasing during the discharge, even though plasma current was constant;
- EFIT reconstruction of poloidal cross-section depicted plasma shrinkage towards inner side, due to the increasing magnetic pressure coming from the above-mentioned vertical field;
- Plasma pressure in terms of its ratio to the poloidal magnetic field pressure (i.e., poloidal beta β_{pol}), estimated from EFIT showed unrealistic large values.

All aforementioned information provides us with enough evidence to conclude that RE beam co-existed with the bulk plasma. For more detailed discussion on the measurements, the interested reader is referred to Section 5.1 in Ref. [10].

4. Method: RE Current Calculation

The overestimate of the β_{pol} is exploited here for the estimate of the RE current I_{RE}. The analysis is based on work done by Fujita et al. [3]. Even though the basic principle used here is the same as Fujita's, the final approach to the calculation is different. While the original paper estimates all necessary parameters theoretically, we use direct measurements from the Thomson scattering and calculations from EFIT reconstruction. In this section, we only address our changes to the method. The main idea behind the Fujita's model is an estimate of RE pressure P_{RE} from the difference between total plasma pressure p_{tot} (coming from EFIT β_{pol})

$$\langle p_{tot} \rangle_V = \frac{\beta_{pol} B_{pol}^2}{2\mu_0} \approx 5 \times 10^{-8} \frac{\beta_{pol}}{\pi} \left(\frac{I_p}{a_p} \right)^2 \tag{1}$$

and pure bulk plasma pressure p_{pl} (coming from the Thomson Scattering measured radial T_e profile)

$$\langle p_{pl} \rangle_V = \frac{2}{a_p^2} \int_0^{a_p} p_{pl}(r) r \, dr, \tag{2}$$

where r is the radial position along the minor radius a_p. Examples of $\langle p_{tot} \rangle_V$ and $\langle p_{pl} \rangle_V$ are presented in Figure 1.

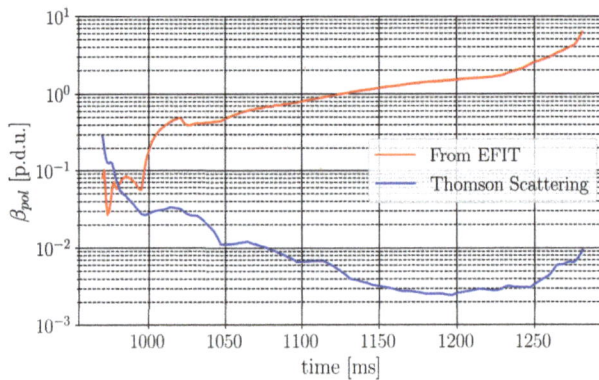

Figure 1. Poloidal beta β_{pol} estimated by EFIT (red) and from Thomson Scattering (blue) for discharge #7298.

The $P_{RE} = \langle p_{tot} \rangle_V - \langle p_{pl} \rangle_V$ is calculated as an average from obtained P_{RE} values corresponding to results when following conditions $\langle p_{tot} \rangle_V > 1.4 \langle p_{pl} \rangle_V$ and $\langle p_{tot} \rangle_V > 2 \langle p_{pl} \rangle_V$ are satisfied simultaneously, to reduce the propagation of error given by this approach. The first condition specifies the accuracy of Thomson Scattering measurements and thereby defines the minimal threshold necessary to distinguish between $\langle p_{tot} \rangle_V$ and $\langle p_{pl} \rangle_V$ and eliminates the error of Thomson Scattering measurements. The second condition is always valid in the case of non-negligible RE pressure and it is artificially taken as a conservative limit.

Furthermore, P_{RE} is connected to the RE density

$$P_{RE} = \langle p_{RE} \rangle = \frac{1}{2} m_e \langle n_{RE} \rangle \langle \gamma v_\parallel^2 \rangle + \frac{1}{4} m_e \langle n_{RE} \rangle \langle \gamma v_\perp^2 \rangle, \tag{3}$$

where $\langle \ \rangle$ represents average in both spatial and momentum coordinates, γ is relativistic Lorentz factor, and $v_\parallel$ and $v_\perp$ denote the RE velocity in the parallel and perpendicular direction to the magnetic field $\vec{B}$, respectively. By estimating the RE energy, one can estimate the RE density $\langle n_{RE} \rangle$, from which RE current I_{RE} can be approximately derived using the following expression:

$$I_{RE} = ec \langle n_{RE} \rangle \langle \beta \rangle A_{RE}, \tag{4}$$

where A_{RE} is the area of the RE beam cross-section and β is normalized relativistic velocity.

4.1. RE Energy Calculation

The model for maximum obtainable RE energy W_{max} is based on Ref. [11], where radiation losses are subtracted from the acceleration. It is a 0D model, as there is no dimensional dependency, but only time evolution. Particularly, the lost power by synchrotron radiation P_{synch} and bremsstrahlung radiation P_{brems} are subtracted from the power gained by the RE thanks to the electric field P_E

$$\frac{dW_{max}}{dt} = P_E - P_{synch} \tag{5}$$

to obtain the time dependence of W_{max}. The power gain by the toroidal electric field E_{tor} is simply calculated as

$$P_E = ecE_{tor} = \frac{ecV_{plasma}}{2\pi R_p}, \tag{6}$$

with R_p standing for the plasma major radius. The synchrotron power loss

$$P_{synch} = \frac{2r_e m_e c^3 \beta^4 \gamma^4}{3R_c^2}, \tag{7}$$

where r_e and m_e are the classical electron radius and mass, respectively. The curvature radius R_c is calculated using Equation (9) from Ref. [12]:

$$R_c = \frac{R_p}{1 + \eta^2 + 2\eta \times 2/\pi} \tag{8}$$

where $2/\pi$ at the end of the denominator comes from the average taken over the gyro-motion angle, and where $\eta = \langle v_\perp \rangle / v_{dr}$ and v_{dr} is the drift velocity

$$v_{dr} = \frac{\langle \gamma \rangle \langle v_\parallel^2 \rangle}{\Omega R_p} \tag{9}$$

with Ω being the fundamental gyro-frequency. Finally, the angle brackets are making the whole calculation dependent on the RE distribution function (REDF).

4.2. RE Distribution Functions

With mentioning REDF, we need to address its influence on the I_{RE} calculation. The REDFs used by Fujita et al. and utilized here are monoenergetic f_{mono}, uniform f_{uni}, and linear f_{lin}. The definitions of the first three distributions as functions of RE energy w are as follows:

$$f_{mono}(w) = \delta(w - W_{max}),$$
$$f_{uni}(w) = \frac{1}{W_{max}}, \tag{10}$$
$$f_{lin}(w) = \frac{2}{W_{max}} \left(1 - \frac{w}{W_{max}}\right),$$

with $\delta()$ being the Dirac delta function. The coefficients come from setting the REDF maximum value to 0 at W_{max} and then normalizing the REDF to 1.

Furthermore, the exponential REDF

$$f_{exp}(w; \epsilon) = \epsilon e^{-\epsilon w}, \tag{11}$$

is introduced to present a distribution with a steep decrease that could be important if the avalanche mechanism were to dominate. Additionally, two more REDFs are considered here: skewed Gaussian f_{sG} [13]

$$f_{sG}(w; l, \zeta, \alpha) = \frac{2}{\zeta} f_{Gpdf}(s) f_{Gcdf}(s\alpha) \tag{12}$$

$$f_{Gpdf}(s) = \frac{1}{\sqrt{2\pi}} e^{-s^2/2} \tag{13}$$

$$f_{Gcdf}(s\alpha) = \frac{1}{2} \left[1 + \text{erf}\left(\frac{s\alpha}{\sqrt{2}} \right) \right] \tag{14}$$

with $s = (w - l)/\zeta$ and Maxwell–Jüttner f_{MJ} [14]

$$f_{MJ}(w; \mathcal{W}) = \frac{\beta \gamma^2}{\mathcal{W} K_2(1/\mathcal{W})} e^{-\gamma/\mathcal{W}} \tag{15}$$

distributions. The (negatively) skewed Gaussian represents a more realistic case of f_{mono}, as it is not a delta-function. In other words, the negatively skewed Gaussian represents REDF where almost all RE tend towards W_{max} but they never reach it. The Maxwell–Jüttner distribution function f_{MJ} is selected as one of the most commonly relativistic distribution functions used in the literature. As all three latter functions have asymptotic behavior towards zero and/or W_{max}, normalization has to be different than for the former three. Their parameters are set in such a way that asymptotic edge is smaller than 10^{-5}, while f_{sG} maximum is located just over $w = 0.95 W_{max}$.

4.3. Pitch Angle

Equation (3) indicates that the pressure separation through the velocities $v_{\parallel}$ and $v_{\perp}$ implies knowing the pitch angle θ. Fujita et al. omitted the $\langle \gamma v_{\perp}^2 \rangle$ term and approximated the other one $\langle \gamma v_{\parallel}^2 \rangle$ by $c^2 \langle \gamma \beta^2 \rangle$. However, we consider the θ influence to be consistent with the estimation of W_{max} reported in Section 4.1, where θ plays a role through the parameter $\eta \propto \langle \beta_{\perp} \rangle / \left(\langle \gamma \rangle \langle \beta_{\parallel}^2 \rangle \right)$.

Table 1 shows the influence of different REDFs on essential parameters for I_{RE} calculation. Firstly, it is important to state that averages presented in Table 1 are done for energy span from 0 to W_{max} using the mean of the variable, i.e., neglecting the RE limit at lower energies. The error is negligible due to the high W_{max} reached in the COMPASS discharges. The average energy $\langle w \rangle$ is calculated for illustration, while $\langle \beta \rangle$ can be seen in Equation (4).

Table 1. Comparison of parameters (RE energy w, relativistic β, the term from Equation (3) $\langle \beta^2 \gamma \rangle = \langle \beta_{\parallel}^2 \gamma \rangle + 0.5 \langle \beta_{\perp}^2 \gamma \rangle$ and drift term η for W_{max} estimation) averaged over w for different REDFs.

REDF	$\langle w \rangle$ [MeV]	$\langle \beta \rangle$	$\langle \beta^2 \gamma \rangle$		$\langle \eta \rangle$	
			$\theta = 0.0$	$\theta = 0.3$	$\theta = 0.1$	$\theta = 0.3$
f_{mono}	10.00	0.9988	20.5	19.6	4.91	15.8
f_{uni}	5.00	0.9721	10.6	10.2	9.55	30.7
f_{lin}	3.33	0.9497	7.30	6.98	13.9	44.7
f_{exp}	0.86	0.8141	2.18	2.09	43.8	140
f_{sG}	8.02	0.9980	16.6	15.9	6.05	19.4
f_{MJ}	1.58	0.9301	3.78	3.62	26.21	84.2

From Equation (4), one would expect that the exponential REDF has the smallest I_{RE} for the same P_{RE}, as it has the far smallest $\langle \beta \rangle$. However, due to the term $\langle \beta_{\parallel}^2 \gamma \rangle + 0.5 \langle \beta_{\perp}^2 \gamma \rangle$ in Equation (3) used for the $\langle n_{RE} \rangle$ estimation (by knowing P_{RE} in Equation (3)), the result is opposite. Namely, difference in $\beta^2 \gamma$ from Table 1 of an order of magnitude creates much larger difference in estimated n_{RE} between

REDFs than $\langle\beta\rangle$ does in Equation (4). Therefore, even though the exponential has by far the smallest $\langle\beta\rangle$, it has the largest I_{RE}.

Furthermore, even though two extreme values are taken for θ to investigate maximum possible effect from the corresponding variations, the difference due to $\langle\beta^2\gamma\rangle$ is smaller than a few percent for every REDF. Furthermore, the aforementioned η is calculated for $\theta = \{0.1, 0.3\}$ rad, as it is zero at $\theta = 0$ rad. Here, a significant influence of the pitch angle on the η is observed. Henceforth, θ is implemented in η that is relevant for W_{max} calculation. Another expected dependence that can be noted from Table 1 is that the most similar REDF to the monoenergetic one is the skewed Gaussian f_{sG} REDF, as their parameters are close to each other.

5. Result: COMPASS Discharge 7298

The pitch angle θ is taken to be 0.2 rad as a mean value found in Ref. [15]. The RE minor radius a_{RE} for the RE area A_{RE} calculation is assumed to be constant and equal to 5 cm. The METIS (with the effective ion charge $Z_{eff} = 2.5$) data for the electric field and plasma pressure $\langle p_{pl} \rangle_V$ from Thomson scattering are interpolated and extrapolated to the EFIT time scale. As EFIT measures before the first and after the last Thomson scattering measurement, the first and the last point of the extrapolated $\langle p_{pl} \rangle_V$ are approximated to be 95% (arbitrarily taken) of the first and the last Thomson scattering measurement, respectively.

The main result is presented in Figure 2. The estimate of the RE current I_{RE} for different RE distribution functions are presented in Figure 2c and their maximum values are listed in Table 2. The plasma current I_p is plotted in Figure 2b for comparison purposes. In the same figure, Shielded HXR and Standard HXR are also plotted.

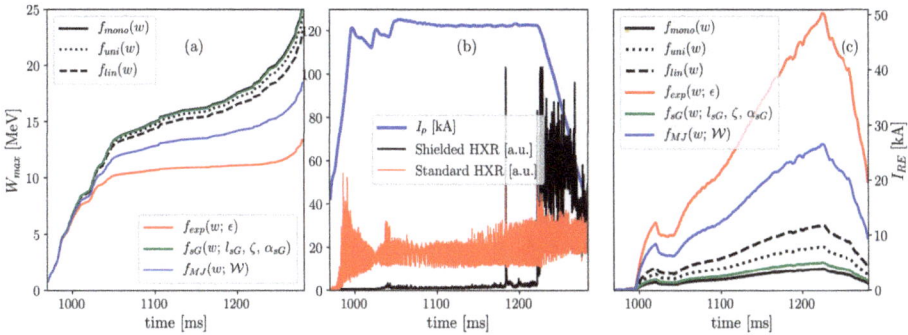

Figure 2. (**a**) The maximum kinetic energy W_{max} of discharge #7298 obtained for the different RE distribution functions: mono-energetic f_{mono} (black solid), uniform f_{uni} (black dotted), linear f_{lin} (black dashed), exponential f_{exp} (red), skewed Gaussian f_{sG} (green) and Maxwell–Jüttner f_{MJ} (blue). (**b**) Time traces of the plasma current I_p (blue), Shielded HXR (black) and Standard HXR (red). (**c**) Estimated RE current I_{RE} corresponding to each RE distribution function with the same labeling as on (**a**).

The comparison between the W_{max} estimate coming from different RE distribution functions described in Section 4.2 is shown in Figure 2a and their maximum values are tabulated in Table 2.

In Figure 2a, one can see that only W_{max} from the exponential f_{exp} and the Maxwell–Jüttner f_{MJ} distribution function are significantly different, as could be expected from the $\langle\eta\rangle$ values in Table 1.

Even though β_{pol} from EFIT increases significantly during the ramp-down phase of the discharge, a relatively strong decrease in RE current I_{RE} can be seen in Figure 2c. This is a consequence of the current drop in the total plasma pressure $\langle p_{tot}\rangle_V$ calculation from EFIT in Equation (1). It is interesting that the I_{RE} drop coincides with the RE losses seen from Shielded HXR signal, even though RE losses are not implemented in the calculation. Henceforth, the I_{RE} decrease can be seen as numerical artifact.

However, physical interpretation could be an additional RE energy limit due to the RE beam drift, which is reported in Section 7. Next, notice that I_{RE} values span over two orders of magnitude—from a few kA to tens of kA.

Even though f_{mono} is the most unrealistic of all the used RE distribution functions, it gives minimum I_{RE} as β, γ and η have extreme values. Therefore, we conclude that at least a few kA of current should definitely be driven by the REs in the COMPASS discharge #7298. However, it is not possible to determine which REDF gives the most accurate RE current I_{RE} estimate from Figure 2.

The above experimental results are compared here with the Kruskal–Bernstein theory [16] and NORSE simulation [17] in Figure 3. Kruskal–Bernstein theory is the analytical solution for growth rate of the Dreicer mechanism of RE generation, whose last shape was derived by Connor and Hastie [16]. The Kruskal–Bernstein equation is a strong function of the E_{tor}. NORSE calculateS only Dreicer mechanism of RE generation and n_e, T_e and E_{tor} are taken as time-varying parameters, while Z_{eff} and B_{tor} are constant. Finally, the n_{RE} is computed by multiplying the total electron density from measurements $n_{e,c}$ with the RE fraction calculated by NORSE.

When applied to the COMPASS discharge #7298, the Kruskal–Bernstein theory predicts RE density n_{RE} larger by 1–3 orders of magnitude than the total n_e. This is a non-physical result of course and we consider this theory not relevant for this particular case. Furthermore, NORSE reaches slide-away regime (i.e., when all electrons runaway) at the very beginning of the discharge. Such early slide-away regime is not supported by the experimental results.

Table 2. The maxima of W_{max} and I_{RE} over the time domain of the whole discharge, plotted in Figure 2.

REDF	max (W_{max}) [MeV]	max (I_{RE}) [kA]
f_{mono}	25.61	3.9
f_{uni}	24.49	7.9
f_{lin}	23.05	11.9
f_{exp}	13.39	50.3
f_{sG}	25.37	5.0
f_{MJ}	18.54	26.5

The possible reasons for the overestimating results from Kruskal–Bernstein theory and NORSE simulation could result from overestimating the electric field E_{tor} by METIS or from too high sensitivity of the theories on the E_{tor} parameter. Moreover, RE energy calculation used here is also simplistic and EFIT results could be misleading in the presence of the RE.

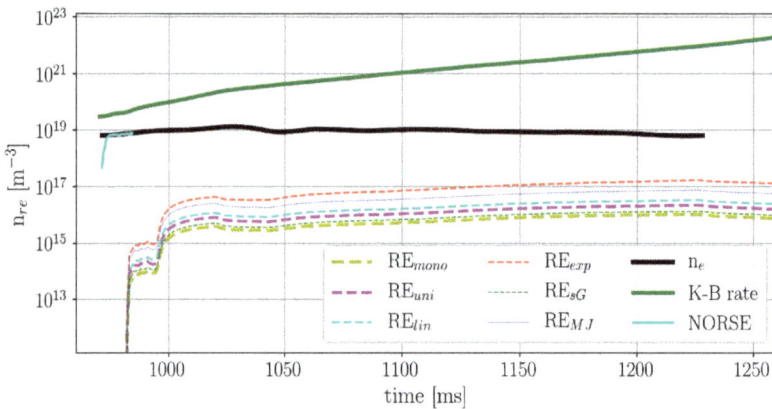

Figure 3. Time traces of total electron density (black) measured with interferometer, RE density from Kruskal–Bernstein theory (green), RE density from NORSE simulation (cyan) and RE densities corresponding to RE currents from Figure 2c (dashed lines).

6. Application: RE Influence on Ramp-Up

REs are frequently generated during the current ramp-up phase in the COMPASS tokamak. Here, the influence of the RE generation on the plasma current I_p at the end of the ramp-up phase is investigated using the reported model for the RE current I_{RE}.

The density scan in the range 1–5×10^{19} m^{-3} was done during the second RE campaign. The scan consisted of 10 COMPASS discharges from #8552 to #8561. Plasma current I_p was feedback controlled to 130 kA during the flat-top phase and the toroidal magnetic field B_{tor} was 1.15 T for all discharges.

Figure 4a presents an estimate of n_{RE}, which is more suitable for comparison of the method reported here with Kruskal–Bernstein and NORSE theories. The three discharges were chosen to cover the standard tokamak discharge (#8553), the slide-away regime (#8559) and the transient between the previous two (#8555). To avoid redundant lines, n_{RE} was calculated only for the two most extreme REDF—the monoenergetic f_{mono} and the exponential f_{exp} ones.

The discharge #8553 has n_{RE} lower by an order of magnitude during the ramp-up phase than the other two discharges (see Figure 4). Surprisingly, discharges #8555 and #8559 have similar n_{RE} at the beginning of the discharge. However, n_{RE} drops for the former discharge as expected from the missing β_p rise from EFIT and density drop, as observed by the Thomson Scattering system. Finally, the corresponding RE current I_{RE} at the end of the ramp-up phase for discharges #8553, #8555 and #8559 are 0.3–2.0 kA, 2.3–17.8 kA and 2.4–19.6 kA, respectively. Therefore, the expected trend is observed. Note that the non-continuous line for discharge #8553 comes from the defined thresholds of the method.

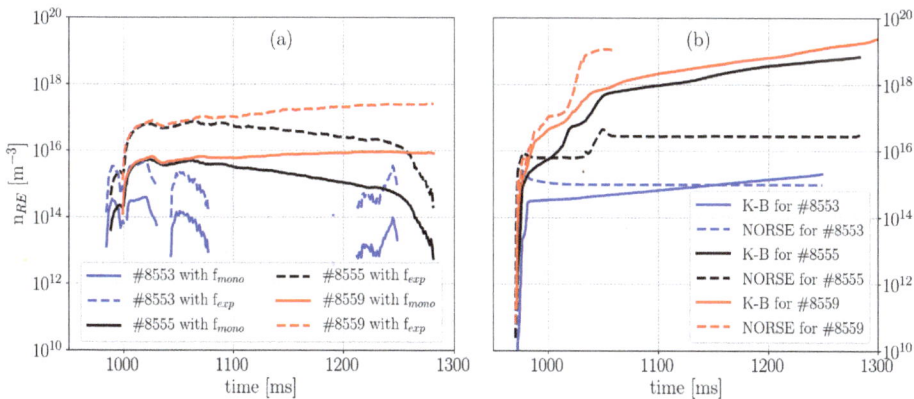

Figure 4. Time traces of the estimated n_{RE} for the three typical discharges: #8553 (blue, standard discharge), #8555 (black, intermediate case) and #8559 (red, slide-away regime). (**a**) Estimates from measurements where for each discharge the two most extreme REDF: the monoenergetic f_{mono} (full line) and the exponential f_{exp} (dashed line). (**b**) Estimates from Kruskal–Bernstein theory (solid line) and NORSE simulation (dashed line).

To complete the analyses, Kruskal–Bernstein theory and NORSE code were used to theoretically estimate n_{RE}. The results are presented in Figure 4b.

First to notice from Figure 4 is that both theories have the expected trend in density of RE rising from #8553 to #8559. Differently from discharge #7298 in Figure 3, the Kruskal–Bernstein theory does not give n_{RE} over n_e, while n_{RE} do rise towards n_e at the end of discharge for #8555 and #8559. For these two discharges, the Kruskal–Bernstein theory does not show a significant difference in n_{RE}. On the other hand, NORSE estimates that those two discharges are quite different—according to this code #8559 reaches slide-away regime already around 1060 ms, which is probably too early.

Interestingly, in the case of standard discharge #8553, both codes seem to approximately agree in the order of magnitude with the experimental n_{RE} (between 10^{14} and 10^{15} m^{-3}). The same is valid when the NORSE simulation is compared with experimental n_{RE} using f_{exp} for discharge #8555. However, the NORSE code is not made to predict the drop of n_{RE}, clearly observed in Figure 4a and from β_{pol} from the EFIT reconstruction.

7. Application: RE Localisation

In this section, we investigate the RE energy limit from the RE outward drift through the analytically derived calculation reported by Zehrfeld [18].

Zehrfeld's analysis provides the W_{drift} as a function of the normalized minor radius $\rho = r/a_p$. One of the main messages from this analysis is that W_{drift} profile can have (depending on machine and plasma parameters) maximum value inside plasma (i.e., $\rho_{max} < 1$—dashed coinciding colored lines in Figure 5), which indicates that REs of certain energy can loose confinement before they reach the plasma edge (i.e., the last closed flux surface) that is represented by R_{out} in Figure 5. Namely, peaking parameter of the plasma current profile m^1 proved to be the most significant parameter for RE confinement in respect to the drift. Therefore, knowledge of the $W_{drift}(\rho)$ profile can be used for localization of the RE beam. Knowing W_{max} at a given time of the discharge, one can find the minimum minor radius ρ_{min} (solid colored lines in Figure 5) where $W_{drift}(\rho_{min}) = W_{max}$, corresponding to the minor radius below which REs with energy W_{max} cannot be confined.

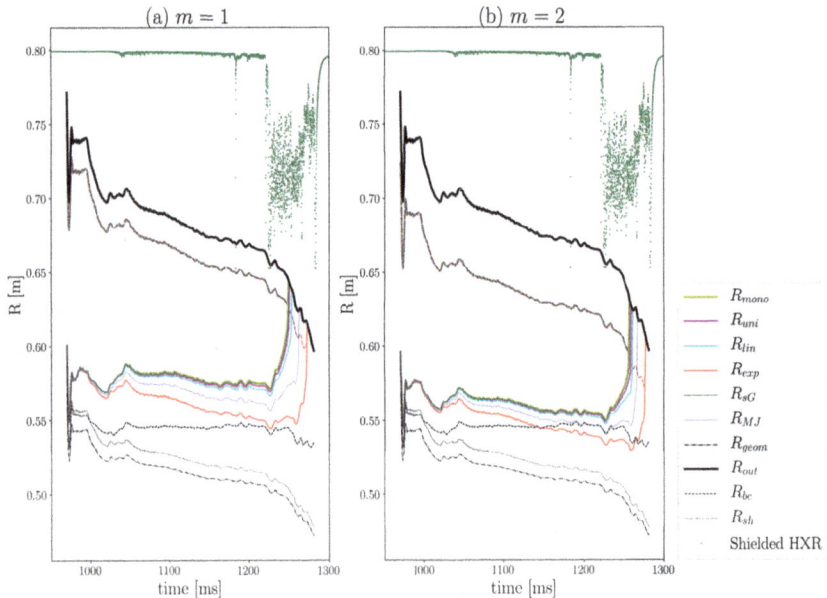

Figure 5. Time traces for lower ρ_{min} (solid colored lines) and upper ρ_{max} (dashed coinciding colored lines) limits of RE major radii for all six REDFs are presented for discharge #7298. For orientation, major radius of plasma geometrical center R_{geom} (black dash-dotted line), outer plasma major radius R_{out} (black solid line), plasma current barycenter R_{bc} (black dashed line) and theoretical bulk plasma barycenter from Shafranov shift R_{sh} (black dotted line) are added. Beside all major radii, the Shielded HXR signal (green points) is presented. Figure is plotted for two of current profile factor m values: (a) $m = 1$; and (b) $m = 2$.

---

[1] m is plasma current profile $I(\rho)$ peaking factor, defined as: $I(\rho) = I_p(1 - (1 - \rho^2)^{m+1})$.

Generally, the major radius of plasma geometrical center R_{geom} and bulk plasma barycenter R_{sh} do not coincide in the tokamak plasma due to the Shafranov shift [19]. Additionally, there is RE beam present in the observed discharges shifting the plasma current barycenter R_{bc} more outwards. All these radii R_{geom}, R_{sh} and R_{bc} are taken from EFIT and shown in Figure 5 for orientation purposes.

EFIT predicts $m = 1.5$ while Zehrfeld's method assumes m to be an integer. Accordingly, we used both rounding integers values: $m = 1$ and $m = 2$.

One can notice from Figure 5 that RE orbits are predicted in minor radius of 10–15 cm, which is not very limiting knowing that the COMPASS minor plasma radius is around 20 cm. RE orbits are localized more inside the plasma for higher m, as could be expected due to the higher peaking of W_{drift} profile. For the same reason, later significant losses of high energy RE are expected for higher m, which can be seen by slightly delayed equalization of the RE radius with R_{out} (for 5–15 ms) for factor of current profile $m = 2$ than in the $m = 1$ case. However, both timings for significant high energy RE losses are a few tens of milliseconds after the Shielded HXR observes the losses, showing that these losses are probably due to the RE outward drift as I_p is decreasing and connection of strong Shielded HXR signal and I_p ramp-down phase is indeed observed in Figure 2b. On the other hand, this does not explain the interaction on the high-field side seen by the visible cameras.

8. Conclusions

Understanding RE physics gives us better prediction towards understanding RE generation and mitigation in ITER. One basic RE parameter to compare model and experiment is the RE current. Therefore, we have presented here one method for estimating the RE current in the COMPASS tokamak.

The pitch angle showed to be relevant for the RE energy calculation, which plays an internal role in the RE current estimate method. Implementation of the method on the COMPASS discharge #7298 shows that at least a few kA of the total plasma current is driven by the RE. The method is then used for n_e-scan and showed the different amount of RE current present in the plasma at the expected trend. Namely, discharges with lower density have higher RE current. Commonly used Kruskal–Bernstein theory does not appear suitable for high RE current discharges. NORSE, a code built for such discharges, showed to be more sensitive to plasma parameters, but further development is necessary for better prediction of the RE dynamics. Applying the knowledge of RE current to the RE drift energy limit, a crucial role of the limit at the final stage of the COMPASS plasma discharge, is reported.

At the first instance, an implementation of the RE current into the equilibrium simulation is left for future work on the COMPASS tokamak. Furthermore, theories used here are also the most commonly used model basis in the RE research community. However, they obviously overestimate the number of REs. Their further development is far beyond the scope of this article. The results obtained from theories are here to show the issue is oversimplified and demonstrate the need for a better modeling of the RE generation.

Author Contributions: Conceptualization, M.V.; methodology, M.V.; software, M.V., O.F., E.M. and COMPASS IT Team; validation, M.V.; formal analysis, M.V.; investigation, M.V., O.F. and J.M.; resources, J.M.; data curation, M.V. and J.M.; writing—original draft preparation, M.V.; writing—review and editing, M.V., O.F., J.M, and E.M; visualization, M.V. and O.F.; supervision, J.M.; project administration, J.M. and COMPASS department; and funding acquisition, J.M., EUROfusion and COMPASS department.

Funding: This research was funded by Ministry of Education, Youth and Sport of the Czech Republic (MEYS) grant number LM2011021 that supported COMPASS operation and by the Euratom research and training programme 2014-2018 under grant agreement No 633053 with the Co-fund by MEYS project number 8D15001.

Acknowledgments: Erasmum Mundus Joint Doctoral Collage in Fusion Science and Engineering (often abbreviated as "Fusion-DC") is acknowledged for providing the necessary infrastructure to the first author for his studies. Moreover, we would like to thank to the whole COMPASS team of scientist, engineers and technicians for their fruitful collaboration.

Abbreviations

The following abbreviations are used in this manuscript:

RE   Runaway Electron
REDF   Runaway Electron Distribution Function
HXR   Hard X-ray

References

1. Hender, T.; Wesley, J.; Bialek, J.; Bondeson, A.; Boozer, A.; Buttery, R.; Garofalo, A.; Goodman, T.; Granetz, R.; Gribov, Y.; et al. Chapter 3: MHD stability, operational limits and disruptions. *Nucl. Fusion* **2007**, *47*, S128. [CrossRef]

2. Pánek, R.; Adámek, J.; Aftanas, M.; Bílková, P.; Böhm, P.; Cahyna, P.; Cavalier, J.; Dejarnac, R.; Dimitrova, M.; Grover, O.; et al. Status of the COMPASS tokamak and characterization of the first H-mode. *Plasma Phys. Control. Fusion* **2016**, *58*, 014015. [CrossRef]

3. Fujita, T.; Fuke, Y.; Yoshida, Z.; Inoue, N.; Tanihara, T.; Mori, K.; Fukao, M.; Tomita, Y.; Mohri, A. High-Current Runaway Electron Beam in a Tokamak Plasma. *J. Phys. Soc. Jpn.* **1991**, *60*, 1237–1246. [CrossRef]

4. Havlíček, J.; Hronová, O. Magnetic Diagnostics of COMPASS Tokamak. 2010. Available online: http://www.ipp.cas.cz/vedecka_struktura_ufp/tokamak/tokamak_compass/diagnostics/magneticka-diagnostika/index.html (accessed on 16 January 2019).

5. Bilkova, P.; Aftanas, M.; Bohm, P.; Weinzettl, V.; Sestak, D.; Melich, R.; Stockel, J.; Scannell, R.; Walsh, M. Design of new Thomson scattering diagnostic system on COMPASS tokamak. *Nucl. Instrum. Methods Phys. Res. Sect. A* **2010**, *623*, 656–659. [CrossRef]

6. Mlynar, J.; Ficker, O.; Vlainic, M.; Weinzettl, V.; Imrisek, M.; Paprok, R.; Rabinski, M.; Jakubowski, M.; the COMPASS Team. Effects of Plasma Control on Runaway Electrons in COMPASS Tokamak. In Proceedings of the 42nd EPS Conference on Plasma Physics, Lisbon, Portugal, 22–26 June 2015; Number P4.102 in ECA; Volume 39E.

7. Ficker, O.; Mlynar, J.; Vlainic, M.; Macusova, E.; Vondracek, P.; Weinzettl, V.; Urban, J.; Cerovsky, J.; Cavalier, J.; Havlicek, J.; et al. Long slide-away discharges in the COMPASS tokamak. In Proceedings of the 58th Annual Meeting of the APS Division of Plasma Physics, San Jose, CA, USA, 31 October–4 November 2016; Volume 61.

8. Artaud, J.; Imbeaux, F.; Garcia, J.; Giruzzi, G.; Aniel, T.; Basiuk, V.; BÂcoulet, A.; Bourdelle, C.; Buravand, Y.; Decker, J.; et al. Metis: a fast integrated tokamak modelling tool for scenario design. *Nucl. Fusion* **2018**, *58*, 105001. [CrossRef]

9. Vlainić, M.; Mlynář, J.; Weinzettl, V.; Papřok, R.; Imrísek, M.; Ficker, O.; Vondráček, P.; Havlíček, J. First Dedicated Observations of Runaway Electrons in COMPASS Tokamak. *Nukleonika* **2015**, *60*, 249–255. [CrossRef]

10. Vlainic, M. Study of Runaway Electrons at COMPASS Tokamak. Ph.D. Thesis, Ghent University, Ghent, Belgium, 2017.

11. Yu, J.H.; Hollmann, E.M.; Commaux, N.; Eidietis, N.W.; Humphreys, D.A.; James, A.N.; Jernigan, T.C.; Moyer, R.A. Visible imaging and spectroscopy of disruption runaway electrons in DIII-D. *Phys. Plasmas* **2013**, *20*, 042113. [CrossRef]

12. Pankratov, I.M. Analysis of the synchrotron radiation spectra of runaway electrons. *Plasma Phys. Rep.* **1999**, *25*, 145–148.

13. Azzalini, A. *The Skew-Normal and Related Families*; Institute of Mathematical Statistics Monographs, Cambridge University Press: Cambridge, UK, 2013.

14. Jüttner, F. Das Maxwellsche Gesetz der Geschwindigkeitsverteilung in der Relativtheorie. *Ann. Phys.* **1911**, *339*, 856–882. [CrossRef]

15. Vlainic, M.; Vondracek, P.; Mlynar, J.; Weinzettl, V.; Ficker, O.; Varavin, M.; Paprok, R.; Imrisek, M.; Havlicek, J.; Panek, R.; et al. Synchrotron Radiation from Runaway Electrons in COMPASS Tokamak. In Proceedings of the 42nd EPS Conference on Plasma Physics, Lisbon, Portugal, 22–26 June 2015; Number P4.108 in ECA; Volume 39E.

16. Connor, J.; Hastie, R. Relativistic limitations on runaway electrons. *Nucl. Fusion* **1975**, *15*, 415. [CrossRef]
17. Stahl, A.; Landreman, M.; Embréus, O.; Fülöp, T. NORSE: A solver for the relativistic non-linear Fokker–Planck equation for electrons in a homogeneous plasma. *Comput. Phys. Commun.* **2017**, *212*, 269–279. [CrossRef]
18. Zehrfeld, H.P.; Fussmann, G.; Green, B.J. Electric field effects on relativistic charged particle motion in Tokamaks. *Plasma Phys.* **1981**, *23*, 473. [CrossRef]
19. Shafranov, V.D. Equilibrium of a toroidal pinch in a magnetic field. *Soviet Atomic Energy* **1962**, *13*, 1149–1158. [CrossRef]

Review

Influence of Nitrogen Admixture on Plasma Characteristics in a dc Argon Glow Discharge and in Afterglow

Nikolay A. Dyatko [1,2,*], Yury Z. Ionikh [3] and Anatoly P. Napartovich [1]

[1]   Troitsk Institute for Innovation and Fusion Research, ul. Pushkovykh, vladenie 12, Troitsk,
      Moscow 108840, Russia; apn@triniti.ru
[2]   Pushkov Institute of Terrestrial Magnetism, Ionosphere and Radio Wave Propagation Russian Academy of
      Sciences, Kaluzhskoe Hwy 4, Troitsk, Moscow 108840, Russia
[3]   Department of Physics, St. Petersburg State University, Universitetskaya Emb., St. Petersburg 199034, Russia;
      yionikh@gmail.com
*   Correspondence: dyatko@triniti.ru; Tel.: +7-495-851-0450

Received: 27 November 2018; Accepted: 14 January 2019; Published: 19 January 2019

Abstract: The present paper is based on the materials of the Invited Lecture presented at 29th Summer School and International Symposium on the Physics of Ionized Gases (28 August 2018–1 September 2018, Belgrade, Serbia). In the paper, the effect of nitrogen admixture on various characteristics of a dc glow discharge in argon (the volt-ampere characteristic, rate of plasma decay in the afterglow, discharge constriction condition, and formation of a partially constricted discharge) is considered.

Keywords: glow discharge; argon; nitrogen admixture; discharge voltage; diffuse discharge; constricted discharge

1. Introduction

It is well-known that an addition of a molecular gas to a rare gas can significantly change characteristics of a discharge. The degree and tendency of these changes depend on the discharge conditions: the sort of a rare gas, sort and percentage of the admixture, gas pressure, etc. The argon/ nitrogen gas mixture is one of the mixtures most commonly used in discharge studies. Over the past decades, many experimental and theoretical works have been carried out in which the effect of nitrogen admixture on the characteristics of an electric discharge in argon has been investigated. Various types of discharges have been studied: microwave [1–3], RF (magnetron, inductive, and capacitive) [4–9], surface [10], barrier [11–13], dc [14–24], and pulse-periodic [25–27] discharges. It is shown that nitrogen impurity can significantly change the electrical and spectral characteristics of the discharge, the plasma ion composition, and the population of argon energy levels. In this paper, we restrict ourselves to examining the effect of nitrogen admixture on the characteristics of a dc glow discharge (maintained in a tube) in argon and on the characteristics of afterglow plasma. The presented brief review is mainly based on the results of our works that have been performed in the last decade. The following effects are discussed: influence of N_2 admixture on the current–voltage characteristic of a diffuse glow discharge; influence of N_2 admixture on the rate of plasma decay in the afterglow; influence of N_2 admixture on discharge constriction conditions and on the characteristics of constriction process; formation of partially constricted discharge in $Ar:N_2$ mixtures at intermediate gas pressures.

2. Influence of N_2 Admixture on the Current–Voltage Characteristic of a Diffuse Glow Discharge in Ar

In glow discharges in pure rare gases the main ionization process is stepwise ionization from the lower metastable states (except for the case of very low pressures and discharge currents). The effect of molecular admixture on discharge characteristics depends, in particular, on the ratio of the ionization energy of the admixture molecules and the energy of the lower metastable state of rare gas atoms. If the ionization energy of molecules is lower than the energy of atoms, then a small admixture of a molecular gas can lead to a reduction in both the breakdown voltage and the discharge operating voltage. Such a situation occurs in discharges in He and Ne (see, e.g., [28–30] and references therein), because the energy of the lower metastable states of He* (19.8 eV) and Ne* (16.6 eV) is higher than the ionization energy of any molecular admixture. The effect of discharge voltage and breakdown voltage reduction is partially due to the low ionization energy of molecules, but is mainly caused by the processes of the Penning ionization (He* + M → He + M+ + e, Ne* + M → Ne + M+ + e). If the percentage of admixture is relatively high, the situation is more complicated. On the one hand, the admixture can appreciably change the electron energy balance in the plasma due to losses of the electron energy going to the excitation of vibrational and lower electronic levels of molecules. This effect leads to an increase in the voltage required to sustain the discharge. On the other hand, as the admixture concentration increases, a new ionization mechanism can occur (related directly to molecules), which may lead to a reduction in the discharge voltage. In general, the resulting change in the discharge characteristics depends on the experimental conditions: the sort of rare gas, the percentage of admixture, the sort of molecular gas and the gas pressure. This section may be divided by subheadings. It should provide a concise and precise description of the experimental results, their interpretation as well as the experimental conclusions that can be drawn.

If the rare gas is argon and the molecular admixture is nitrogen, the Penning ionization mechanism is absent, since the energy of the lower metastable level of Ar atom (11.6 eV) is lower than the ionization energy of nitrogen molecules (15.58 eV). The ionization energy of Ar atoms (15.76 eV) is close to the ionization energy of nitrogen molecules, therefore nitrogen is not an easily ionized additive. Cross sections for the excitation of vibrational levels of N_2 molecules by electron impact are rather high in the energy range 2–3.5 eV (i.e., at energies below the energy of the lower argon metastable state), so in Ar:N_2 discharge plasma the electron energy losses due to excitation of vibrational levels can be significant. Moreover, in Ar:N_2 plasma metastable states of Ar atoms are effectively quenched in collisions with N_2 molecules [31]

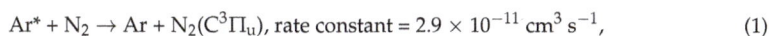

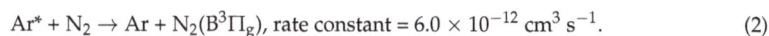

$$Ar^* + N_2 \rightarrow Ar + N_2(C^3\Pi_u), \text{ rate constant} = 2.9 \times 10^{-11} \text{ cm}^3 \text{ s}^{-1}, \tag{1}$$

$$Ar^* + N_2 \rightarrow Ar + N_2(B^3\Pi_g), \text{ rate constant} = 6.0 \times 10^{-12} \text{ cm}^3 \text{ s}^{-1}. \tag{2}$$

In addition, it is also well known that, for the same conditions (discharge tube geometry, gas pressure, and discharge current) the electric field in the positive column (or the discharge voltage) required to sustain a dc diffuse glow discharge in pure nitrogen is much higher than that in pure argon [32].

Given the above arguments, one would expect that the addition of nitrogen would lead to an increase in the electric field in the positive column of the dc discharge in argon. Indeed, this is exactly what happens at relatively low gas pressures (P~1–2 Torr for a tube radius of R~1–2 cm; see, for example, [14,33]). This is illustrated in Figure 1, which shows current–voltage characteristic of discharges in pure argon and in Ar + 0.075%N_2 and Ar + 1%N_2 gas mixtures at gas pressure $P = 2$ Torr ($R = 1.4$ cm, interelectrode distance ≈ 75 cm). One can see that the discharge voltage increases with a factor 2 when 1% of nitrogen is added to argon.

However, at intermediate pressures (tens of Torr), the situation changes dramatically. It was observed that at intermediate pressures a small additive of nitrogen to argon resulted in a noticeable decrease in the discharge voltage of a diffuse dc discharge. This effect was discovered in experiments with partially constricted discharges in Ar:N$_2$ mixtures [15,16,34]. Figure 2 shows current–voltage characteristics measured in pure argon and in Ar + 0.75%N$_2$ and Ar + 1%N$_2$ gas mixtures at gas pressure P = 40 Torr (R = 1.4 cm, interelectrode distance $\approx$ 75 cm). As can be seen from Figure 2, the discharge voltage of a diffuse discharge decreases by 2–1.5 times (depending on discharge current value) when 0.75% of nitrogen is added to argon. In the case of 1% of nitrogen the effect is slightly less pronounced.

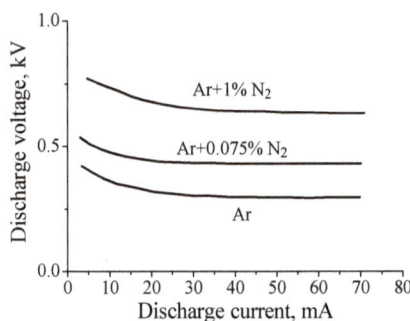

Figure 1. Current–voltage characteristic of discharges in pure argon and the Ar/N$_2$ mixtures at gas pressure P = 2 Torr [33].

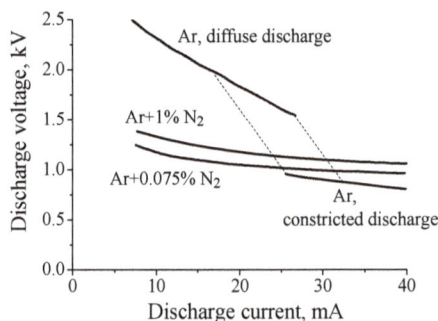

Figure 2. Current–voltage characteristic of discharges in pure argon and the Ar/N$_2$ mixtures at gas pressure P = 40 Torr [15,34].

The explanation of the observed effect was given in paper [17], in which characteristics of discharge plasma in pure argon and Ar + 1%N$_2$ mixture were studied both experimentally and theoretically (R = 1.4 cm, interelectrode distance $\approx$ 75 cm). To reproduce in the calculations the current–voltage characteristics measured at various pressures (2 Torr, 40 Torr and 80 Torr, see Figure 3), a rather complex (complete) zero-dimensional kinetic model was elaborated [17]. The model included kinetics of excited states of Ar atoms (four lower levels and the higher states combined into three lumped levels), the kinetics of electronic levels of N$_2$ (N$_2$(A$^3\Sigma_u^+$), N$_2$(B$^3\Pi_g$), N$_2$(B$'^3\Sigma_u^-$), N$_2$(a$'^1\Sigma_u^-$), N$_2$(a$^1\Pi_g$), N$_2$(w$^1\Delta_u$), and N$_2$(C$^3\Pi_u$)), the vibrational kinetics of nitrogen molecules in the ground state N$_2$(X$^1\Sigma_g^+$, v) (45 vibrational levels), the kinetics of electronic states of N atoms (N(^{4}S), N(^{2}D), and N(^{2}P)), and the kinetics of electrons and Ar$^+$, Ar$_2^+$, N$^+$, N$_2^+$, N$_3^+$, and N$_4^+$ ions. Electron transport coefficients and rate constants for electron induced processes were calculated by solving the electron Boltzmann equation in parallel with the system of kinetic equations. Besides, the model included the equation for the electric circuit.

The gas temperature on the tube axis was evaluated from the experimental data and used in the model as a parameter. Naturally, the radial temperature profile was not taken into account within 0D model. The temperature value that was used in the calculations corresponded to the gas temperature on the tube axis. It was estimated as follows. First, for given experimental data on the electric field strength, E, and the discharge current, I, the power deposited per unit length of the discharge, $Q = IE$, was calculated. Then, a one-dimensional (along the tube radius) thermal balance equation with a given heat source was solved numerically and the radial profile of the gas temperature was calculated. It was assumed that the wall temperature was 300 K and the radial profile of the deposited energy was Besselian.

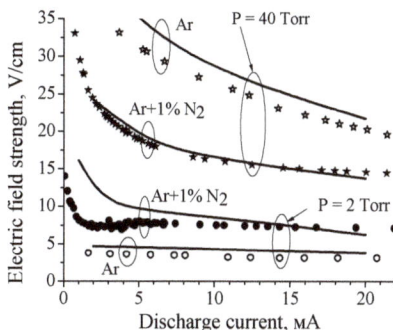

Figure 3. Calculated (curves) and measured (symbols) dependences of the electric field in the positive columns of discharges in pure argon and the Ar + 1%N$_2$ mixture on the discharge current for P = 2 Torr and 40 Torr [17].

The gas temperature in a glow discharge under the conditions close to the experimental conditions [17] was measured in [35] (Ar, Ar + 1%N$_2$, pressure of 50 Torr, tube diameter of 3.8 cm, and currents of 5–50 mA). In that paper, the gas temperature was also estimated using the procedure described above. It was shown that for the case of gas mixture the calculated values of the gas temperature agreed quite well with the measured ones. For pure Ar discharge the estimated gas temperatures were lower than measured. This was explained by the fact that, under considered conditions, the discharge in argon contracted at approximately 35 mA whereas the discharge in mixture remained diffuse even at 200 mA. In pure Ar, at discharge currents relatively close to 35 mA the radial profile of the deposited energy is noticeably narrower than the Besselian profile and, as a result, the measured temperature profile is narrower than the calculated one (for example, at I = 20 mA, see comments in [35]). Naturally, with decreasing current, the radial profile of the discharge current density becomes wider, and the theoretical estimate of the gas temperature becomes more reliable. In [17], the relatively low discharge currents were considered ($\leq$20 mA), so it can be expected that the performed estimation of the gas temperature was rather correct.

Gas temperatures at the axis of the discharge tube calculated in pure argon and the Ar + 1%N$_2$ mixture for P = 2 Torr and 40 Torr are shown in Figure 4. According to calculations, at P = 40 Torr, the gas temperature in pure Ar discharge is higher than in Ar + 1%N$_2$ mixture discharge, which is consistent with experimental data [35].

The calculated $E(I)$ characteristics were in a reasonable agreement with the experimental data (see Figure 3). As for the reduced electric field (E/N, N is the gas number density), in the case of P = 40 Torr the calculated E/N value varies from 3.9 Td to 3.3 Td (in pure Ar) and from 2.1 Td to 1.6 Td (in Ar + 1%N$_2$ mixture) with discharge current increase from 2 mA to 20 mA. That is, in argon, the reduced electric field is two times higher than in the mixture. At low pressure (2 Torr) the E/N value varies from 7.59 Td to 6.6 Td (in pure Ar) and from 22.5 Td to 10.9 Td (in Ar + 1%N$_2$ mixture) with discharge current increase from 2 mA to 20 mA.

The performed analysis showed that, in pure argon at intermediate pressures (e.g., 40 Torr) the stepwise ionization from the lower excited electronic states of Ar atoms was the dominant ionization process. The excitation of these states was provided by electron impact from the ground state, the energy of the lower state is about 11.6 eV. Due to ion conversion reaction the main ion in plasma was

Ar_2^+. Losses of electrons were mainly provided by recombination with Ar_2^+ ions and, partially, by ambipolar diffusion.

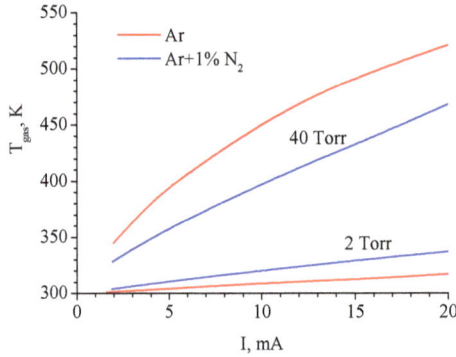

Figure 4. Gas temperature at the axis of the discharge tube calculated in pure argon and the Ar + 1%N_2 mixture for P = 2 Torr and 40 Torr [17].

According to calculations [17], in the Ar + 1%N_2 mixture at intermediate pressures, a very effective ionization mechanism was realized. The ionization was mainly provided by processes of associative ionization of excited nitrogen atoms

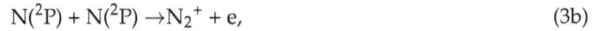

$$N(^2P) + N(^2D) \rightarrow N_2^+ + e, \tag{3a}$$

$$N(^2P) + N(^2P) \rightarrow N_2^+ + e, \tag{3b}$$

and (to a lesser extent) by processes of associative ionization of excited nitrogen molecules

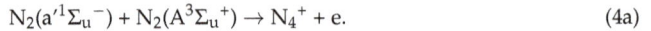

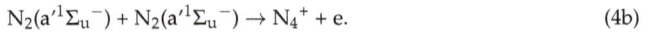

$$N_2(a'^1\Sigma_u^-) + N_2(A^3\Sigma_u^+) \rightarrow N_4^+ + e. \tag{4a}$$

$$N_2(a'^1\Sigma_u^-) + N_2(a'^1\Sigma_u^-) \rightarrow N_4^+ + e. \tag{4b}$$

It was also shown that, at intermediate pressures, the degree of vibrational excitation of nitrogen was very high and the processes involving vibrationally excited molecules substantially contributed to the production of N atoms

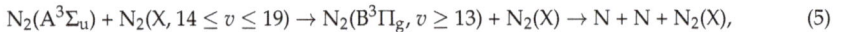

$$N_2(A^3\Sigma_u) + N_2(X, 14 \leq v \leq 19) \rightarrow N_2(B^3\Pi_g, v \geq 13) + N_2(X) \rightarrow N + N + N_2(X), \tag{5}$$

and $N_2(a')$ molecules

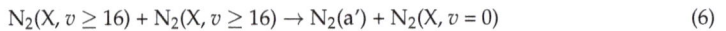

$$N_2(X, v \geq 16) + N_2(X, v \geq 16) \rightarrow N_2(a') + N_2(X, v = 0) \tag{6}$$

Due to charge transfer and ion conversion processes

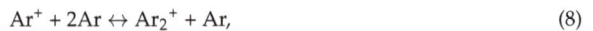

$$N_2^+ + Ar \leftrightarrow N_2 + Ar^+, \tag{7}$$

$$Ar^+ + 2Ar \leftrightarrow Ar_2^+ + Ar, \tag{8}$$

the major ion in discharge appeared to be Ar_2^+, i.e., the same as in pure Ar discharge.

Note that the energies of $N(^2D)$ and $N(^2P)$ states are 2.38 eV and 3.57 eV, respectively. Nitrogen atoms are produced in processes (5) with the participation of vibrationally excited molecules $N_2(X, 14 \leq v \leq 19)$ and electronically excited $N_2(A)$ molecules. Lower vibrational levels of N_2 molecules ($v \leq 8$) are excited by electron impact, the threshold for the excitation of the first vibrational level is about 0.29 eV. The upper vibrational levels are populated due to V–V exchange processes. The lower metastable state $N_2(A^3\Sigma_u)$ is excited by electron impact from the ground electronic state $N_2(X, v)$, the energy of the $N_2(A^3\Sigma_u)$ state is about 6.17 eV. One can see that the energies of key species, which provide ionization in Ar + 1%N$_2$ mixture at intermediate pressures, are essentially lower than 11.6 eV. For this reason, the ionization mechanism in Ar + 1%N$_2$ mixture is realized at E/N values noticeably lower than those in pure Ar.

Therefore, at intermediate gas pressures, the addition of N$_2$ to Ar leads to more effective ionization processes, while the major ion in plasma and, thus, the mechanisms of electron losses remain the same. As a result, the electric field in Ar + 1%N$_2$ discharge (and the reduced electric field) is lower than that in the pure argon discharge.

At low gas pressures (e.g., 2 Torr) the losses of electrons and ions in discharges in pure argon and in Ar + 1%N$_2$ mixture are due to ambipolar diffusion, the rate of losses is noticeably higher than that at intermediate pressures (e.g., 40 Torr). For this reason, the reduced electric field E/N in the plasma increases [17], because, in order to preserve ionization balance in the discharge plasma, it is necessary that the ionization rate be sufficiently high. In pure argon discharge, the required rate of ionization is provided, as before, mainly by stepwise ionization processes. In Ar + 1%N$_2$ discharge at low pressures the Processes (3) and (4) cannot provide the required rate of ionization and the ionization processes involving argon atoms (ionization by electron impact from the ground state, stepwise ionization and chemoionization) contribute substantially to electron production [17]. At that, the value of E/N (and, accordingly, the value of E) in Ar + 1%N$_2$ mixture is appreciably higher than that in pure argon. This is related, in particular, to the fact that, in the mixture, excited argon atoms are efficiently quenched by nitrogen molecules (Processes (1) and (2)).

Naturally, the results of the calculations depend on the used values of the rate constants. The choice of the rate constants for the processes included in the model is discussed in detail in paper [17]. Here we briefly discuss the situation with rate constants for the key ionization Processes (3) and (4). The rate constant of Process (3a) is known only approximately. Estimate performed in [36] yield a value of ~10^{-12} cm^3 s^{-1}, which is lower by one order of magnitude than the estimate given in [37]. The rate constant used in the model [17] was taken from [36]. As for the Process (3b), it was shown in [38] that the use of this process in the model of the afterglow nitrogen plasma allows one to adequately describe specific features of the experimentally observed plasma decay dynamics. According to estimates [38], the rate constant of this process is 2×10^{-11} cm^3 s^{-1}, it is this value that was used in the model [17]. The rate constants of Reactions (4a) and (4b) used in different studies differ by one to two orders of magnitude. In the model [17] rate constants of these processes were chosen in accordance with recommendations made in [39]: 10^{-11} cm^3 s^{-1} and 5×10^{-11} cm^3 s^{-1}, respectively.

It is also worth noting that, in contrast to argon, the addition of nitrogen (1%, for example) to neon or helium leads to an increase in the electric field (discharge voltage) in a dc glow discharge even at intermediate gas pressures. The explanation of this effect is as follows [18]. Addition of N$_2$ to Ne also leads to changing the ionization mechanism in dc discharge plasma, ionization processes in Ne + 1%N$_2$ discharge are similar to that in Ar + 1%N$_2$ discharge. On the other hand, in contrast to argon atoms, the ionization energy of Ne atoms is essentially higher than that of N$_2$ molecules, therefore the charge transfer process from N$_2^+$ ion to Ne atom is absent Ne + 1%N$_2$ plasma. For this reason, the addition of N$_2$ to Ne leads to the replacement of Ne$_2^+$ (major ion in plasma in pure Ne) with N$_4^+$ in Ne + 1%N$_2$ mixture. The rate constant for the process of electron recombination with N$_4^+$ ion is one order of magnitude higher than that with Ne$_2^+$ ion. The increase in the rate of electron losses due to recombination with N$_4^+$ ions appears to be more significant factor than the new ionization mechanism, so the electric field needed for the glow discharge maintenance increases with the addition of N$_2$ to Ne. In the case of He, the situation is similar to that in Ne.

3. Effect of Nitrogen Addition to Argon on the Rate of Plasma Decay in the Afterglow

It is shown in [17] that, in a dc glow discharge in Ar + 1%N$_2$ mixture, a very high degree of vibrational excitation of nitrogen molecules is achieved. In the afterglow of such a discharge, the electrons gain energy in superelastic (second kind) collisions with vibrationally excited molecules, so the effective electron temperature ($T_e = 2/3u_m$, where u_m is the mean electron energy) in the afterglow plasma can be quite high for some time (about the relaxation time of the vibrational distribution function). In turn, the high electron temperature provides the high rate of plasma decay due to ambipolar diffusion process.

It should be noted that the evolution of the electron energy distribution function (EEDF) in nitrogen afterglow plasma was a subject of a large body of studies within the past 30 years (see [40,41] and references therein). It was ascertained that, in the afterglow plasma, the EEDF form was governed by heating of electrons in superelastic collisions with vibrationally excited molecules. As a consequence, there was a strong coupling between the degree of the vibrational excitation of nitrogen and the electron temperature. The EEDF and the effective electron temperature in pure nitrogen afterglow plasma were studied in a large number of papers (see [40,41] and references therein), though there are a few works, in which similar studies were performed in gas mixtures of a rare gas with nitrogen [3,26,42–44].

In particular, in paper [43] the EEDF in the afterglow of a pulsed direct current discharge was measured in an Ar + 1%N$_2$ mixture. The vibrational temperature, T_v, of N$_2$ molecules was also experimentally estimated. The discharge was maintained in a cylindrical glass tube of 3 cm internal diameter at gas pressures of 0.5 Torr and 1 Torr. The pulse duration was 40 μs and the pulse repetition frequency was 1 kHz. The estimated vibrational temperature was about 4000–5000 K, and the electron temperature, calculated by the measured EEDFs, was varied in the range 4000–6000 K (depending on the gas pressure and the discharge current). In [3], the electron temperature in the afterglow of Ar:N$_2$ power-pulsed microwave plasma was measured. Measurements were performed in a quartz tube with an inner radius of 0.3 cm in mixtures with N$_2$ percentage of 1–20% at a total gas pressure of 8–30 Torr. It was shown that in mixtures with a relatively low N$_2$ percentage (for example, 1%) the electron temperature in the afterglow plasma can be as high as 0.8 eV, which indicates that the degree of vibrational excitation of nitrogen is high.

Actually, the vibrational distribution function in the afterglow plasma (as well as in the discharge plasma) is not the Boltzmann function, i.e., is not characterized by only the temperature T_v. Calculations of the vibrational distribution function and the EEDF in the afterglow of a dc glow discharge in Ar + 1%N$_2$ were presented in conference paper [44]. In calculations the kinetic model [17] was used, and the procedure of simulation was as follows. Firstly, time–evolution of plasma parameters was calculated up to approaching steady-state discharge conditions which were characterized by the discharge current value. Then, the applied voltage (the electric field in plasma) was switched off, and the time-variation of plasma parameters in the post-discharge was calculated. The following conditions were considered: discharge tube radius $R = 1.5$ cm, gas pressure $P = 5$ Torr, gas temperature $T_{gas} = 350$ K, discharge current $I = 20$ mA.

According to simulations in [44], the reduced electric field in the steady-state discharge is about $E/N = 4.7$ Td, the electron number density is $n_e = 2.0 \times 10^{10}$ cm^{-3}. Vibrational distribution functions in discharge and afterglow plasma are shown in Figure 5. The shape of the distribution function (and the degree of the vibrational excitation) can be characterized by 'local' vibrational temperatures, $T_v^{i,i+1}$, calculated using the populations of two successive vibrational levels, i and $i + 1$. In the discharge plasma the vibrational temperature $T_v^{0,1}$ is as high as 10,580 K and in the afterglow, it decreases down to 3300 K during 50 ms (see Table 1). Note also that $T_v^{i,i+1}$ values ($i = 1, 2, 3, 4$, see Table 1) are noticeably higher than $T_v^{0,1}$ values.

The high degree of the vibrational excitation of nitrogen molecules leads to high effective electron temperatures in the afterglow plasma (see Table 1). At $t = 10$ ms, the electron temperature is as high as $T_e \approx 8520$ K, and the further decrease in T_e value is explained by the decrease in the degree of vibrational excitation of nitrogen molecules, which is illustrated in Figure 5 by growth of N$_2$(X, $v = 0$) population. At that, even at $t = 30$ ms the electron temperature is rather high $T_e \approx 6590$ eV. It is worth noting that the calculated T_e values agree with those measured in the afterglow of a pulsed microwave discharge in the Ar + 1%N$_2$ mixture [3].

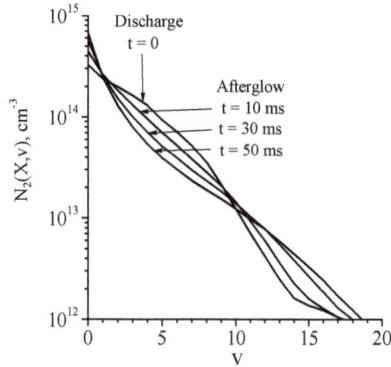

Figure 5. Ar + 1%N_2. Calculated vibrational distribution functions in discharge ($t = 0$) and afterglow ($t > 0$) plasma [44]. $R = 1.5$ cm, $P = 5$ Torr, $T_{gas} = 350$ K, $I = 20$ mA.

Table 1. Ar + 1%N_2. Calculated 'local' vibrational temperatures ($T_v^{i,i+1}$, see comments in the text) and effective electron temperatures (T_e), in discharge ($t = 0$) and afterglow ($t > 0$) plasma [44]. $R = 1.5$ cm, $P = 5$ Torr, $T_{gas} = 350$ K, $I = 20$ mA.

	Discharge	Afterglow		
	$t = 0$	10 ms	30 ms	50 ms
$T_v^{0,1}$, K	10,580	6300	4110	3300
$T_v^{1,2}$, K	17,260	9150	6350	5100
$T_v^{2,3}$, K	16,560	9990	7900	6760
$T_v^{3,4}$, K	14,980	10,530	9410	8610
$T_v^{4,5}$, K	9560	10,190	10,070	9870
T_e, K	28,900	8520	6590	6050

It was mentioned in [44] that, under the considered conditions, the rate of plasma decay due to ambipolar diffusion process was high because of the high electron temperature in the afterglow. However, no comparison with plasma decay rate in pure argon afterglow was performed. Such a comparison was presented in paper [26], in which the time variation of the electron concentration in the afterglow of dc glow discharges in pure Ar and Ar + 1%N_2 mixture was numerically studied. As in work [44], the kinetic model [17] was used in simulations. The following conditions were considered: discharge tube radius $R = 1.4$ cm, gas pressures $P = 1$, 2, and 5 Torr, discharge currents $I = 20$ mA and 56 mA. Time variation of the effective electron temperature in Ar + 1%N_2 afterglow calculated at fixed discharge current $I = 20$ mA and various gas pressures is shown in Figure 6. As follows from the calculations, the lower the gas pressure, the faster the electron temperature decreases in the discharge afterglow. In the afterglow of a discharge in pure argon, electrons are heated in superelastic collisions with electronically excited atoms. In addition, fast electrons appear in chemoionization processes Ar* + Ar* = Ar + Ar$^+$ + e (≈ 7.6 eV). In [26], the electron temperature in the pure argon afterglow was not calculated because of the problem of accounting for the letter process in the Boltzmann equation. On the other hand, it was observed in [3,45] that the electron temperature in the argon afterglow plasma dropped rapidly (during ~100 μs at pressures of 6–30 Torr) to T_e~1200 K. Therefore, in calculations [26], the electron energy distribution function in pure argon afterglow was assumed to be Maxwellian with the temperature $T_e = 1000$ K or $T_e = T_{gas}$ (for comparison).

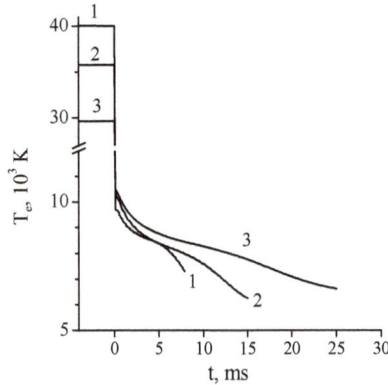

Figure 6. Calculated effective electron temperature in a dc discharge in the Ar + 1%N$_2$ mixture ($t < 0$) and its time evolution in the discharge afterglow ($t > 0$). $R = 1.4$ cm, $I = 20$ mA. $P = 1$ Torr (1), 2 Torr (2), and 5 Torr (3) [26].

Electron densities in the discharge and their time evolution in the afterglow calculated for pure Ar and Ar + 1%N$_2$ mixture at a discharge current of $I = 20$ mA are shown in Figure 7a–c. From Figure 7 it follows that in some cases an increase in the electron concentration is observed at the very beginning of the afterglow. According to calculations, in steady state Ar + 1%N$_2$ discharge plasma under considered conditions, production of electrons is mainly due to associative ionization (3) and (4) and chemoionization (Ar* + Ar* → Ar$^+$ + Ar + e) processes and losses of electrons are due to ambipolar diffusion process. After turning off the electric field, the mean electron energy instantaneously (within the model used) decreases. The decrease in the mean electron energy leads to the decrease in the rate of electron losses due to ambipolar diffusion process while the rate of electron production remains near the same, since the rates of the ionization processes mentioned above do not depend on the electron temperature. As a result, this leads to an increase in the electron concentration at the very beginning of the afterglow.

In pure Ar discharge under considered conditions losses of electrons are also due to ambipolar diffusion process. At that, at pressure 1 Torr, production of electrons is partially due to chemoionization processes. As a result, an increase in the electron concentration at the very beginning of the afterglow is observed in simulations. It follows from simulations that, with increasing pressure, the contribution of chemoionization processes to ionization rate decreases and the production of electrons is almost completely provided by stepwise ionization processes. In this case, the decrease in the electron temperature in the afterglow leads to significant decrease in the rate constant of stepwise ionization and, accordingly, the rate of electron production. As a result, the concentration of electrons monotonously decreases in the afterglow.

As Figure 7a shows, at $P = 1$ Torr, the afterglow plasma of a discharge in the Ar + 1%N$_2$ mixture decays noticeably faster than that of a discharge in pure Ar. In this case, the decay of plasma (both in pure argon and in a gas mixture) is governed by the ambipolar diffusion process, the rate of which in Ar + 1%N$_2$ afterglow plasma is high due to the high electron temperature (see Figure 6).

However, the situation changes significantly with the pressure increase. At $P = 5$ Torr (Figure 7c), at the beginning of the afterglow (during 1 ms) the electron density in the Ar + 1%N$_2$ mixture decreases much slower than in pure argon, although the rate of ambipolar diffusion in argon is significantly lower. As a result, even at $t = 13$ ms after the end of the discharge, the electron density in Ar + 1%N$_2$ mixture remains slightly higher than that in pure Ar. In contrast, at $t > 15$ ms, the electron density in pure argon is significantly higher with respect to that in Ar + 1%N$_2$ mixture. This result is explained by two main effects [26]. The first effect consists in the following: at the beginning of the afterglow in pure argon, the Ar$^+$ ions (dominant ions in the discharge plasma) are quickly (~0.3 ms) converted in molecular ions Ar_2^+ (see comments in [26]). At the beginning of the afterglow, the rate of plasma decay due to recombination of electrons with molecular ions is higher than that due to the ambipolar diffusion process. The rate of the recombination process decreases with the electron (and ion) concentration decrease, so that, at $t > 15$ ms, the plasma decay is governed by the ambipolar diffusion only.

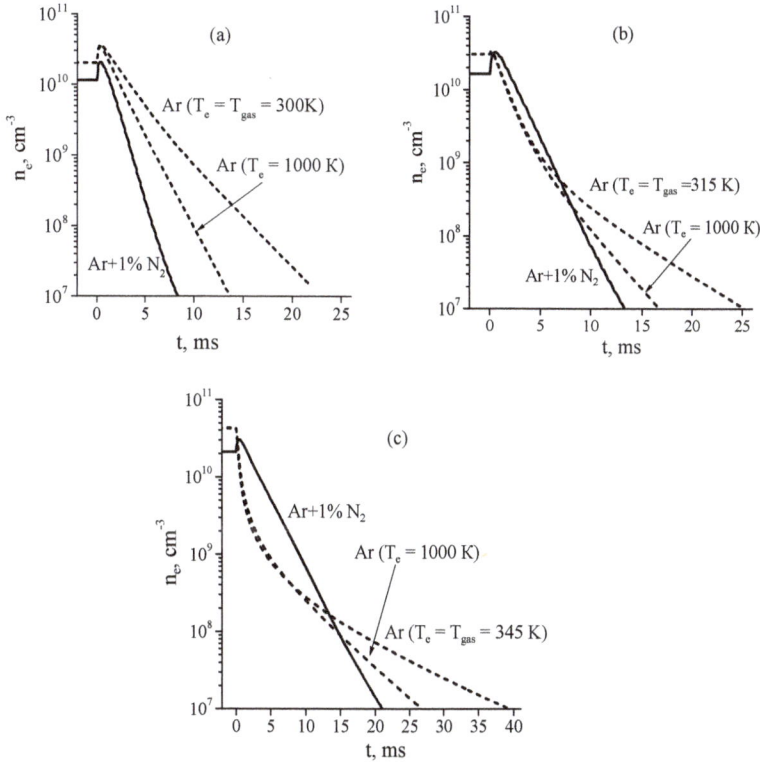

Figure 7. Calculated electron densities in the discharge ($t \leq 0$) and in the afterglow ($t > 0$) at a discharge current of $I = 20$ mA and gas pressures of $P = 1$ (**a**), 2 (**b**), and 5 Torr (**c**), respectively [26].

The second reason is that, according to calculations, in Ar + 1%N$_2$ afterglow the high rate of plasma decay due to ambipolar diffusion process is balanced (to a large extent) by the high rate of electron production via associative ionization of excited nitrogen atoms and molecules (3) and (4). As a result, the plasma decay rate during 1 ms after the end of the discharge is appreciably lower than in the Ar afterglow plasma.

The above described results of simulations were used in [26] for the qualitative explanation of the experimental data on the influence of a nitrogen admixture on the anomalous memory effect in the breakdown of low-pressure argon in a long discharge tube.

4. Effect of Nitrogen Addition to Argon on Discharge Constriction Conditions

The constriction of a glow discharge is the transformation of discharge from the diffuse form where plasma fills up the whole tube cross-section more or less uniformly, to the constricted one, where the plasma cord is narrower (sometimes much narrower) than the tube diameter [32]. The constriction is a result of the onset of plasma instability against transverse perturbations of the electron density (see, for example, comments in the review [46]). Depending on experimental conditions the instability can be caused by different physical mechanisms. In discharges in noble gases at intermediate pressures (except for helium), the most important is the supernlinear dependence the excitation rates of electronic states and the rate of stepwise ionization on the electron density due to the influence of electron–electron (e–e) collisions on the electron energy spectrum [46].

The simplest way to observe the constriction of a dc discharge is to gradually enlarge the discharge current by increasing a power supply voltage. When the current reaches some critical value I_C, the discharge positive column sharply constricts to a narrow cord. Simultaneously the sharp bend of the discharge volt-amp characteristic (VAC) occurs (Figure 8). Current values for the transition from the diffuse form to the constricted one and, vice versa, from the constricted form to the diffuse differ, so that hysteresis occurs.

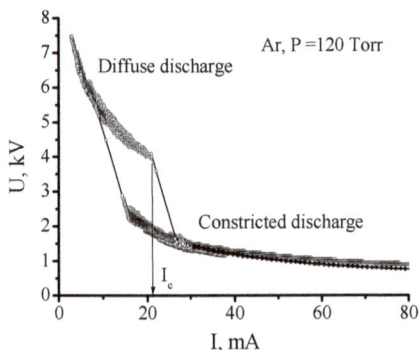

Figure 8. Volt-amp characteristic of the discharge under constriction [47]. (© IOP Publishing. Reproduced with permission. All rights reserved.)

For the given noble gas, the I_C value depends strongly on the gas pressure and discharge tube diameter. In the context of this paper, it is important that even a small nitrogen admixture to argon also affects essentially the I_C and the whole VAC. This effect is illustrated by Figure 9, where the set of data for pure argon and argon with various nitrogen admixtures is depicted. The discharge tube inner diameter is 2.8 cm, the distance between electrodes is 75 cm. One can see that as low as 0.02 and 0.075 percentage of nitrogen admixtures shifts the critical current from 17 mA to 45 mA and to even 100 mA, respectively. Increasing N_2 concentration up to 1 percent makes it impossible to reach constricting for the given electrical scheme. The qualitative explanation of this effect is as follows. As it was mentioned above, it is e–e collisions that provide the conditions necessary for the stepwise discharge constriction. The degree of the influence of e–e collisions on electron energy spectrum depends, in particular, on the ratio of the e–e collision frequency and the frequency of electron energy losses in elastic and inelastic collisions with atoms and molecules. The higher the percentage of nitrogen admixture, the higher the rate (frequency) of electron energy losses due to excitation of vibrational and electronic levels of nitrogen molecules and, consequently, the higher electron concentration (discharge current) is needed to provide the conditions for the discharge constriction.

As Figure 9 shows, N_2 addition also diminishes drastically the discharge voltage in the diffuse region. This issue was discussed in Section 2. At the same time, VACs in the constricted region go approximately along the same curve. As is shown in [16], it can be explained by the fact that, due to a very high electron density in a constricted discharge, N_2 addition has very little influence on ionization mechanism which in Ar and Ar-N_2 cases is mainly stepwise ionization of Ar metastables.

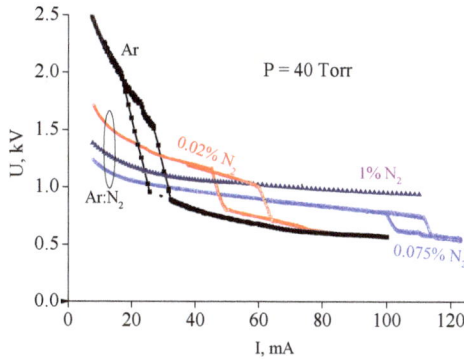

Figure 9. Volt–amp characteristics of a glow discharge in pure argon and in argon with nitrogen admixture. Percentages of admixture are indicated [16]. (© IOP Publishing. Reproduced with permission. All rights reserved.)

5. Effect of Nitrogen Addition to Argon on the Characteristics of Constriction Process

Visually the process of discharge constriction in pure noble gases usually occurs instantly and simultaneously along the whole tube. More detailed study shows that if the current exceeds I_C value only slightly, then the constriction begins near one of the electrodes and then its front propagates toward the other electrode with a finite speed. It seems that this speed can correlate with a group velocity of moving strata whose appearance accompanies the discharge constriction [46]. Experiments with Ar-N_2 mixtures showed that a small nitrogen addition affected the constriction front speed.

In Figure 10, small portions of VAC in the region of transition from the diffuse to constricted part are shown for pure argon and argon with 0.02 percent nitrogen admixture.

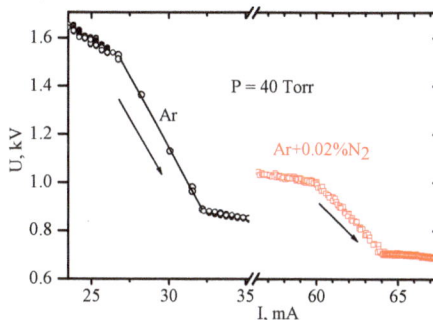

Figure 10. Volt–amp characteristics for Ar and Ar-N_2 mixture in a region of constricting [16]. (© IOP Publishing. Reproduced with permission. All rights reserved.)

The experimental data were collected in continuous mode with a sample time of 10 ms. It can be seen that, the process of constricting in Ar lasts 50 ms, while in the mixture it lasts 600 ms. In the course of this time, the boundary between diffused and constricted parts moves from one electrode to the other. For argon, this movement is too fast to be noticeable by eye, but in the mixture, it is easily distinguished. In this case, the process is so sluggish that it is possible to interfere with its course by varying the power supply voltage and to obtain rather peculiar VACs. In Figure 11, after the beginning of the contraction the power supply voltage was first decreased (from point A to point B) and then increased (after B).

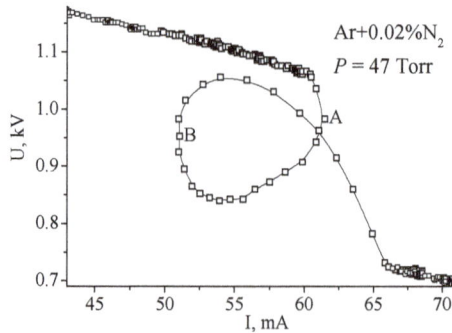

Figure 11. Volt–amp characteristic recorded at lowering (from point A) and increasing (from B) of power supply voltage [16]. (© IOP Publishing. Reproduced with permission. All rights reserved.)

6. Formation of Partially Constricted Discharge in Ar:N₂ Mixtures at Intermediate Gas Pressures

It turned out that, during the slow move of the boundary it was possible by proper variation of power supply voltage to make the movement still slower and finally to bring it to a complete stop at some position between the electrodes, i.e., to get the steady-state partially constricted discharge (PCD) (Figure 12).

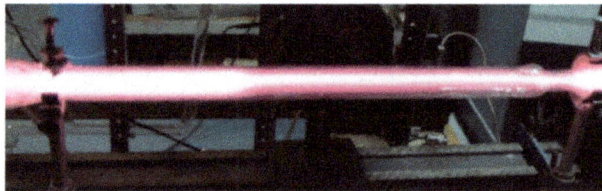

Figure 12. Steady-state partially constricted glow discharge in the Ar + 0.075%N_2 mixture, $P = 40$ Torr [19] (© 2011 IEEE).

The PCD, formed in such a way, was stable and could exist for a long time without any further adjustment of the electric circuit parameters. It was also shown that the point on the *U-I* plane which corresponded to the formed PCD was placed inside the hysteresis loop.

The constricted part of the positive column can touch either the cathode or the anode. For these two cases, the interface between the diffuse and constricted parts was different (Figure 12). In the former case, the boundary between two modes was sharp, while in the latter case, one mode gradually transformed into the other. Striations moving through the constricted part from the anode to the cathode were observed (Figure 13, the lower panel). The typical value of the phase velocity of striations in a constricted discharge is several tens of m/s, the phase velocity decreases with increasing discharge current [16].

The boundary between the constricted and diffuse parts could be obtained in different positions between the electrodes. It influenced the discharge voltage value but the current remained at approximately the same level (Figure 14).

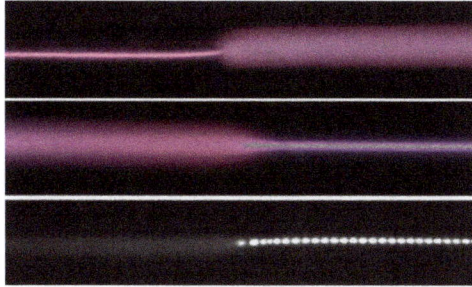

Figure 13. Two upper panels are the photos of the transition region between diffuse and constricted parts of discharge. The lower panel is a 120 µs exposure CMOS camera image. In all cases, the cathode is on the left. Ar + 0.075%, P = 80 Torr [19] (© 2011 IEEE).

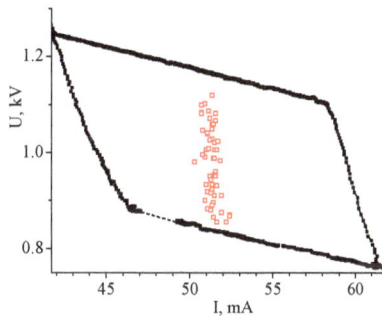

Figure 14. Manifold of PCD points (open squares) for the discharge in the Ar + 0.02%N_2 mixture at P = 50 Torr [16]. (© IOP Publishing. Reproduced with permission. All rights reserved.)

At constant parameters of the electrical circuit, the discharge could remain partially constricted for a few minutes. For the constricted part adjacent to the anode, the boundary was immovable during this period. Then, probably due to small uncontrolled changes in the parameters of the plasma or of the external circuit, the boundary shifted a little. After ~10 s, a new shifting occurred with the same displacement. Furthermore, these jumps repeated with a gradually diminishing period until the discharge became homogeneous (diffuse or constricted).

If the constricted part was adjacent to the cathode, then from the very beginning, the boundary performed irregular oscillations with amplitude of a several centimeters and a period of several seconds. Such a regime could exist for a few minutes. Then, at some moment, the amplitude of oscillations enhanced and they turned into the movement of the boundary toward one of the electrodes.

7. Conclusions

The effect of nitrogen admixture on various characteristics of a dc glow discharge (maintained in a tube) in argon has been reviewed. The following specific effects have been discussed.

At relatively low pressures (several Torr) the addition of nitrogen leads to an increase in the discharge voltage, while at intermediate gas pressures (tens of Torr) it leads to a noticeable decrease in the discharge voltage. A high degree of a vibrational excitation of molecules is achieved in the Ar:N_2 discharge, which provides a high electron temperature in the afterglow due to superelastic collisions of electrons with vibrationally excited molecules. This, in turn, leads to an increase in the rate of plasma decay due to ambipolar diffusion process.

At intermediate gas pressures, a stepwise transition of the positive column from the diffuse to the constricted form is observed when the discharge current exceeds a certain critical value. The addition of nitrogen to argon leads to a noticeable increase in the critical current value. According to observations, the constriction starts near one of the electrodes and then the constricted region boundary propagates towards the other electrode. For the Ar:N$_2$ mixture, the transition time is considerably longer than in pure argon. By proper variation of power supply voltage during the transition, in the case of the Ar:N$_2$ mixture, there can be formed a steady-state partially constricted discharge in which the constricted and diffuse forms of the positive column simultaneously exist in the discharge tube.

Author Contributions: All authors have contributed equally to the writing of this review manuscript.

Funding: This research received no external funding.

Conflicts of Interest: The authors declare no conflicts of interest.

References

1. Sá, P.A.; Loureiro, J. A time-dependent analysis of the nitrogen afterglow in N$_2$ and N$_2$–Ar microwave discharges. *J. Phys. D Appl. Phys.* **1997**, *30*, 2320–2330. [CrossRef]
2. Henriques, J.; Tatarova, E.; Dias, F.M.; Ferreira, C.M. Spatial structure of a slot-antenna excited microwave N$_2$–Ar plasma source. *J. Phys. D Appl. Phys.* **2008**, *103*, 103304. [CrossRef]
3. Hübner, S.; Carbone, E.; Palomares, J.M.; van der Mullen, J. Afterglow of argon plasmas with H$_2$, O$_2$, N$_2$, and CO$_2$ admixtures observed by Thomson scattering. *Plasma Process. Polym.* **2014**, *11*, 482–488. [CrossRef]
4. Tochikubo, F.; Petrović, Z.L.; Nakano, N.; Makabe, T. Influence of Ar Metastable on the Discharge Structure in Ar and N$_2$ Mixture in RF Discharges at 13.56 MHz. *Jpn. J. Appl. Phys.* **1994**, *33*, 4271–4275. [CrossRef]
5. Fiebrandt, M.; Hillebrand, B.; Spiekermeier, S.; Bibinov, N.; Boke, M.; Awakowicz, P. Measurement of Ar resonance and metastable level number densities in argon containing plasmas. *J. Phys. D Appl. Phys.* **2017**, *50*, 355202. [CrossRef]
6. Britun, N.; Gaillard, M.; Ricard, A.; Kim, Y.M.; Kim, K.S.; Han, J.G. Determination of the vibrational, rotational and electron temperatures in N$_2$ and Ar–N$_2$ rf discharge. *J. Phys. D Appl. Phys.* **2007**, *40*, 1022–1029. [CrossRef]
7. Moravej, M.; Yang, X.; Barankin, M.; Penelon, J.; Babayan, S.E.; Hicks, R.F. Properties of an atmospheric pressure radio-frequency argon and nitrogen plasma. *Plasma Sources Sci. Technol.* **2006**, *15*, 204–210. [CrossRef]
8. Fritsche, B.; Chevolleau, T.; Kourtev, J.; Kolitsch, A.; Möller, W. Plasma diagnostic of an RF magnetron Ar/N$_2$ discharge. *Vacuum* **2003**, *69*, 139–145. [CrossRef]
9. Kim, Y.-C.; Lee, H.-C.; Kim, Y.-S.; Chung, C.-W. Correlation between vibrational temperature of N$_2$ and plasma parameters in inductively coupled Ar/N$_2$ plasmas. *Phys. Plasmas* **2015**, *22*, 083512. [CrossRef]
10. Bravo, J.A.; Rincón, R.; Muñoz, J.; Sánchez, A.; Calzada, M.D. Spectroscopic characterization of argon–nitrogen surface-wave discharges in dielectric tubes at atmospheric pressure. *Plasma Chem. Plasma Process.* **2015**, *35*, 993–1014. [CrossRef]
11. Becker, K.H.; Masoud, N.M.; Martus, K.E.; Schoenbach, K.H. Electron-driven processes in high-pressure plasmas. *Eur. Phys. J. D* **2005**, *35*, 279–297. [CrossRef]
12. Barkhordari, A.; Ganjovi, A.; Mirzaei, I.; Falahat, A.; Rostami Ravari, M.N. A pulsed plasma jet with the various Ar/N$_2$ mixtures. *J. Theor. Appl. Phys.* **2017**, *11*, 301–312. [CrossRef]
13. Masoud, N.; Martus, K.; Becker, K. VUV emission from a cylindrical dielectric barrier discharge in Ar and in Ar/N$_2$ and Ar/air mixtures. *J. Phys. D Appl. Phys.* **2005**, *38*, 1674–1683. [CrossRef]
14. Kimura, T.; Akatsuka, K.; Ohe, K. Experimental and theoretical investigations of DC glow discharges in argon-nitrogen mixtures. *J. Phys. D Appl. Phys.* **1994**, *27*, 1664–1671. [CrossRef]
15. Ionikh, Y.Z.; Meshchanov, A.V.; Petrov, F.B.; Dyatko, N.A.; Napartovich, A.P. Partially constricted glow discharge in an argon–nitrogen mixture. *Plasma Phys. Rep.* **2008**, *34*, 867–878. [CrossRef]
16. Ionikh, Y.Z.; Dyatko, N.A.; Meshchanov, A.V.; Napartovich, A.P.; Petrov, F.B. Partial constriction in a glow discharge in argon with nitrogen admixture. *Plasma Sources Sci. Technol.* **2012**, *21*, 055008. [CrossRef]
17. Dyatko, N.A.; Ionikh, Y.Z.; Meshchanov, A.V.; Napartovich, A.P.; Barzilovich, K.A. Specific features of the current–voltage characteristics of diffuse glow discharges in Ar:N$_2$ mixtures. *Plasma Phys. Rep.* **2010**, *36*, 1040–1064. [CrossRef]

18. Dyatko, N.A.; Ionikh, Y.Z.; Meshchanov, A.V.; Napartovich, A.P. Theoretical and experimental study of the influence of nitrogen admixture on characteristics of dc diffuse glow discharge in rare gases at intermediate pressures. *J. Phys. D Appl. Phys.* **2013**, *46*, 355202. [CrossRef]
19. Dyatko, N.A.; Ionikh, Y.Z.; Meshchanov, A.V.; Napartovich, A.P. Steady-state partially constricted glow discharge. *IEEE Trans. Plasma Sci.* **2011**, *39*, 2532–2533. [CrossRef]
20. Dyatko, N.; Napartovich, A. Ionization mechanisms in Ar:N$_2$ glow discharge at elevated pressures. In Proceedings of the 41st Plasmadynamics and Lasers Conference, Chicago, IL, USA, 28 June–1 July 2010. Paper AIAA 2010-4884.
21. Isola, L.M.; López, M.; Cruceño, J.M.; Gómez, B.J. Measurement of the Ar(1sy) state densities by two OES methods in Ar–N$_2$ discharges. *Plasma Sources Sci. Technol.* **2014**, *23*, 015014. [CrossRef]
22. Reyes, P.G.; Torres, C.; Martinez, H. Electron temperature and ion density measurements in a glow discharge of an Ar–N$_2$ mixture. *Radiat. Eff. Defects Solids* **2014**, *169*, 285–292. [CrossRef]
23. Zhovtyansky, V.A.; Anisimova, O.V. Kinetis of plasma chemical reactions producing nitrogen atoms in the glow discharge in a nitrogen-argon gas mixturte. *J. Phys.* **2014**, *59*, 1155–1163.
24. Bogaerts, A. Hybrid Monte Carlo-Fluid model for studying the effects of nitrogen addition to argon glow discharges. *Spectrochim. Acta Part B* **2009**, *64*, 126–140. [CrossRef]
25. Jackson, G.P.; King, F.L. Probing excitation/ionization processes in millisecond-pulsed glow discharges in argon through the addition of nitrogen. *Spectrochim. Acta Part B* **2003**, *58*, 185–209. [CrossRef]
26. Dyatko, N.A.; Ionikh, Y.Z.; Meshchanov, A.V.; Napartovich, A.P. Influence of a nitrogen admixture on the anomalous memory effect in the breakdown of low-pressure argon in a long discharge tube. *Plasma Phys. Rep.* **2018**, *44*, 334–344. [CrossRef]
27. Qayyum, A.; Zeb, S.; Naveed, M.A.; Rehman, N.U.; Ghauri, S.A.; Zakaullah, M.J. Optical emission spectroscopy of Ar–N$_2$ mixture plasma. *J. Quant. Spectrosc. Radiat. Transf.* **2007**, *107*, 361–371. [CrossRef]
28. Martens, T.; Bogaerts, A.; Brok, W.J.M.; Dijk, J.V. The dominant role of impurities in the composition of high pressure noble gas plasmas. *Appl. Phys. Lett.* **2008**, *92*, 041504. [CrossRef]
29. Wang, Y.; Wang, D. Influence of impurities on the uniform atmospheric-pressure discharge in helium. *Phys. Plasmas* **2005**, *12*, 023503. [CrossRef]
30. Sasaki, N.; Shoji, M.; Uchida, Y. Capacitively Coupled RF Discharge Breakdown in Gas Mixtures. *IEEJ Trans. Fundam. Mater.* **2007**, *127*, 714–718. [CrossRef]
31. Zhiglinskii, A.G. (Ed.) *Handbook of Constants for Elementary Atomic, Ionic, Electronic, and Photonic Processes*; PGU: St. Petersburg, Russia, 1994. (In Russian)
32. Raizer, Y.P. *Gas Discharge Physics*; Nauka: Moscow, Russia, 1987; Springer: Berlin, Germany, 1991.
33. Ionikh, Y.Z.; Chernysheva, N.V. Radiation of gas-discharge plasma in mixtures of inert and molecular gases. In *Encyclopedia of Low Temperature Plasma*; Fortov, V.E., Ed.; Fizmatlit: Moscow, Russia, 2008; Series B; Volume III 2, pp. 427–443. (In Russian)
34. Dyatko, N.A.; Ionikh, Y.Z.; Meshchanov, A.V.; Napartovich, A.P.; Petrov, F.B. Volt-ampere characteristics of the partially constricted glow discharge in Ar:N$_2$ mixtures. In Proceedings of the XIX European Sectional Conference on Atomic and Molecular Physics of Ionized Gases, Granada, Spain, 15–19 July 2008; Poster 2.36. Available online: http://www.escampig2008.csic.es/PosterSessions/135.pdf (accessed on 19 January 2019).
35. Yalin, A.P.; Ionikh, Y.Z.; Miles, R.B. Gas temperature measurements in weakly ionized glow discharges with filtered Rayleigh scattering. *Appl. Opt.* **2002**, *41*, 3753–3762. [CrossRef]
36. Mnatskanyan, A.K.; Naidis, G.V. Processes of production and loss of charged particles in a nitrogen-oxygen plasma. In *Plasma Chemistry*; Smirnov, B.M., Ed.; Energoatomizdat: Moscow, Russia, 1987; Volume 14, pp. 227–255. (In Russian)
37. Berdichevskii, M.G.; Marusin, V.V. Nonequilibrium and ionization mechanism of the nitrogen plasma of an electrodeless RF capacitive discharge at medium pressures. *Proc. Sib. Branch USSR Acad. Sci.* **1979**, *8*, 72–79. (In Russian)
38. Popov, N.A. Associative ionization reactions involving excited atoms in nitrogen plasma. *Plasma Phys. Rep.* **2009**, *35*, 436–449. [CrossRef]
39. Guerra, V.; Loureiro, J. Electron and heavy particle kinetics in a low-pressure nitrogen glow discharge. *Plasma Sources Sci. Technol.* **1997**, *6*, 361–372. [CrossRef]
40. Capitelli, M.; Colonna, G.; De Pascale, O.; Gorse, C.; Hassouni, K.; Longo, S. Electron energy distribution functions and second kind collisions. *Plasma Sources Sci. Technol.* **2009**, *18*, 014014. [CrossRef]

41. Dyatko, N.A.; Ionikh, Y.Z.; Kolokolov, N.B.; Meshchanov, A.V.; Napartovich, A.P. Experimental and theoretical studies of the electron temperature in nitrogen afterglow. *IEEE Trans. Plasma Sci.* **2003**, *31*, 553–563. [CrossRef]
42. Dilecce, G.; Benedictis, S.D. Relaxation of the electron energy in the post-discharge of an He-N_2 mixture. *Plasma Sources Sci. Technol.* **1993**, *2*, 119–122. [CrossRef]
43. Dyatko, N.A.; Ionikh, Y.Z.; Kolokolov, N.B.; Meshchanov, A.V.; Napartovich, A.P. Jumps and bi-stabilities in electron energy distribution in Ar–N_2 post discharge plasma. *J. Phys. D Appl. Phys.* **2000**, *33*, 2010–2018. [CrossRef]
44. Dyatko, N.A.; Napartovich, A.P. Theoretical Study of Plasma Parameters in a dc Glow Discharge and Postdischarge in Argon-nitrogen Mixtures. In Proceedings of the 23rd Europhysics Conference on Atomic and Molecular Physics of Ionized Gases, Bratislava, Slovakia, 12–16 July 2016; EPS ECA (Europhysics Conference Abstracts); European Physical Society: Bratislava, Slovakia, 2016; p. 109.
45. Hübner, S.; Palomares, J.M.; Carbone, E.A.D.; van der Mullen, J.J.A.M. A power pulsed low-pressure argon microwave plasma investigated by Thomson scattering: Evidence for molecular assisted recombination. *J. Phys. D Appl. Phys.* **2012**, *45*, 055203. [CrossRef]
46. Golubovskii, Y.B.; Nekuchaev, V.; Gorchakov, S.; Uhrlandt, D. Contraction of the positive column of discharges in noble gases. *Plasma Sources Sci. Technol.* **2011**, *20*, 053002. [CrossRef]
47. Dyatko, N.A.; Ionikh, Y.Z.; Kochetov, I.V.; Marinov, D.L.; Meschanov, A.V.; Napartovich, A.P.; Petrov, F.B.; Starostin, S.A. Experimental and theoretical study of the transition between diffuse and contracted forms of the glow discharge in argon. *J. Phys. D Appl. Phys.* **2008**, *41*, 055204. [CrossRef]

atoms

MDPI

Review

The Equivalent Circuit Approach for the Electrical Diagnostics of Dielectric Barrier Discharges: The Classical Theory and Recent Developments

Andrei V. Pipa * and Ronny Brandenburg

Leibniz Institute for Plasma Science and Technology (INP), Felix-Hausdorff-Straße 2, 17489 Greifswald, Germany; brandenburg@inp-greifswald.de
* Correspondence: avpipa@gmail.com

Received: 30 November 2018; Accepted: 15 January 2019; Published: 23 January 2019

check for updates

Abstract: Measurements of current and voltage are the basic diagnostics for electrical discharges. However, in the case of dielectric barrier discharges (DBDs), the measured current and voltage waveforms are influenced by the discharge reactor geometry, and thus, interpretation of measured quantities is required to determine the discharge properties. This contribution presents the main stages of the development of electrical diagnostics of DBDs, which are based on lumped electrical elements. The compilation and revision of the contributions to the equivalent circuit approach are targeted to indicate: (1) the interconnection between the stage of development, (2) its applicability, and (3) the current state-of-the-art of this approach.

Keywords: electrical theory of DBDs; QV-plot; instantaneous power

1. Introduction

Dielectric barrier discharges (DBDs) are a well-established method to generate non-thermal plasmas at atmospheric pressure [1]. Their main peculiarity is the presence of at least one dielectric between the electrodes. It prevents the transition into a spark discharge. DBDs have many industrial applications, e.g., ozone generation, exhaust gas cleaning, surface activation, and light sources. They are further under investigation for plasma medicine, surface deposition, and flow control.

The non-thermal plasma produces highly-reactive species without an extensive heating of the gas. The application of non-thermal plasma generated by DBDs is of continuous interest for and permanently expanded to new technological areas such as conversion of carbon dioxide [2] or removal of odors [3]. DBD can also initiate very specific chemical processes, which are hardly accessible by other technologies, like ozone generation [4] or polymerization of D-ribose [5].

The parameters of the plasma are influenced by the discharge geometry and profound knowledge about the dissipated energy and power, the current and voltage in the gas gap are crucial for the characterization, comparability, and up-scaling of DBD reactors. Therefore, current, voltage, and charge measurements are used and interpreted based on equivalent circuits.

However, the growing interest in DBDs and the exploration of new applications has led to a great variety of DBD designs, reactor configurations, power excitation schemes, and discharge regimes, which differ from the DBD in the classical ozonizer [6]. Thus, the interpretation of electrical measurements is not straightforward, and equivalent circuits have been reviewed and further developed. This contribution aims to summarize the state-of-the-art of the stages of development of this approach.

It does not pretend to be a comprehensive review about electrical diagnostics of DBDs, but is focused on the principle stages of development of equivalent circuits, which gives information about a DBD's properties directly from measured current/charge and voltage waveforms.

We will follow the historical development of the approach, starting with "the classical electrical theory of ozonisers" formulated by Manley [7]. First, the main principle of discharge operation and the challenges for the electrical diagnostics will be given (Section 2). The work [7] provides the basic principles of the DBD electrical characteristics and will be discussed in detail in Section 3. The development of the experimental techniques led to new insights, which could not be explained within the framework of the classical theory. The equivalent circuit approach was suggested by three groups independently [8–10] to overcome these difficulties. This will be reviewed in Section 4. Section 5 will revise the works [7–9] taking into account the previous revision described in [11–13]. Sections 3–5 deal with the volume DBD where the capacitances of the reactor do not depend on the operation conditions; in other words, the charges deposited on the dielectric surfaces cover the whole cross-section of the electrodes uniformly. Section 6 will describe the electrical diagnostic of DBDs when discharge expansion on the dielectric surfaces is influenced by the amplitude of the applied voltage [14–16]. The concluding Section 7 will summarize the state-of-the-art and remaining challenges for the equivalent circuit approach for DBDs.

2. Basics of DBD Operation and Challenges for the Electric Diagnostics

The basic design of a DBD is schematically shown in Figure 1. The discharge is ignited at sufficiently high applied voltage $V(t)$ (range of a few kV) with frequencies in the range of 50 Hz–1 MHz. Electrical charges are deposited on the dielectric surfaces during discharge operation. The deposited charges shield the external electric field, and thus, the discharge is self-extinguished. In other words, the dielectric limits the charge transfer through the gas gap and restricts the heating of the gas. For the next discharge ignition, the applied voltage must be increased further or the polarity must be changed. Thereby, DBD has active and passive phases within one period of the applied voltage, with and without active charge transfer through the gas gap, respectively.

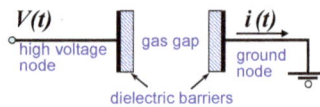

Figure 1. Schematic presentation of the dielectric barrier discharge cell. Reprinted from [12] with the permission of ©AIP Publishing.

Since the charges deposited on the dielectric surfaces strongly influence the electric field in the gas gap, the gas gap voltage $U_g(t)$ significantly differs from the applied voltage $V(t)$. The term $U_g(t)$ is not defined neatly, as it assumes implicitly an equipotential internal dielectric surface. This is fulfilled only in the case homogeneous (or uniform/diffuse) discharge regimes, but obviously not for the more common filamentary modes. $U_g(t)$ can be understood as an effective characteristic, averaged over the whole dielectric surface, with the following properties: the active DBD phase starts when $U_g(t)$ overcomes the breakdown potential U_b, and the passive phase begins when $U_g(t)$ falls to the discharge extinguish voltage U_{ext}. This can be expected at the moment t', when $U_g(t') = U_{ext}$ is reached near the maximum of the applied voltage, i.e., the increase of $V(t)$ does not compensate the screening of the electric field by the deposited charges. The value of U_{ext} should be between zero and the breakdown potential U_b ($0 < U_{ext} \leq U_b$). The direct measurement of $U_g(t)$ is impossible, but sometimes, it can be obtained via an interpretation of the externally-measured voltage $V(t)$ and current $i(t)$. The methods to infer $V(t)$ and $i(t)$ are revised in the present work.

The measured current $i(t)$ contains the current associated with charge transfer in the gas gap (the discharge current $j_R(t)$) and the displacement current in the gas gap. At the low amplitude of the alternating applied voltage $V(t)$, the discharge is not ignited, and there is no charge transfer in the gas gap. In this situation, the reactor cell behaves as an ideal capacitor C_{cell}, and the measured current will be determined by the relation:

$$i_{off}(t) = C_{cell} \frac{dV(t)}{dt}. \tag{1}$$

The index *off* in this equation emphasis that the current is measured without discharge ignition (passive or plasma-off phase). The difference of currents measured with and without discharge ignition reflects the discharge current, but is not entirely equal to this due to the following reason. The displacement current in the gas gap is proportional to the derivative of $U_g(t)$, which does not coincide with the derivative of $V(t)$ in the active DBD phase. Therefore, $i_{off}(t)$ is not always equal to the displacement current in the gas gap, and additional efforts are required to establish the relation between measured $i(t)$ and discharged $j_R(t)$ current.

The total power dissipated in a DBD, averaged over the voltage period T, can be deduced directly from the measured $i(t)$ and $V(t)$:

$$P = \frac{1}{T} \int_0^T i(t)V(t)dt. \tag{2}$$

The product of measured current and voltage gives the power dissipated in the discharge and stored on the dielectric surfaces. Thus, only the power averaged over the whole discharge period is available from directly-measured $i(t)$ and $V(t)$. However, the knowledge of $U_g(t)$ and $j_R(t)$ can provide instantaneous power:

$$P(t) = U_g(t)j_R(t). \tag{3}$$

The relations between $V(t)$ and $U_g(t)$, as well as $i(t)$ and $j_R(t)$ depend on the DBD reactor geometry. The comparison of DBDs with different geometries should be based on internal discharge characteristics $U_g(t)$ and $j_R(t)$, which can be obtained by means of the equivalent circuit approach. It allows one to infer the measured waveform comprehensively, but the applicability of the approach should be accurately examined. This is the main aim of the present work.

3. The Classical Electrical Theory of Ozonizers

DBD was introduced by Siemens in 1857 [4] as a low-temperature discharge for ozone generation. The principle ideas for electrical characterization of DBD were formulated much later by Manley in 1943 [7], now referred to as "The classical electrical theory of ozonisers". It is the base for the equivalent circuit approach. Other approaches to infer electrical characteristics can be found in [17], but they are not further discussed here.

The main results of the classical theory are based on measurements of three quantities: applied voltage $V(t)$, current $i(t)$ waveform, and charge as a function of the applied voltage $Q(V)$. The charge $Q(t)$ can be obtained as an integral of the measured current waveform:

$$\int_0^t i(\tau)d\tau = Q(t) + const, \tag{4}$$

or can be measured as a voltage drop $V_0(t)$ across a given capacitor C_0 inserted in series into the reactor cell:

$$Q(t) = C_0 V_0(t). \tag{5}$$

If the charge is measured via capacitance, it can be shown as a function of the applied voltage directly on the oscilloscope. Examples of such oscilloscope screen shots [7] are reproduced in Figure 2.

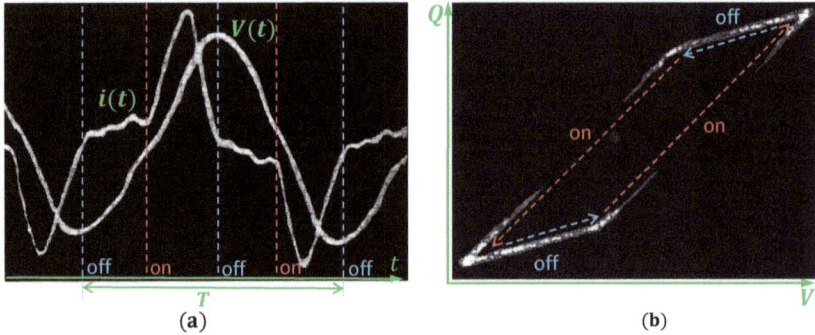

Figure 2. Oscilloscope screen shots of the measured electrical characteristics of DBD from [7]. (a) Measured voltage $V(t)$ and current waveforms $i(t)$. Vertical lines indicate moments of switching between active (discharge on) and passive (discharge off) phases of DBD. T is the discharge period. (b) Charge voltage characteristics $Q(V)$ (QV-plot). The arrows indicate temporal development. The screen shots are reprinted from [7] with the permission of ©Electrochemical Society.

The discharge cell was driven by sinusoidal voltage $V(t)$, whereas the current waveform is more complex. The active discharge phase is associated with a hump on the current waveform, which ends when the voltage reaches its maximum. The hump can be well identified by the variation of the voltage amplitude V_{max}. It appears when V_{max} is high enough and rapidly grows with V_{max}. When $V(t)$ reaches its maximum, the discharge turns to the passive phase. The switching between passive and active phases can be verified with synchronous measurements of the light emission [7]. The phases of the discharge can be seen more distinctly in the charge-voltage characteristics (QV-plot). It appears as a parallelogram, and each side corresponds to one of the discharge phases. The discharge power averaged over period T is determined by the integration of the product $i(t)V(t)$ (see Equation (2)) or as the area of the QV-plot multiplied with the frequency $1/T$:

$$P = \frac{1}{T} \oint_T Q(V)\mathrm{d}V. \tag{6}$$

Further conclusions can be drawn from detailed discussions of the QV-plot, which is schematically shown in Figure 3. When the amplitude of the applied voltage is too low for the discharge being ignited, the reactor cell behaves as a capacitor C_{cell}. C_{cell} can be represented as a serial connection of the capacitances associated with a gas gap C_g and dielectric barriers C_d:

$$C_{cell} = \frac{C_d C_g}{C_d + C_g}. \tag{7}$$

Then, the QV-plot is a straight line with slope C_{cell} (i.e., $Q(t) = C_{cell}V(t)$). In the case of the passive discharge phase, the measured charge is shifted by $\pm Q_0$ due to the charges deposited on the dielectric surfaces; see Figure 3a. Therefore, the measured charge can be described as:

$$Q(t) \pm Q_0 = C_{cell}V(t). \tag{8}$$

This can be inferred in terms of the equivalent circuit for the passive discharge phase (off); see Figure 3b. Q_0 is part of the internal node charge, which is present on the plate of the capacitor C_d; see [12] for details.

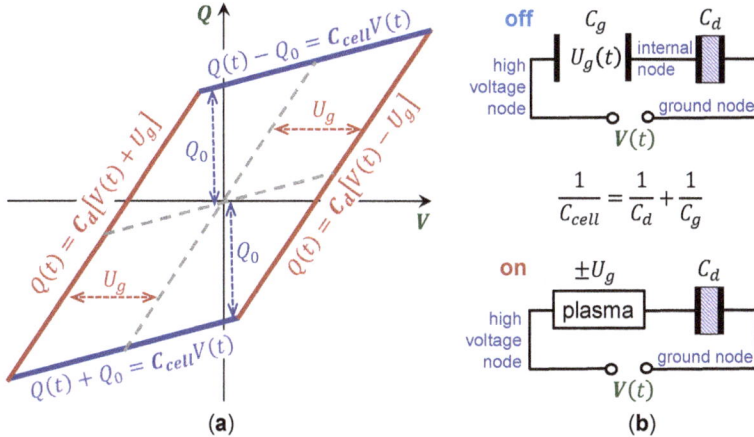

Figure 3. Interpretation of the classical charge-voltage characteristics of sinusoidal voltage-driven ozonizers. (**a**) Schematic presentation of the QV-plot. (**b**) Equivalent circuits corresponding to passive (plasma-off) and active (plasma on) discharge phases.

In the active part of the discharge, the plasma connects the electrodes, and the reactor capacitance is only determined by the dielectric barriers as represented in the equivalent circuit for the active part of the discharge (on); see Figure 3b. The corresponding part of the measured QV-plot is a straight line with the slope of C_d shifted by the gas gap voltage U_g; see Figure 3a. U_g does not depend on the applied voltage and is constant over the whole active discharge phase ($U_g = U_b = U_{ext}$):

$$Q(t) = C_d \left[V(t) \pm U_g \right]. \tag{9}$$

The work of Manley [7] contains an additional important result, which is not often mentioned. The gas gap voltage of the DBD in the active phase, measured for different gas gaps distances (d) and pressures (p), was compared with the breakdown voltage of the gas (air) between the parallel plate electrodes. The influence of the gas temperature was accounted for by an extrapolation of measured values to the zero discharge power. U_g appears as a function of pd, similar to the breakdown voltage in a homogeneous electric field. However, the absolute values were somehow lower. The breakdown voltage strongly depends on pre-ionization of the gas; thus, the residual charges, left from a previous DBD cycle, are responsible for the reduction of the U_g. The influence of the residual charges is also confirmed by the dependence of the breakdown voltage on the time between external re-ignitions [18]. Nevertheless, the observation, that U_g is a function of pd, supports the use of U_g also in the case of filamentary DBDs, as investigated in [7], in spite of the implicit assumption about the equipotential barrier surfaces.

The classical electrical theory of ozonizers can be summarized as follows: (a) DBD can be represented by two equivalent circuits, which correspond to passive and active discharge phases (see Figure 3); (b) during the active discharge phase, U_g is constant and does not depend on the voltage amplitude, but it is a function of the inter-electrode distance and the gas density (as assumed by Paschen's law); (c) geometrical parameters of the discharge cell can be determined from the QV-plot, namely capacitances C_{cell} and C_d; (d) there is no conclusion about discharge current; however, it is assumed that the total current, measured in the active phase, corresponds to the charge transfer through the gas gap (as seen from the equivalent circuit for active discharge phase; see Figure 3b (on)). This means that Equation (1) could not be used for the determination of the displacement current in the active discharge phase.

4. Suggestions for the Equivalent Circuit Approach

With the improvement of experimental techniques for the generation of high voltages, as well as for the measurement of current and voltage waveforms, new types of QV-plots were reported, which differ significantly from a parallelogram and, thus, cannot be explained by the classical theory in a straightforward manner. Examples of such QV-plots are presented in Figure 4. The most exotic QV-plots are obtained under pulsed excitation (Figure 4b,c), but even the QV-plot for a sinusoidal operated DBD, in Figure 4a, requires additional explanations.

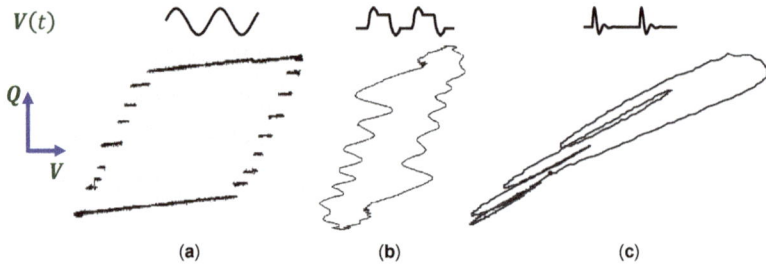

Figure 4. Schematic presentation of different types of the voltage waveforms (upper line) and the corresponding QV-plots (lower line). (a) Staircase-shaped QV-plot measured for a sinusoidal operated DBD [19]. ©Penerbit UTM Press. (b) QV-plot measured for bipolar pulsed operated DBD [8]. ©V.E. Zuev Institute of Atmospheric Optics SB RAS, reproduced with permission. (c) QV-plot measured for pulsed operation in the form of damped oscillations [20] ©IOP Publishing. Reproduced with permission. All rights reserved. The figures are reproduced with the kind permission of the authors.

It was inferred that the gas gap voltage U_g is varying during the active discharge phase. For more detailed interpretation of the DBD electrical characteristics, an equivalent circuit approach was presented in 2001 by Lomaev [8], as well as Liu and Neiger [9]; see Figure 5. The approach was also used by Bibinov et al. at the same time [10]. These equivalent circuits contain a capacitor associated with dielectric barriers C_d in serial connection to the gas gap. The latter is represented as a parallel connection of a gas gap capacitor C_g and a time-dependent current source or resistor R. This resistor/current-source is a "black box" approximation of the discharge, i.e., the plasma is just characterized by the current $j_R(t)$.

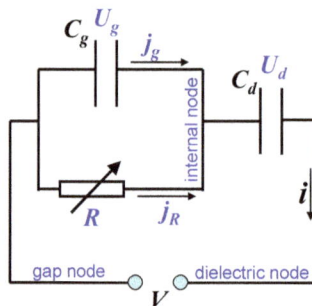

Figure 5. Simplest equivalent circuit of a DBD. Reprinted from [12] with the permission of ©AIP Publishing.

Four coupled equations based on the definition of capacitance and Kirchhoff's laws can be derived from Figure 5:

$$U_d(t) = \frac{Q(t)}{C_d},$$

(10)

$$U_g(t) = V(t) - U_d(t), \tag{11}$$

$$j_g(t) = C_g \frac{dU_g(t)}{dt}, \tag{12}$$

$$j_R = i(t) - j_g(t), \tag{13}$$

where $U_d(t)$ is the voltage across dielectric barriers and $j_g(t)$ is the current through the gas gap capacitance C_g. Note that these equations do not contain any information or definition for R; thus, they are valid for any element R. This approach strongly differs from circuits with linear elements, where all circuit parameters can be obtained from a specified applied voltage $V(t)$. The element R can be nonlinear, and its properties are unknown. However, the measured current $i(t)$ is an input parameter, in addition to $V(t)$.

Substituting Equation (10) into (11) gives the following expression for the gas gap voltage:

$$U_g(t) = V(t) - \frac{Q(t)}{C_d}. \tag{14}$$

Substituting Equation (12) into (13) gives an expression for the discharge current, as already suggested in [8,21]:

$$j_R = i(t) - C_g \frac{dU_g(t)}{dt}. \tag{15}$$

Equation (15) indicates that the measured current is the sum of discharge current and displacement current through the gas gap. It can be seen that the expression (12) differs from Equation (1) for current i_{off}. Therefore, the difference of the measured current with and without discharge ignition is not equal to the discharge current in this approach either.

Substituting Equation (14) into (15) and taking into account that the derivative of the measured charge is the measured current $dQ(t)/dt = i(t)$, an expression for the discharge current $j_R(t)$ can be obtained in the form as suggested in [8–10]:

$$j_R(t) = \left[1 + \frac{C_g}{C_d}\right] i(t) - C_g \frac{dV(t)}{dt}. \tag{16}$$

Expressions (16) and (14) enable instantaneous power determination as given in Equation (3) if the capacitances C_g and C_d are known.

5. Revision of the Equivalent Circuit Approach and the Classical Electrical Theory of Ozonizers

The equivalent circuit approach provides new possibilities to infer measured electrical characteristics. However, two important questions remain: (1) Why is the equivalent circuit appropriate, or why does it reflect the properties of DBDs? The replacement validity of the reactor cell (see Figure 1) by the equivalent circuit in Figure 5 is not obvious. (2) How does one determine the capacitances C_{cell}, C_d, C_g? Capacitances can be calculated in the case of simple geometries, but it is difficult to account for edge effects and non-uniform gaps, especially for small-sized laboratory reactors. These questions were in focus in [11–13] and are revised in this section.

5.1. Validity of the Equivalent Circuit Approach

The work [12] discusses the circuit in Figure 5 as the simplest equivalent circuit for the interpretation of classical QV-plots. Namely, the circuit (i) describes the change of the DBD capacitances by varying the resistance of the "black box" R and (ii) provides the same Equations (8) and (9) for the charge measured in passive and active phases of the discharge as in the classical theory. The charge Q_0 in Equation (8) can be related to the charge transferred through the gas gap. The relation is the same, whether the classical theory or the equivalent circuit is applied. The absence of the contradictions between the theories was concluded. Two antiparallel Zener diodes can be considered as a specific case of the "black box" R for the description of the classical QV-plot, as was suggested by Kogelschatz [22].

Here, we demonstrate the close relation between the theories in a slightly different way. The classical theory [7] implicitly suggests two equivalent circuits for active and passive discharge phases, respectively; see Figure 3. When there is no current flow through the "black box" R, the equivalent circuit in Figure 5 coincides with the suggestion of the classical theory for the passive discharge phase. When current flows through the "black box" R, the equivalent circuit will be identical to the circuit of the classical theory for the active discharge phase, if the restriction of the gas gap voltage $U_g = constant$ is applied. If $U_g = constant$, then the displacement current through C_g is zero (see Equation (12)), and the total measured current is associated with the charge transferred through the gas gap (see Equation (15)). Thus, the equivalent circuit approach (see Figure 5) generalizes the classical theory: (i) it uses one circuit instead of two, and (ii) it does not require the condition $U_g = constant$. The replacement of the box "plasma" with "black box" R does not increase the complexity of the description, nor does it introduce any additional assumptions. The main concerns about the applicability of the equivalent circuit approach are related to the term "gas gap voltage", which assumes an equipotential surface of the dielectric barriers. This term was introduced in the classical theory.

5.2. Determination of C_{cell} and C_d

The work [11] proposes to observe the point of maximal charge Q_{max} at different voltage amplitudes V_{max} for the determination of the capacitances C_{cell} and C_d; see Figure 6. Figure 6a presents examples for volume DBDs operated by square voltage pulses from QV-plots. In the presented examples, the applied voltage oscillates around its amplitude value during the active part of the discharge, and an extreme value of $V(t)$ does not correspond to the maximum charge. Point 5 on the QV-plot indicates the moment just before the beginning of the $V(t)$ falling slope and the $Q_{max}(V_{max})$ point. The $Q_{max}(V_{max})$ point is at the upper right corner in the QV-plots, as shown by the arrows in Figure 6a. All $Q_{max}(V_{max})$ points measured for different amplitudes of the applied voltages are displayed in Figure 6b, and a linear slope is obtained. Without discharge ignition, the slope of the line represents C_{cell}, and when the discharge is ignited, the slope of the line represents C_d. In the example in Figure 6b, square voltage pulses with two different rise times (20 ns and 75 ns) were used for the same reactor. The resulting straight lines have slightly different slopes, but this uncertainty is included in the error bars of the values shown in Figure 6b.

Without discharge ignition, the reactor cell acts as a capacitance C_{cell}, and its determination is obvious. For explanation of the C_d determination, the measured charge can be expressed by Equation (14):

$$Q(t) = C_d \left[V(t) - U_g(t) \right]. \tag{17}$$

This is similar to Equation (9) (classical theory); however, Equation (14) is valid for the whole discharge period, and $U_g(t)$ is not necessarily constant. Due to the dependence of the gas gap voltage on time in the active discharge phase, the corresponding part of the QV-plot is not linear; see the part between Points 1 and 5 in Figure 6a. The gas gap voltage at Moment 5, at the $Q_{max}(V_{max})$ point, can be denoted as the residual gas gap voltage U_{res}. Then, Equation (17) can be written as follows:

$$Q_{max} = C_d \left[V_{max} - U_{res} \right]. \tag{18}$$

If U_{res} does not depend on applied voltage amplitude V_{max}, $Q_{max}(V_{max})$ appears as a straight line with the slope C_d. Therefore, the method suggested in [11], based on the assumption $U_{res}(V_{max}) = const$, is valid for volume discharges of different geometries and various applied voltage waveforms [11].

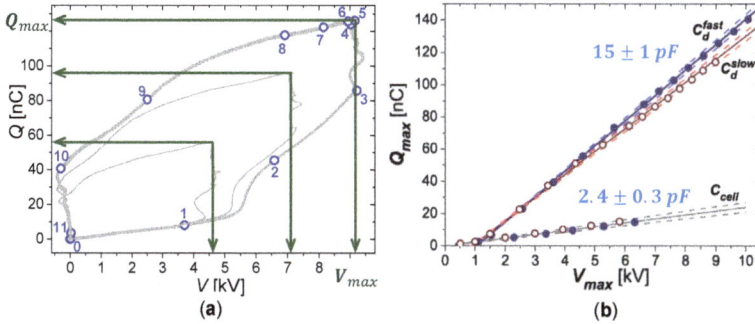

Figure 6. Determination of the reactor capacitances, based on experimental data from [13]. (a) Examples of QV-plots for DBD operated by square voltage pulses. Arrows indicate values of Q_{max} and V_{max}. Digits enumerate selected moments in the QV-plot with the largest amplitude of the applied voltage. (b) An example of the $Q_{max}V_{max}$ plot, reproduced from [13] with the permission of ©2013 WILEY-VCH Verlag GmbH & Co. KGaA, Weinheim. The points are the measured values of Q_{max} and V_{max} for DBD operated by square voltage pulses with fast (solid circles) and slow (open circles) rise times. Solid straight lines have slopes corresponding o the capacitance, and dashed lines indicate linear fit uncertainties.

5.3. Discharge Current $j_R(t)$

The expression for the discharge current (16) contains the gas gap capacitance C_g, which can be expressed by Equation (7):

$$C_g = \frac{C_d C_{cell}}{C_d - C_{cell}}. \tag{19}$$

In order to avoid an increase of the experimental uncertainty due to the direct use of Equation (19), this is substituted in (16). After re-arrangement, the form of the equation suggested in [12] is obtained:

$$j_R(t) = \frac{1}{1 - \frac{C_{cell}}{C_d}} \left[i(t) - C_{cell} \frac{dV(t)}{dt} \right]. \tag{20}$$

Equation (20) contains only directly-measurable quantities and allows the interpretation of the measured current $i(t)$. The derivative of the applied voltage $V(t)$ scaled on capacitance C_{cell} is the current i_{off}, measurable without discharge ignition; see Equation (1). The difference of the measured current with and without discharge is in brackets in Equation (20). It is proportional (but not equal) to the discharge current. The proportionality coefficient depends on the geometrical properties of the discharge arrangement C_{cell} and C_d.

Note that Expressions (20) and (15) represent the same relation between measured $i(t)$ and discharge $j(t)$ current. A simple physical meaning can be easily seen from Equation (15), namely that the measured current is the sum of the discharge and displacement current in the gas gap. Equation (20) has more practical impact as it contains only measurable quantities.

5.4. Dissipated Energy and Relevance of the Equivalent Circuit

An examples of the total energies derived from measured current $i(t)$ and voltage $V(t)$ and instantaneous energies from discharge current $j_R(t)$ and gas gap voltage $U_g(t)$ are shown in Figure 7. This illustrates the advantage of the equivalent circuit approach for the analysis of DBD electrical characteristics. The amplitude of the applied voltage and the geometry of the discharge cell are the same for the data in Figure 7a,b. The integration of the measured current and voltage product:

$$E_{total}(t) = \int_0^t i(\tau)V(\tau)d\tau, \tag{21}$$

is the total energy, i.e., the dissipated energy and the energy stored on dielectric barriers. During the falling voltage slope, between Moments 5 and 11 (see also Figure 6), the energy stored on dielectric barriers is released. A part of this energy is not dissipated in the discharge and, thus, leads to the decrease of $E_{total}(t)$. The discharge current $j_R(t)$ and the gas gap voltage $U_g(t)$ derived from the equivalent circuit allow one to determine the instantaneous energy dissipated in the DBD:

$$E(t) = \int_0^t j_R(\tau) U_g(\tau) d\tau. \tag{22}$$

Both energies ($E_{total}(t)$ and $E(t)$) merge at the end of the discharge period, confirming that the determination of the energy or power averaged over the discharge period does not require knowledge about the equivalent circuit. However, the determination of the energy dissipated in rising and falling voltage slopes requires knowledge about $j_R(t)$ and $U_g(t)$. The equivalent circuit allows concluding that (i) a larger portion of energy is consumed during the rising voltage slope and (ii) the energy dissipated in the falling voltage slope drops much more strongly with the decrease of the voltage rise time.

The decrease of the total energy dissipated over the period can be compensated by the increase of the voltage amplitude; however, the discharges with fast and slow voltage pulses will not be the same. The fast voltage pulses generate two comparable discharge pulses during one period, whereas slow voltage pulses couple the major energy during the rising edge of the voltage. Thus, the equivalent circuit approach is an important tool for electrical characterization of DBDs, which provides the instantaneous discharge power.

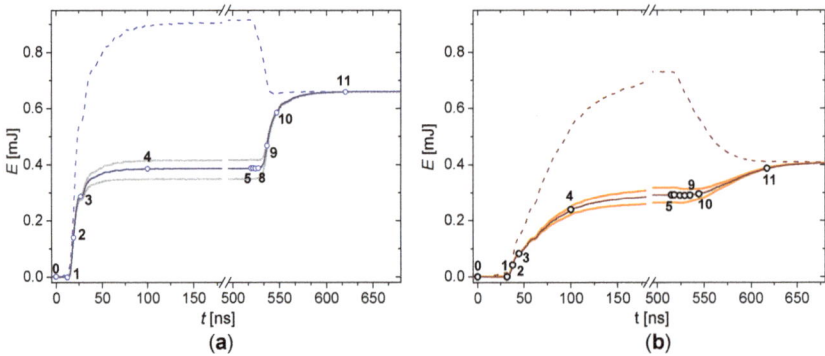

Figure 7. Instantaneous energy dissipated in DBD excited by square voltage pulses of 9 kV in amplitude with a rise time of 20 ns (**a**) and 70 ns (**b**). Experimental data from [13]. Dashed lines are the total energy defined as the integral of the product $i(t) \times V(t)$. Solid lines are the discharge energy defined as the integral of the products $j_R(t) \times U_g(t)$. Grey lines correspond to the extreme values of the energy caused by uncertainties in capacitance C_{cell} and C_d displayed in Figure 6b. Enumerated moments for fast switching (**a**) are the same as in Figure 6a.

6. Further Development of the Equivalent Circuit Approach and Open Questions

The equivalent circuit approach is reliable for characterization of DBD geometries, where the capacitances of the reactor do not depend on the operation conditions. The simplest equivalent circuit (Figure 5) contains the non-linear element R, the "black box". Current measurements in addition to the measurements of the applied voltage compensate the lack of knowledge about this element. The introduction of additional nonlinear elements, such as variable capacitances, is problematic. A large variety of the DBD geometries has been used in applications [6], and the capacitance might depend on the operation conditions for some of them. Examples of such DBD arrangements are shown in Figure 8 schematically and discussed in the following subsections.

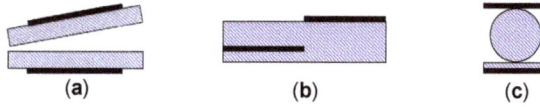

Figure 8. Schematic presentation of DBD arrangements with variable capacitance. (**a**) Tilted electrodes, (**b**) surface discharge, and (**c**) packed bed reactor.

6.1. Discharge with Tilted Electrodes and Partial Discharging

Peeters and van de Sanden [14] investigated a discharge with tilted electrodes under excitation by sinusoidal voltage. The measured charge-voltage characteristics are presented in Figure 9. The QV-plot for one amplitude of the applied voltage (Figure 9a) is nearly a classical parallelogram. Therefore, it could be assumed that reactor capacitances are constant over the discharge period. However, the corresponding non-linear $Q_{max}V_{max}$-plot in Figure 9b could not be inferred in the framework of the classical theory or the simplest equivalent circuit. The authors [14] concluded that the discharge volume grows with the amplitude of the applied voltage, and thus, the reactor capacitance grows, leading to the non-linearity of the $Q_{max}V_{max}$-plot.

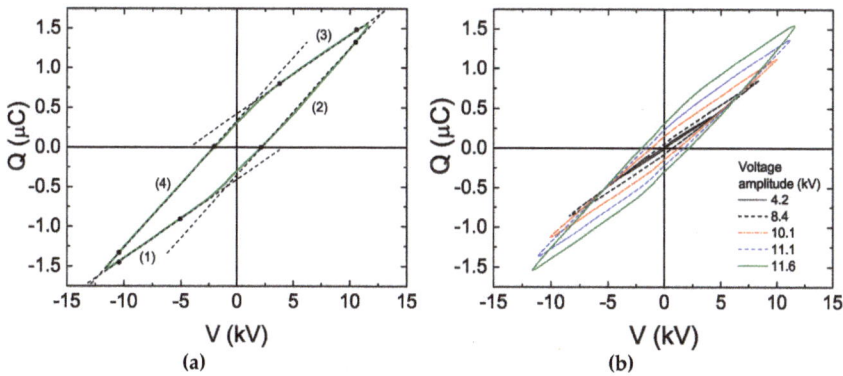

Figure 9. QV-plots measured in [14] for DBD arrangement with tilted electrodes. (**a**) QV-plot for a single amplitude of the applied voltage. Dashed lines and black circles emphasize the linear parts of the plot. (**b**) QV-plots for different voltage amplitudes. ©IOP Publishing. Reproduced from [14] with permission of the authors and IOP Publishing. All rights reserved.

For interpretation of the charge-voltage characteristics, the authors suggested an equivalent circuit where the reactor cell is divided into a discharging (β) and non-discharging (α) part; see Figure 10a. The sum of the reactor parts is: $\alpha + \beta = 1$. This representation has an additional physical meaning. The non-discharging part can be considered as a parasitic capacitance C_p, and for simplicity, the equivalent circuit in Figure 10b can be considered.

The parasitic capacitance will influence the measured slopes of the QV-plot. The measured slopes will be $C_{cell} + C_p$ for the passive phase and $C_d + C_p$ for the active phase of the discharge. It can be seen from the circuit in Figure 10b and was discussed also in [14,23]. In order to illustrate the role of C_p equations for gas gap voltage $U_g(t)$ and discharge current $j_R(t)$, they are re-derived here.

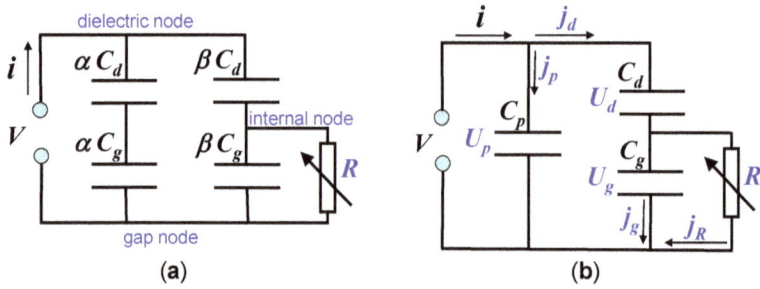

Figure 10. The equivalent circuit for partial discharging (**a**) suggested in [14] and the equivalent circuit accounting for parasitic capacitance (**b**).

Note that Equations (14) and (20) obtained for the simplest equivalent circuit (Figure 5b) are valid for the circuit with parasitic capacitance (Figure 10b) if the measured current is replaced by a current through the dielectric capacitance: $i(t) \longrightarrow j_d(t)$ and the same for the charge $Q(t) \longrightarrow Q_d(t)$. Taking into account the relations from the circuit in Figure 10b, it follows:

$$j_d(t) = i(t) - C_p \frac{dV(t)}{dt} \quad \text{and} \quad Q_d(t) = Q(t) - C_p V(t), \tag{23}$$

and introducing notations ζ_{cell} and ζ_d for capacitance values measurable from the QV plot gives:

$$\zeta_{cell} = C_{cell} + C_p \quad \text{and} \quad \zeta_d = C_d + C_p. \tag{24}$$

After re-arrangement, the relations for $U_g(t)$ and $j_R(t)$ can be obtained in the following form, similar to Equations (14) and (20):

$$U_g(t) = \frac{1}{\left\{1 - \frac{C_p}{\zeta_d}\right\}} \left[V(t) - \frac{Q(t)}{\zeta_d}\right], \tag{25}$$

$$j_R(t) = \left\{1 - \frac{C_p}{\zeta_d}\right\} \frac{1}{1 - \frac{\zeta_{cell}}{\zeta_d}} \left[i(t) - \zeta_{cell} \frac{dV(t)}{dt}\right]. \tag{26}$$

Both equations depend on the parasitic capacitance C_p via the factor in the curly brackets. The value of C_p often cannot be determined. If the capacitance values measured from the charge-voltage characteristics ζ_{cell} and ζ_d are used for the determination of $U_g(t)$ and $j_R(t)$, then the unknown factor in curly brackets may introduce a significant error. This error grows with the value of C_p. Surprisingly, the instantaneous power, the product of $U_g(t)$ and $j_R(t)$, does not depend on this factor and can still be evaluated accurately.

Besides the change of the slopes of the QV-plot, the parasitic capacitance has one more effect on the interpretation of the QV-plot. In classic theory, the measured charge during the active phase is shifted by the value U_g; see Figure 3 and Equation (9). Expressing the measured charge from Equation (25), the equation analogous to Equation (9) can be obtained:

$$Q(t) = \zeta_d \left[V(t) - \left\{1 - \frac{C_p}{\zeta_d}\right\} U_g(t)\right]. \tag{27}$$

Equation (27) indicates that the shift of the charge is proportional to $U_g(t)$, but it is smaller by the factor in the curly brackets, which depends on the value C_p.

It is important to note the difference between the situations with parasitic capacitance (Figure 10b) and partial discharging (Figure 10a). In Figure 10b, the contribution of C_d to $U_g(t)$ and $j_R(t)$ could not be evaluated due to the unknown value of C_p; whereas the partial discharging (Figure 10a) assumes experimental conditions when the whole reactor cell is discharged, which enables one to determine C_d and thereafter the coefficients α and β and all parameters of the circuit for arbitrary experimental conditions.

The corresponding equations for the circuit of the partial discharging (Figure 10a) can be obtained by substitutions $C_{cell} \longrightarrow \beta C_{cell}$, $C_d \longrightarrow \beta C_d$, $C_p \longrightarrow \alpha C_{cell}$, and $\zeta_{cell} \longrightarrow C_{cell}$ in Equations (24)–(26). The results for the coefficients α and β were already suggested in [14]:

$$\alpha = \frac{C_d - \zeta_d}{C_d - C_{cell}} \quad \text{and} \quad \beta = \frac{\zeta_d - C_{cell}}{C_d - C_{cell}}. \tag{28}$$

The equation for the discharge current corresponds to the simplest equivalent circuit [14]; see Equation (20). For the equation for the gas gap voltage, we will use the directly-measured values ζ_d instead of coefficients α and β in contrast to [14]:

$$U_g(t) = \frac{1 - \frac{C_{cell}}{C_d}}{1 - \frac{C_{cell}}{\zeta_d}} \left[V(t) - \frac{Q(t)}{\zeta_d} \right]. \tag{29}$$

This form is more similar to the equation for $U_g(t)$ from the simplest equivalent circuit; see Equation (14). In the case $\zeta_d = C_{cell}$, Equation (29) is undetermined [14] because $U_g(t)$ is the voltage on capacitance βC_g (see Figure 10a) and $\beta = 0$.

The idea of the partial discharging is an important stage of development for the equivalent circuit approach. It is applicable when (i) the parasitic capacitance is negligible, (ii) the experimental condition of complete discharging ($\beta = 1$) is reachable (i.e., the results of the simplest equivalent circuit can be used to determine C_d), and (iii) the measured QV-plot resembles a parallelogram, indicating constant capacitances during the active discharge phase, as is observed, e.g., for sinusoidal applied voltage with constant gas gap voltage.

6.2. Surface Discharge

The surface discharge is a type of DBD, and the plasma expansion over the dielectric surface depends on the applied voltage amplitude. An example of the experimental data obtained from electrical diagnostics of a surface DBD [15] are presented in Figure 11. The measured QV-plot of a surface discharge has an almond-like shape (Figure 11a), and the authors suggest to use the charge derivative $C(t) = dQ(t)/dV(t)$ as a measure of the effective capacitance of the reactor. The examples of obtained $C(t)$ values correlated with the voltage waveform are shown in Figure 11b.

In the passive discharge phase, $C(t)$ equals C_{cell}, which is indicated as C_0 in Figure 11a,b. The passive phase starts just after the moment when the applied voltage reaches the amplitude and is characterized by a straight line with minimal slope in the QV-plot. During the active discharge phase, $C(t)$ increases until it saturates as $C(t) = C_{eff}$. The value C_{eff} measured for different amplitudes and frequencies of the applied voltage is shown in Figure 11c as a function of the visual plasma expansion Δx. The Δx values were determined as the length of the luminescent area from photographs taken form the discharge at the different experimental conditions.

However, the charge derivative $C(t) = dQ(t)/dV(t)$ can be associated with a capacitance only in the classical theory where the gas gap voltage and dielectric capacitance are constant over the discharge cycle. There is no evidence that this assumption is valid if the measured QV-plot differs from a parallelogram. The alternation of the reactor capacitance during the active discharge phase can be represented as a variable capacitance C_d in the simplest equivalent circuit (see Figure 5), and the charge derivative can be deduced from Equation (17):

$$\frac{dQ(t)}{dV(t)} = C_d(t) + [V(t) - U_g(t)] \frac{dC_d(t)}{dV(t)} - C_d \frac{dU_g(t)}{dV(t)}. \tag{30}$$

The variation of the gas gap voltage alone can make the charge derivative undetermined; see the QV-plot in Figure 6a between Moments 3 and 4. A variable capacitance $C_d(t)$ makes this consideration even more complex. Note that the expressions for the discharge current are not valid with variable $C_d(t)$. The substitution of Equation (14) into Equation (15) accounts for additional terms related to the derivative of the dielectric capacitor. In any case, the introduction of a new "black box" into the equivalent circuit makes the evaluation of the discharge properties impossible without additional measurable parameters.

Nevertheless, the charge derivative $C(t) = \mathrm{d}Q(t)/\mathrm{d}V(t)$ in Figure 11b has a plateau where the derivative is constant $C(t) = C_{eff}$. In this time period, just before the voltage maximum, it can be assumed that $C_d(t)$ and $U_g(t)$ are constant and C_{eff} is associated with the dielectric capacitance C_d. The linear dependence of C_{eff} on the discharge expansion Δx (see Figure 11) is an argument in favor of this assumption.

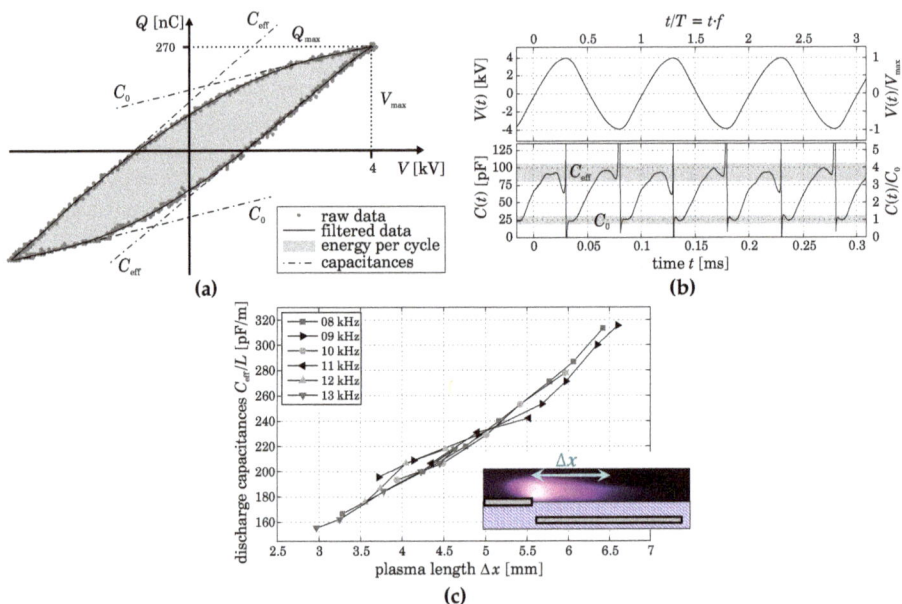

Figure 11. Experimental data for surface DBD [15]. (**a**) QV-plot, (**b**) applied voltage $V(t)$ (top) and charge derivative $C(t) = \mathrm{d}Q(t)/\mathrm{d}V(t)$ (bottom) waveforms, and (**c**) effective capacitance as a function of discharge expansion Δx for different frequencies of the applied voltage, reprinted from [15] with the kind permission of authors and ©AIP Publishing. The added color inset in (**c**) shows Δx schematically.

6.3. Packed Bed Reactor

The electrode arrangement of the packed bed reactor (see Figure 8c) is similar to a DBD with tiled electrodes (see Figure 8a), as well as to the surface discharge (see Figure 8b). These similarities were pointed out in [16], which emphasized that the QV-plots of packed bed reactors are far away from the classical parallelogram, namely an almond-like shape (similar to surface discharge). The suggested scheme of the QV-plot formation is presented in Figure 12.

The left side of Figure 12 depicts the region of contact between the dielectric pellet and the top electrode. The numerated areas are associated with the moments of discharge expansion over the pellet surface within the voltage period. The discharge expansion should lead to an increase of the effective reactor capacitance, and the selected moments might correlate with parts of the QV-plots, as shown in Figure 12. Similar to the surface discharge, the charge derivative $C(t) = \mathrm{d}Q(t)/\mathrm{d}V(t)$ near the maximum of the voltage amplitude reflects the dielectric capacitance for this voltage amplitude.

At higher voltage amplitudes, the discharge occupies a larger surface, and consequently, the effective dielectric capacitance increases. This fact was inferred in terms of the equivalent circuit for partial discharging [14]; see Figure 10a. Thus, the current and voltage waveforms can be used for the estimation of the spread of the discharge, characterized by the coefficients α and β.

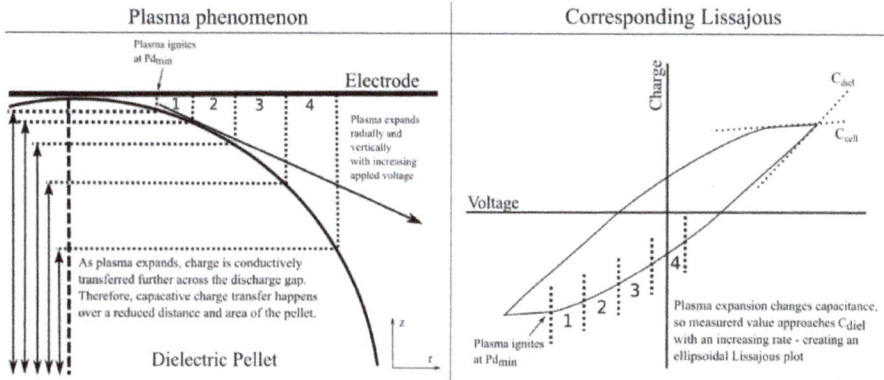

Figure 12. Formation scheme of the QV-plot for a packed bed reactor suggested in [16]. Reprinted from [16] with the kind permission of the authors, used under the terms of the Creative Commons Attribution 3.0 license, https://creativecommons.org/licenses/by/3.0/.

7. Summary

The present work emphasizes that the equivalent circuit approach is a generalization of the classical electric theory of ozonizers [7]. This approach does not introduce additional assumptions, but does not require a constant gas gap voltage during the active discharge phase. This allows one to correlate measured current and voltage waveforms with the discharge current $j_R(t)$ and the gas gap voltage $U_g(t)$. It is applicable to volume DBDs, where the reactor cell capacitance C_{cell} and the dielectric capacitance C_d are independent of the experimental conditions. In the case of a significant (although with an unknown value) parasitic capacitance, $j_R(t)$ and $U_g(t)$ could not be evaluated separately, however, the product $j_R(t) \times U_g(t)$, i.e., the instantaneous power, can still be measured accurately. The instantaneous power allows one to separate the energy dissipated in the discharge from the energy stored on the dielectric surfaces. However, it should not be confused with the dissipated power averaged over the discharge period, which does not require the detailed considerations of the equivalent circuit at all.

The equivalent circuit is applicable for any type of voltage waveform; however, attention should be paid to the technical difficulties of the accurate current measurement. For example, if the discharge current waveform $j_R(t)$ is recorded for sinusoidal applied voltage, a sufficient time interval should be investigated with high sampling rate and resolution of analog-to-digital converter in a large dynamic range. This can be achieved with modern measurement technique (see [24]), but requires significant efforts.

The interpretation of the current and voltage waveform for DBDs with variable capacitance C_d remains challenging, but the equivalent circuit for partial discharging can be used for characterization of the discharge expansion over the dielectric surface.

The range of applicability of the present approach is restricted to the situations where power dissipated in the dielectric is negligible. Dielectric losses strongly depend on the type and temperature of the dielectric, the discharge geometry, as well as the operation frequency. The power losses in the dielectric can be significant even for a glass dielectric and excitation frequency above kilohertz [25]. A further analysis of this point remains for future work.

For completeness, other approaches for electrical diagnostics of DBD should be mentioned: for example, statistical methods based on detailed analysis of the shapes of the current pulses measured in a DBD driven by sinusoidal voltage [24,26,27], modeling of the discharge processes [28–30], the simulation of the measured characteristics by defining the properties of lumped electrical elements of the equivalent circuit in XCOS/Scilab or the Simulink environment [31–34], or optical measurements of the electric field [35–37] and electron densities [38,39].

The peculiarity of the approach discussed here is that it is based on current/charge and voltage measurements and does not need any information about operation gas or discharge processes. The main attention was paid to the determination of instantaneous power dissipated in DBDs. The plasma chemistry is activated by electrons with energies above the threshold of the reaction. Thus, the maximum instantaneous power density can determine the chemical pathway and yield of reactions. The monitoring of the instantaneous power also would enable more reliable comparison of DBDs with different electrode configurations, since it accounts for the influence of the reactor geometry on the measured electrical characteristics. Additionally, the discharge current $j_R(t)$ and the gas gap voltage $U_g(t)$ obtained in the framework of the equivalent circuit can be used for consideration of the current voltage characteristics of the discharge [40,41] to gain more comprehensive analysis of the discharge processes.

We hope that the present work will encourage the use of the equivalent circuit approach and stimulate its further development.

Author Contributions: A.V.P. and R.B. have jointly contributed to the work.

Funding: This research received no external funding.

Acknowledgments: A.V.P. is thankful to Milivoje Ivković for the invitation to give a lecture at the 29th Summer School and International Symposium on the Physics of Ionized Gases, Belgrade, Serbia, 2018, which became the main motivation for this contribution. The authors are also thankful to Tomáš Hoder for useful discussions.

Conflicts of Interest: The authors declare no conflict of interest.

Abbreviations

The following abbreviations are used in this manuscript:

DBD	dielectric barrier discharge
QV-plot	charge voltage characteristics
C_{cell}	capacitance of the reactor cell without discharge
C_d	the capacitance associated with dielectric barriers of the reactor cell, sometimes able to be seen as a reactor capacitance during the active discharge phase
ζ_{cell}, ζ_d	capacitance values obtained from the slopes of the QV-plot, which can coincide with C_{cell} and C_d if a parasitic capacitance C_p is negligible
C_g	the capacitance associated with the gas gap of the reactor cell
$i(t), V(t), Q(t)$	measurable values: external current, applied voltage, and charge
$i_{off}(t)$	current measured without discharge (discharge off)
$j_R(t), U_g(t)$	equivalent circuit parameters: discharge current and gas gap voltage
U_b, U_{ext}	the values of the gas gap voltage corresponding to the ignition (breakdown) and extinguishing of the discharge
Q_{max}	the maximal value of the measured charge
V_{max}	the value of the applied voltage when Q_{max} is reached, often corresponding to the voltage amplitude or the maximum of the applied voltage
U_{res}	the value of the gas gap voltages when Q_{max} is reached, residual voltage
α, β	the relative areas of the reactor cell, normalized on the whole area, which are not influenced and occupied by the discharge, respectively
C_{eff}	the value of the QV-plot derivative near to V_{max} in the case of the sinusoidal applied voltage

References

1. Kogelschatz, U. Advanced Ozone Generation. In *Process Technologies for Water Treatment (Earlier Brown Boveri Symposia)*; Stucki, S., Ed.; Plenum: New York, NY, USA, 1998; pp. 87–118. [CrossRef]
2. Zhou, A.; Chen, D.; Ma, C.; Yu, F.; Dai, B. DBD Plasma-ZrO_2 Catalytic Decomposition of CO_2 at Low Temperatures. *Catalysts* **2018**, *8*, 256. [CrossRef]
3. Xuan, K.; Zhu, X.; Cai, Y.; Tu, X. Plasma Oxidation of H_2S over Non-stoichiometric La_xMnO_3 Perovskite Catalysts in a Dielectric Barrier Discharge Reactor. *Catalysts* **2018**, *8*, 317. [CrossRef]
4. Siemens, W. Ueber die elektrostatische Induction und die Verzögerung des Stroms in Flaschendrähten. *Poggendorffs Ann. Phys. Chem.* **1857**, *102*, 66-122. [CrossRef]
5. Li, Y.; Atif, R.; Chen, K.; Cheng, J.; Chen, Q.; Qiao, Z.; Fridman, G.; Fridman, A.; Ji, H.-F. Polymerization of D-Ribose in Dielectric Barrier Discharge Plasma. *Plasma* **2018**, *1*, 13. [CrossRef]
6. Brandenburg, R. Dielectric barrier discharges: Progress on plasma sources and on the understanding of regimes and single filaments. *Plasma Sources Sci. Technol.* **2017**, *26*, 053001. [CrossRef]
7. Manley, T.C. The electric characteristics of the ozonator discharge. *Trans. Electrochem. Soc.* **1943**, *84*, 83–96. [CrossRef]
8. Lomaev, M.I. Determination of energy input in barrier discharge excilamps. *Atmos. Ocean. Opt.* **2001**, *14*, 1005–1008.
9. Liu, S.; Neiger, M. Excitation of dielectric barrier discharges by unipolar submicrosecond square pulses. *J. Phys. D Appl. Phys.* **2001**, *34*, 1632–1638. [CrossRef]
10. Bibinov, N.K.; Fateev, A.A.; Wiesemann, K. Variations of the gas temperature in He/N_2 barrier discharges. *Plasma Sources Sci. Technol.* **2001**, *10*, 579–588. [CrossRef]
11. Pipa, A.V.; Hoder, T.; Koskulics, J.; Schmidt, M.; Brandenburg, R. Experimental determination of dielectric barrier discharge capacitance. *Rev. Sci. Instrum.* **2012**, *83*, 075111. [CrossRef] [PubMed]
12. Pipa, A.V.; Koskulics, J.; Brandenburg, R.; Hoder, T. The simplest equivalent circuit of a pulsed dielectric barrier discharge and the determination of the gas gap charge transfer. *Rev. Sci. Instrum.* **2012**, *83*, 115112. [CrossRef] [PubMed]
13. Pipa, A.V.; Hoder, T.; Brandenburg, R. On the Role of Capacitance Determination Accuracy for the Electrical Characterization of Pulsed Driven Dielectric Barrier Discharges. *Contrib. Plasma Phys.* **2013**, *53*, 469–480. [CrossRef]
14. Peeters, F.J.J.; van de Sanden, M.C.M. The influence of partial surface discharging on the electrical characterization of DBDs. *Plasma Sources Sci. Technol.* **2015**, *24*, 015016. [CrossRef]
15. Kriegseis, J.; Grundmann, S.; Tropea, C. Power consumption, discharge capacitance and light emission as measures for thrust production of dielectric barrier discharge plasma actuators. *J. Appl. Phys.* **2011**, *110*, 013305. [CrossRef]
16. Butterworth, T.; Allen, R.W.K. Plasma-catalyst interaction studied in a single pellet DBD reactor: Dielectric constant effect on plasma dynamics. *Plasma Sources Sci. Technol.* **2017**, *26*, 065008. [CrossRef]
17. Samojlovich, V.G.; Gibalov, I.V.; Kozlov, K.V. *Physical Chemistry of the Barrier Discharge*; DVS: Düsseldorf, Germany, 1997; pp. 1–261, ISBN 978-3-87155-744-6.
18. Brandenburg, R.; Navratil, Z.; Jansky, J.; St'ahel, P.; Trunec, D.; Wagner, H.-E. The transition between different modes of barrier discharges at atmospheric pressure. *J. Phys. D Appl. Phys.* **2009**, *42*, 085208. [CrossRef]
19. Buntat, Z.; Harry, J.E.; Smith, I.R. Generation of a Homogeneous Glow Discharge in Air at Atmospheric Pressure. *Elektrika* **2007**, *9*, 60–65.
20. Mildren, R.P.; Carman, R.J. Enhanced performance of a dielectric barrier discharge lamp using short-pulsed excitation. *J. Phys. D Appl. Phys.* **2001**, *34*, L1–L6. [CrossRef]
21. Massines, F.; Gherardi, N.; Naude, N.; Segur, P. Glow and Townsend dielectric barrier discharge in various atmosphere. *Plasma Phys. Control Fusion* **2005**, *47*, B577–B588. [CrossRef]
22. Kogelschatz, U. Fundamentals of Dielectric-Barrier Discharges. In *Non-Equilibrium Air Plasmas at Atmospheric Pressure*; Becker, K.H., Kogelschatz, U., Schoenbach, K.H., Barker, R.J., Eds.; Institute of Physics: Bristol, UK, 2004; pp. 68–75, ISBN 9780750309622.
23. Falkenstein, Z.; Coogan, J.J. Microdischarge behavior in the silent discharge of nitrogen-oxygen and water-air mixtures. *J. Phys. D Appl. Phys.* **1997**, *30*, 817–825. [CrossRef]

24. Synek, P.; Zemánek, M.; Kudrle, V.; Hoder, T. Advanced electrical current measurements of microdischarges: Evidence of sub-critical pulses and ion currents in barrier discharge in air. *Plasma Sources Sci. Technol.* **2018**, *27*, 045008. [CrossRef]
25. Cho, G.; Shin, M.; Jeong, J.; Kim, J.; Hong, B.; Koo, J.; Kim, Y.; Choi, E.; Fechner, J.; Letz, M.; et al. Glass tube of high dielectric constant and low dielectric loss for external electrode fluorescent lamps. *J. Appl. Phys.* **2007**, *102*, 113307. [CrossRef]
26. Siliprandi, R.A.; Romana, H.E.; Barni, R.; Riccardi, C. Characterization of the streamer regime in dielectric barrier discharges. *J. Appl. Phys.* **2008**, *104*, 063309. [CrossRef]
27. Tay, W.H.; Yap, S.L.; Wong, C.S. The Electrical Characteristics of a Filamentary Dielectric Barrier Discharge. *AIP Conf. Proc.* **2010**, *1250*, 532. [CrossRef]
28. Takashima, K.; Yin, Z.; Adamovich, I.V. Measurements and kinetic modeling of energy coupling in volume and surface nanosecond pulse discharges. *Plasma Sources Sci. Technol.* **2013**, *22*, 015013. [CrossRef]
29. Akishev, Y.; Aponin, G.; Balakirev, A.; Grushin, M.; Karalnik, V.; Petryakov, A.; Trushkin, N. Spatial-temporal development of a plasma sheet in a surface dielectric barrier discharge powered by a step voltage of moderate duration. *Plasma Sources Sci. Technol.* **2013**, *22*, 015004. [CrossRef]
30. Becker, M.M.; Hoder, T.; Brandenburg, R.; Loffhagen, D. Analysis of microdischarges in asymmetric dielectric barrier discharges in argon. *J. Phys. D Appl. Phys.* **2013**, *46*, 355203. [CrossRef]
31. Pinchuk, M.E.; Stepanova, O.M.; Lazukin, A.V.; Astafiev, A.M. A simple XCOS/SCILAB model of a DBD Plasma jet impinging on a target. In *29th Summer School and International Symposium on the Physics of Ionized Gases*; Poparić, G., Obradović, B., Borka, D., Rajković, M., Eds.; Vinča Institute of Nuclear Sciences: Belgrade, Serbia, 2018; pp. 214–217, ISBN 978-86-7306-146-7.
32. Fang, Z.; Ji, S.; Pan, J.; Shao, T.; Zhang, C. Electrical Model and Experimental Analysis of the Atmospheric-Pressure Homogeneous Dielectric Barrier Discharge in He. *IEEE Trans. Plasma Sci.* **2012**, *40*, 883–891. [CrossRef]
33. Tay, W.H.; Yap, S.L.; Wong, C.S. Electrical Characteristics and Modeling of a Filamentary Dielectric Barrier Discharge in Atmospheric Air. *Sains Malays.* **2014**, *43*, 583–594. Available online: http://www.ukm.my/jsm/pdf_files/SM-PDF-43-4-2014/12%20W.H.%20Tay.pdf (accessed on 18 January 2019).
34. Valdivia-Barrientos, R.; Pacheco-Sotelo, J.; Pacheco-Pacheco, M.; Benítez-Read, J.S.; López-Callejas, R. Analysis and electrical modelling of a cylindrical DBD configuration at different operating frequencies. *Plasma Sources Sci. Technol.* **2006**, *15*, 237–245. [CrossRef]
35. Simeni, M.S.; Tang, Y.; Frederickson, K.; Adamovich, I.V. Electric field distribution in a surface plasma flow actuator powered by ns discharge pulse trains. *Plasma Sources Sci. Technol.* **2018**, *27*, 104001. [CrossRef]
36. Obrusník, A.; Bílek, P.; Hoder, T.; Šimek, M.; Bonaventura, Z. Electric field determination in air plasmas from intensity ratio of nitrogen spectral bands: I. Sensitivity analysis and uncertainty quantification of dominant processes. *Plasma Sources Sci. Technol.* **2018**, *27*, 085013. [CrossRef]
37. Ivković, S.S.; Obradović, B.M.; Cvetanović, N.; Kuraica, M.M.; Purić, J. Measurement of electric field development in dielectric barrier discharge in helium. *J. Phys. D Appl. Phys.* **2009**, *42*, 225206. [CrossRef]
38. Cvetanović, N.; Galmiz, O.; Synek, P.; Zemánek, M.; Brablec, A.; Hoder, T. Electron density in surface barrier discharge emerging at argon/water interface: Quantification for streamers and leaders. *Plasma Sources Sci. Technol.* **2018**, *27*, 025002. [CrossRef]
39. Konjević, N.; Ivković, M.; Sakan, N. Hydrogen Balmer lines for low electron number density plasma diagnostics. *Spectrochim. Acta B* **2012**, *76*, 16–26. [CrossRef]
40. Naudé, N.; Cambronne, J.-P.; Gherardi, N.; Massines, F. Electrical model and analysis of the transition from an atmospheric pressure Townsend discharge to a filamentary discharge. *J. Phys. D Appl. Phys.* **2005**, *38*, 530–538. [CrossRef]
41. Enache, I.; Naudé, N.; Cambronne, J.-P.; Gherardi, N.; Massines, F. Electrical model of the atmospheric pressure glow discharge (APGD) in helium. *Eur. Phys. J. Appl. Phys.* **2006**, *33*, 15–21. [CrossRef]

Article

Quantum Rainbows in Positron Transmission through Carbon Nanotubes

Marko Ćosić [1],*, Srđan Petrović [1] and Nebojša Nešković [2]

[1] Laboratory of Physics, Vinča Institute of Nuclear Sciences, University of Belgrade, P. O. Box 522,
    11001 Belgrade, Serbia; petrovs@vinca.rs
[2] World Academy of Art and Science, Napa, CA 94558, USA; nneskovic49@gmail.com
* Correspondence: mcosic@vinca.rs; Tel.: +381-11-644-7700

Received: 28 November 2018; Accepted: 23 January 2019; Published: 28 January 2019

Abstract: Here we report the results of the theoretical investigation of the transmission of channeled positrons through various short chiral single walled carbon nanotubes (SWCNT). The main question answered by this study is "What are the manifestations of the rainbow effect in the channeling of quantum particles that happens during the channeling of classical particles?" To answer this question, the corresponding classical and quantum problems were solved in parallel, critically examined, and compared with each other. Positron energies were taken to be 1 MeV when the quantum approach was necessary. The continuum positron-nanotube potential was constructed from the thermally averaged Molière's positron-carbon potential. In the classical approach, a positron beam is considered as an ensemble of noninteracting particles. In the quantum approach, it is considered as an ensemble of noninteracting wave packages. Distributions of transmitted positrons were constructed from the numerical solutions of Newton's equation and the time-dependent Schrödinger equation. For the transmission of 1-MeV positrons through 200-nm long SWCNT (14; 4), in addition to the central maximum, the quantum angular distribution has a prominent peak pair (close to the classical rainbows) and two smaller peaks pairs. We have shown that even though the semiclassical approximation is not strictly applicable it is useful for explanation of the observed behavior. In vicinity of the most prominent peak, i.e., the primary rainbow peak, rays interfere constructively. On one of its sides, rays become complex, which explains the exponential decay of the probability density in that region. On the other side, the ray interference alternates between constructive and destructive, thus generating two observed supernumerary rainbow peaks. The developed model was then applied for the explanation of the angular distributions of 1-MeV positrons transmitting through 200 nm long (7, 3), (8, 5), (9, 7), (14, 4), (16, 5) and (17, 7) SWCNTs. It has been shown that this explains most but not all rainbow patterns. Therefore, a new method for the identification and classification of quantum rainbows was developed relying only on the morphological properties of the positron wave function amplitude and the phase function families. This led to a detailed explanation of the way the quantum rainbows are generated. All wave packets wrinkle due to their internal focusing in a mutually coordinated way and are concentrated near the position of the corresponding classical rainbow. This explanation is general and applicable to the investigations of quantum effects occurring in various other atomic collision processes.

Keywords: rainbow scattering; positron channeling effect; time-dependent Schrödinger equation; chiral single wall carbon nanotubes

---

1. Introduction

Let us consider a single perfect graphene sheet shown in the Figure 1a. The primitive vectors of the graphene lattice are denoted as a_1, and a_2. Single wall carbon nanotubes (SWCNTs) can be seen as a graphene sheet rolled-up to form a cylinder [1]. However, rolling up of the graphene sheet is possible only in certain directions. To form a nanotube it is necessary that its circumference be equal to the length of the vector, called the chiral vector $C_h = ma_1 + na_2$ specifying possible distances between atoms of the sheet. The resulting SWCNT is made of the infinite number of nanotube unit cells containing N carbon atoms translationally repeating itself in the direction orthogonal to the vector C_h, defining the nanotube axis. Chiral indices (m, n) uniquely determine the structure of nanotube [1], and are used for the identification of the nanotubes. Depending on the direction of the vector C_h, all SWCNTs can be classified in the three classes: zig-zag $C_h = (0, n)$, armchair $C_h = (n, n)$, and generic nanotubes also called chiral $C_h = (m, n)$. Views in the direction of the axis of the zig-zag, armchair, and chiral SWCNT are shown in the Figure 1a–c, respectively. Nanotubes have extraordinary elastic, electronic and thermal properties. A good overview of nanotube properties which are important for the potential applications can be found in the references [2–4].

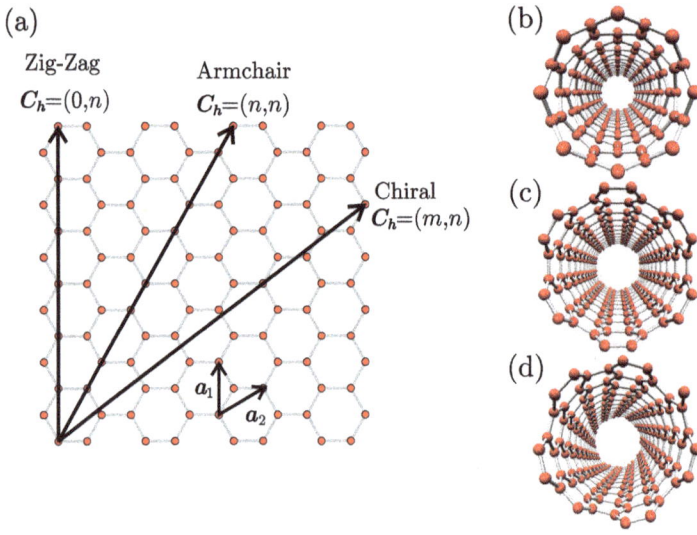

Figure 1. (**a**) Section of the graphene sheet. Small arrows labeled a_1 and a_2 represent primitive vectors of the graphene lattice. Large arrows show chiral vectors C_h of zig-zag, armchair, and generic chiral single wall carbon nanotubes (SWCNT). Views in direction of axis in the case of: (**b**) zig-zag, (**c**) armchair, and (**d**) chiral SWCNT.

Viewed in the direction of their axes, SWCNTs can be described as an arrangement of atomic strings (see Figure 1b–d). Let us now examine the scattering of a positively charged particle by an atomic string. A schematic representation of this process is shown in Figure 2a. If the positively charged particle is directed towards the atomic string at a small angle, then it will be reflected back by the correlated series of small angle scatterings on atoms of the string. The particle does not fall under the influence of individual atoms, rather it behaves as if being scattered by the atomic string itself. To deflect the particle trajectory, the potential energy of the atomic string U, at the distance of the closest approach, must be equal to the particle transverse kinetic energy. Lindhard has shown that minimal approach distance is approximately equal to the atom screening radius a_{sc} [5]. Consequently, the maximal incident angle Θ_c, called the critical angle, is defined by relation

$$\Theta_c \approx \sqrt{\frac{U(a_{sc})}{E_k}}, \tag{1}$$

where E_k is the kinetic energy of the incoming particle.

Figure 2b show the schematic representation of the particle bounded motion in the potential of the SWCNT. Such a motion occurs if the angle between SWCNT axis and ion velocity vector in the entrance plane of the nanotube is smaller than Θ_c. The subsequent series of scatterings by atomic strings then gently guides the particle trajectory through the regions of low electron density. At all times an angle measuring the deflection of the particle from the SWCNT axis remains small. The described mode of particle motion is called channeling.

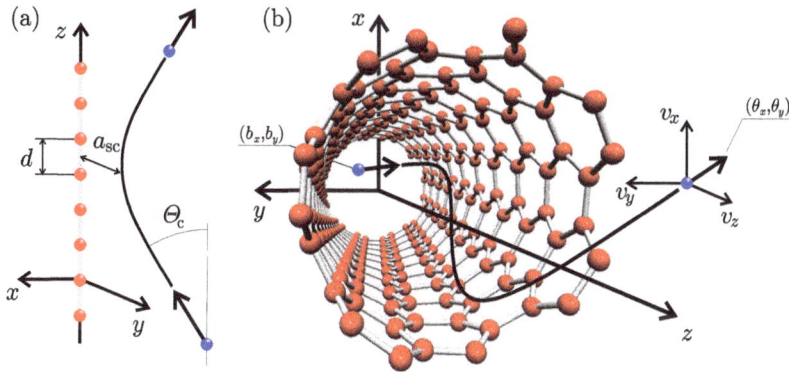

Figure 2. (**a**) Schematic representation of the ion Scattering by the atomic string. The interatomic distance of the string is d. The maximal ion incident angle Θ_c and its minimal approach distance a_{sc} are indicated (**b**) Schematic representation of the ion channeling process. The deflection angles (θ_x, θ_y) at the exit of the SWCNT are smaller than the critical angle.

Nanotubes were discovered in 1991 by Iijima [6]. Soon after their discovery, Klimov and Letokhov demonstrated that SWCNT can be used for the channeling of positively charged particles [7]. The same authors predicted that the motion of channeled particles would generate X-ray and γ-ray radiation [7,8]. A lot of subsequent studies were devoted to the investigation of the possibility to use nanotube for ion guiding and the construction of nanotube based undulators. A good review devoted to particle channeling in the SWCNT can be found in the Ref. [9].

Rainbow scattering occurs if the neighboring sections of the impact parameter plane are scattered to the same section of the scattering angle plane. As a consequence, the differential cross-section becomes infinite along certain lines, called rainbows. The best-known example of the rainbow scattering is the scattering of the light rays by the droplet of water generating the meteorological rainbow [10]. Rainbow scattering happens in nucleus-nucleus collisions [11], elastic scattering [12], electron-molecule collisions [13], particle scattering form the surfaces [14], and ion channeling in crystals [15].

Petrović et al. have shown that the rainbow effect appears also in ion channeling through SWCNTs [16]. It was shown that the theory of rainbows, developed for the explanation of the rainbow channeling in crystals [17], can also be applied for the explanation of the most important features of the rainbow channeling in SWCNTs. A summary of the most important findings of the mentioned group can be found in the Ref. [18].

Besides its theoretical significance, the rainbow effect has a number of practical uses. It was used to extract the correct proton-Si interaction potential [19], and there is also suggestion to be employed for production of the ion beams focused to the subatomic precision [20]. A new method for characterization of the short SWCNTs proposed in the Refs. [21,22] is based on the rainbow effect. It has been shown that the quantum rainbow channeling effect is even more sensitive to the variation of the SWCNT radius [23,24]. Therefore, it is reasonable to expect that quantum rainbow channeling is useful for investigation of the nanotubes and other nanostructured materials.

It should be noted that the classical approach is usually sufficient for the description of the channeling effects of energetic charged particles. Recently, Takabayashi et al. reported on the first experimental observation of the rainbow effect in the planar channeling of 1GeV electrons in the Si crystal [25]. In their experiments, no wave-features of the transmitted electron beam were observed. According to the classical theory, the density of ion trajectories on the rainbow line is infinite. The classical particle density is a strictly additive quantity; therefore, on the rainbow line it is also infinite. However, any particle also behaves as a wave. Due to the interference individual contributions to the wave function amplitude of the wave trains moving approximately along the classical trajectories can be additive or subtractive. The net result is a finite particle density of the rainbow peak, and in a number of additional smaller peaks called supernumerary rainbows. Therefore, even for classical particles it is not possible to understand the true nature of the rainbow effect without quantum mechanics. However, for particles of high energy, the rainbow pattern is so fine that it is difficult to observe it even using detectors of very high resolution.

For light particles whose energies are in the MeV range the quantum description becomes mandatory. Recent theoretical publications of the Kharkov group were devoted to the investigation of the quantum rainbow channeling of electrons in ultra-thin crystals [26,27]. They explained observed wave features as a result of the electron diffraction on the periodic arrangement of the atomic planes or strings. They did not provide any finer classification of the rainbow peaks. Schüller and Winter experimentally observed supernumerary rainbows in scattering of fast atoms by the LiF(001) surface [28]. Their interpretation of the results was based on the semiclassical approach.

The problem of the form of the quantum wave function in the vicinity of the rainbow line and classification of rainbow peaks is well known. It is usually treated in the framework of the semiclassical approach employing uniform approximation [11,29]. However, this approach is applicable only when longitudinal energy is so large that asymptotic approximations are applicable [30]. Another approach is to treat the motion of quantum channeled particles using the formalism of dynamical diffraction [31–34]. However, expanding the incoming wave function in the Bloch state basis is unable to describe the propagation of the evanescent waves, which are shown to be important for the description of the interference in the vicinity of the rainbow line [35]. It is in principle possible to introduce complex interaction potentials which would generate, required imaginary branches of the dispersion relations, but it is difficult to obtain parametrization of such a potential which reproduces observable results. In both approaches, the incoming particle beam is represented as a plane wave. This assumption is perfectly adequate for the description of the wave diffraction. However, there are two main reasons why this is not desirable in particle channeling. Firstly, with simple plane wave it is difficult to model the influence of the beam divergence on the resulting distributions. Angular divergence is extremely important quantity. If it is larger than the critical angle Θ_c then there is no channeling at all. Secondly, the plane wave is infinite, it interacts with the whole sample at the same time. The experimentally proven characteristic feature of the channeling effect is that all physical quantities (such as energy loss, dechanneling probability, etc.) are orientational and impact parameter dependent. This means that one needs to consider fine details of the individual scattered waves. Using the plane wave, one immediately gets a wave describing the interference of all scattered waves and such detailed investigation is impossible. Also, it has been found that the difference between angular distributions of the transmitted parallel positron beam represented as the plane wave and diverging beam represented as an ensemble of wave packets can be large (compare Figure 4 against Figures 7 and 8 of the Ref. [23]).

The simplest way to remedy all the mentioned drawbacks is to represent particles as wave packets and to base the analysis on the explicit solution of the time dependent Schödinger equation. In this report, the transmission of 1-MeV positrons will be examined in detail, when quantum treatment is needed. Initially, quantum particles will be represented as Gaussian wave packets. The corresponding classical problem will be examined in parallel, and both results will be compared and critically examined. We start with a brief review of the classical rainbow channeling theory, and give a short description of the developed model of quantum channeling. Next, we show how to interpret obtained exact solutions using the language of the semiclassical approach. At the end, we present a method for classification of prominent peaks of transmitted distributions relaying only on the information contained in the corresponding quantum amplitude and phase functions families.

2. Theory

In this section, a brief review of the theory of rainbow channeling will be given, and a model of quantum rainbow channeling will be presented. The z axis of the adopted coordinate system is aligned with the axis of the nanotube. The x and y axes are vertical and horizontal axes, respectively.

2.1. Interaction Potential

The primitive vectors of the graphene lattice are denoted as a_1, and a_2. Their lengths are $|a_1| = |a_2| = \sqrt{3}l$, where $l = 0.14$ nm stands for the carbon-carbon bond length. The angle between vectors is $\pi/3$. Chiral vector is defined by expression $C_h = m a_1 + n a_2$. Consequently, the radius of the nanotube is given by the expression

$$R = \frac{|C_h|}{2\pi} = \frac{\sqrt{3}l}{2\pi} \left(m^2 + mn + n^2 \right)^{1/2}, \tag{2}$$

and translational vector of SWCNT unit cell is

$$T = \frac{1}{q_{mn}} \left[(2n + m) a_1 + (2m + n) a_2 \right]. \tag{3}$$

where q_{mn} is the greatest common divisor of $2m + n$ and $2n + m$. The number of atoms in the SWCT unit cell

$$N = \frac{4}{q_{mn}} (m^2 + nm + n^2) \tag{4}$$

is equal to the number of graphene atoms contained in a rectangle defined by vectors C_h, and T. Each atom is the starting point of one atomic string forming circumference of the SWCNT.

We assume that the potential describing charged particle carbon interaction is given by the Molière's expression [36]

$$V(\vec{r}) = \frac{Z_1 Z_2 e^2}{4\pi\varepsilon_0 |\vec{r}|} \sum_{k=1}^{3} \alpha_k \exp\left[-\beta_k \frac{|\vec{r}|}{a_{sc}} \right]. \tag{5}$$

where Z_1, Z_2 are charge state of the incoming particle and carbon atomic number ($Z_2 = 6$), respectively; e is the elementary charge; $\vec{r} = (\bar{x}, \bar{y}, \bar{z}) = r - r_o$ represent the distance vector between positions of the particle $r = (x, y, z)$, and carbon atom $r_o = (x_0, y_0, z_0)$; ε_0 is dielectric permittivity of the vacuum; $a_{sc} = [9\pi^2/(128 Z_2)]^{1/3} a_B$ is Thomas-Fermi screening radius, while a_B is the Bohr's radius; $\alpha = [0.35, 0.55, 0.10]$, and $\beta = [0.1, 1.2, 6.0]$ are Molière's fitting parameters. The channeled particle does not feel the influence of the potential of individual atoms $V(\vec{r})$, rather, its trajectory is influenced by the longitudinally averaged atomic potential of the atomic string

$$U(\vec{\rho}) = \frac{1}{|T|} \int_{-\infty}^{\infty} V(\vec{r}) d\bar{z} = \frac{Z_1 Z_2 e^2}{2\pi\varepsilon_0 |T|} \sum_{k=1}^{3} \alpha_k K_0 \left(\beta_k \frac{|\vec{\rho}|}{a_{sc}} \right), \tag{6}$$

where $\vec{\rho} = (\bar{x}, \bar{y}) = \rho - \rho_o$ is distance vector between transverse positions of particle $\rho = (x, y)$, and atomic string $\rho_o = (x_0, y_0)$; K_0 is modified Bessel function of the second kind and 0-th order [37]. Potential of the SWCNT U_{C_h} at the transverse point ρ is the sum of contributions of all atomic strings located at transverse positions ρ_s ($s = 1, \ldots, N$). It can be shown that the potential U_{C_h} is given by the expression [38]:

$$
\begin{aligned}
U_{C_h}(\rho; R) &= \sum_{s=1}^{N} U(\rho - \rho_s) \\
&= \frac{Z_1 Z_2 e^2 R}{3\sqrt{3}l^2 \varepsilon_0} \sum_{k=1}^{3} \left[U_0^k(\rho; R) + 2 \sum_{\mu=1}^{\infty} U_{\mu N}^k(\rho; R) \cos\left(\frac{\mu N}{2} \Delta\phi\right) \cos\left(\frac{\mu N}{2}(\phi - \Delta\phi)\right) \right],
\end{aligned}
\tag{7}
$$

where ρ and ϕ are coordinates of the vector ρ in polar coordinate system; $\Delta\phi = (m + n)\pi/(2m^2 + 2mn + 2n^2)$; while quantities $U_\nu^k(\rho)$ are defined by expression:

$$
U_\nu^k(\rho; R) =
\begin{cases}
\alpha_k I_\nu\left(\beta_k \dfrac{\rho}{a_{sc}}\right) K_\nu\left(\beta_k \dfrac{R}{a_{sc}}\right), & \text{for } \rho \leq R, \\[2mm]
\alpha_k K_\nu\left(\beta_k \dfrac{\rho}{a_{sc}}\right) I_\nu\left(\beta_k \dfrac{R}{a_{sc}}\right), & \text{for } \rho > R.
\end{cases}
\tag{8}
$$

I_ν, and K_ν are modified Bessel functions of the first and second kind and ν-th order [37]. In channeling, thermal effects are introduced by averaging the static potential U_{C_h} over the distribution of atoms thermal vibrations [38,39]

$$
U_{C_h}^{th}(\rho; R) = \int_{\rho'} P_{th}(\rho - \rho') U_{C_h}(\rho'; R) d\rho',
\tag{9}
$$

where $P_{th}(\rho) = \frac{1}{\sqrt{(2\pi)^2 |\det \Sigma|}} \exp[-\frac{1}{2} \rho^T \cdot \Sigma^{-1} \cdot \rho]$ is distribution of carbon transverse thermal motion; Σ is its associate covariance matrix; while ρ^T is transposed vector.

2.2. Theory of Rainbow Channeling

For simplicity, we assume that the incoming particle beam is monochromatic, perfectly collimated and aligned with the nanotube axis. We also assume that the energy of the particle is sufficiently large so that the energy loss and fluctuation of the scattering angle due to the interaction with SWCNT electrons can be neglected. Once the interaction potential is known, the particle trajectories can be found by solving Newton's equations of motion

$$
m\frac{d^2 r}{dt^2} = -\nabla U_{C_h}^{th}(\rho; R),
\tag{10}
$$

where m is particle mass, t denotes the time, and $\nabla = (\partial_x, \partial_y, \partial_z)$. Appropriate initial conditions are $r(t = 0) = (b, 0)$, and $v = (0, 0, v_z)$; $b = (b_x, b_y)$ is the particle impact parameter. The distribution of the incoming, macroscopic particle beam is uniform on the scale of the nanotube; therefore, the impact parameters b should be random samples form uniform distribution. The Equation (10) shows that the motion of the particle in the longitudinal direction is free. It is inertial motion with constant velocity v_z, while motion in the transverse plane satisfies equation

$$
m\frac{d^2 \rho}{dt^2} = -\nabla U_{C_h}^{th}(\rho; R).
\tag{11}
$$

Therefore, the particle trajectory can be parameterized by a value of the longitudinal coordinate z. At the exit of the SWCNT of length L, trajectory end point determines particle exit transverse position $\rho(L)$ and deflection angle $\theta = (\theta_x, \theta_y)$ (see Figure 2b). Angular and spatial distributions of transmitted particles Y_θ and Y_ρ are constructed by counting the number of particles detected at the specific angle and at the specific position. It should be noted that in principle spatial yield Y_ρ

is measurable. To observe it a position sensitive detector of picometer resolution is required, which still does not exist.

Particle trajectories define two mappings: a mapping of the impact parameter plane to the final transmission position plane $b \rightarrow \rho$, and a mapping of the impact parameter plane to the final transmission angle plane $b \rightarrow \theta$. Since the initial distribution of particles is uniform, the differential cross-sections describing scattering process are defined by the following expressions

$$\sigma_{\text{diff}}^{\rho}(\rho) = \frac{db_x db_y}{dx\, dy} = \frac{1}{|J_r|}, \quad \sigma_{\text{diff}}^{\theta}(\theta) = \frac{db_x db_y}{d\theta_x d\theta_y} = \frac{1}{|J_\theta|}, \tag{12}$$

where J_ρ and J_θ are determinants of Jacobian matrices associated with mappings $b \rightarrow \rho$, and $b \rightarrow \theta$, respectively. Differential cross-sections are infinite whenever the following equations are satisfied.

$$J_\rho(b) = \frac{\partial x}{\partial b_x}\frac{\partial y}{\partial b_y} - \frac{\partial x}{\partial b_y}\frac{\partial y}{\partial b_x} = 0, \quad J_\theta(b) = \frac{\partial \theta_x}{\partial b_x}\frac{\partial \theta_y}{\partial b_y} - \frac{\partial \theta_x}{\partial b_y}\frac{\partial \theta_y}{\partial b_x} = 0. \tag{13}$$

The solutions of Equations (13) form lines in the impact parameter plane, called spatial, and angular impact parameter rainbow lines, respectively. Their images obtained by the application of the corresponding mapping $b \rightarrow \rho$, and $b \rightarrow \theta$ respectively, are also lines, called spatial and angular rainbow lines, respectively. Note that spatial and angular rainbow lines separate areas of different multiplicities of the mappings $\theta \rightarrow b$, and $\rho \rightarrow b$. The side of higher multiplicity is called the bright side of the rainbow, while the rainbow side of lower multiplicity is called the dark side of the rainbow. Thus singularities (i.e., rainbow lines) and multiplicity of mappings $\theta \rightarrow b$ and $\rho \rightarrow b$ dominantly determine the shape of the observable distributions Y_θ and Y_ρ, respectively.

2.3. Model of Quantum Rainbow Channeling

In the quantum approach particles are represented as wave packets Ψ. Evolution of any individual state in the spatial representation satisfy the time-dependent Schrödiner equation.

$$i\hbar \frac{\partial}{\partial t} \Psi(r, t) = \left[-\frac{\hbar^2}{2m} \nabla^2 + U_{Ch}^{th}(\rho; R) \right] \Psi(r, t), \tag{14}$$

Since the particle is free in the z direction, and the initial particle beam is monochromatic, the wave function Ψ must be an eigenstate of the longitude momentum operator $\hat{p}_z$. Therefore, wave function Ψ can be represented in the form

$$\Psi(r, t) = \psi(\rho, t; b) \exp\left[\frac{i}{\hbar}(p_z z - E_k t) \right], \tag{15}$$

where $p_z = \hbar k_z$ is longitude momentum eigenvalue, k_z is longitudinal wave vector, and E_k is initial kinetic energy, while $\psi_b(\rho, t)$ is the transverse part of the wave function associated with the impact parameter b which satisfies the following equation

$$i\hbar \frac{\partial}{\partial t} \psi_b(\rho, t) = \left[-\frac{\hbar^2}{2m} \nabla_\rho^2 + U_{Ch}^{th}(\rho; R) \right] \psi_b(\rho, t), \tag{16}$$

where $\nabla_\rho^2 = \partial_{xx}^2 + \partial_{yy}^2$. The corresponding wave function in the angular representation φ_b is given by expression

$$\varphi_b(\theta, t) = \frac{k_z}{2\pi} \int \psi_b(\rho, t) \exp\left[-ik_z \theta \cdot \rho \right] d\rho^2. \tag{17}$$

Initially, the wave function is represented as Gaussian wave packets

$$\psi_b(\rho, t=0) = \frac{1}{\sqrt{2\pi}\sigma_\rho} \exp\left[-\frac{(\rho-b)^2}{4\sigma_r^2}\right], \quad \varphi_b(\theta, t=0) = \frac{\exp\left[-ik_z\theta \cdot \rho\right]}{\sqrt{2\pi}\sigma_\theta} \exp\left[-\frac{\theta^2}{4\sigma_\theta^2}\right], \quad (18)$$

here σ_ρ and $\sigma_\theta = 1/(2k_z\sigma_\rho)$ are corresponding standard deviations of the probability distributions in spatial and angular representations, respectively. According to the rules of quantum mechanics, spatial and angular yields of transmitted particles are defined by relations

$$Y_\rho(\rho, t) = \sum_b w_b |\psi_b(\rho, t)|^2, \quad Y_\theta(\theta, t) = \sum_b w_b |\varphi_b(\theta, t)|^2. \quad (19)$$

where expansion coefficients w_b satisfy constrain $\sum_b w_b = 1$. We assume that at the entrance plane of the SWCNT the spatial distribution of the incoming beam is uniform, while its angular distribution is Gaussian normal with standard deviation Δ_θ. It is easy to see that $Y_\theta(\theta, t=0) = \frac{1}{2\pi\sigma_\theta^2} \exp[-\theta^2/2\sigma_\theta^2]$; therefore, $\Delta_\theta = \sigma_\theta$. Expansion coefficients w_b should be determined in such a manner that $Y_\rho(\rho, t=0)$, composed of Gaussian distributions of standard deviations $\sigma_\rho = 1/(2k_z\Delta_\theta)$, is constant in the region of the channel.

3. Results

For simplicity in this section we will focus on the channeling through chiral SWCNT. For arbitrary chiral indices m, n, the greatest common divisor q_{mn} is generally small; therefore, according to the Equation (4) the number of atoms in the unit cell of the chiral nanotube N is large. Consequently, a large number of the atomic strings almost uniformly cover the SWCNT circumference making its potential effectively axially symmetric. Note that the general expression for the SWCNT potential (7) represents the Fourier expansion in the polar angle ϕ. The first term in the square bracket U_0^k gives the axially averaged value of the function, and the remaining terms $U_{\mu N/2}^k$ represent amplitudes of the higher harmonics. Using asymptotic formulas for Bessel functions of the large order [40]

$$I_\nu \sim \frac{1}{2\nu\sqrt{\pi}} \left(\frac{ze}{2\nu}\right)^\nu, \quad I_\nu \sim \frac{\pi}{2\nu} \left(\frac{ze}{2\nu}\right)^{-\nu}, \quad (20)$$

it can be shown that $U_{\mu N/2}^k \sim 1/(\mu N)$ which is negligible compared with U_0^k. The general expression for the SWCNT potential given by Equation (7) reduces to:

$$U_{C_h}(\rho; R) = \frac{Z_1 Z_2 e^2 R}{3\sqrt{3}l^2 \epsilon_0} \sum_{k=1}^{3} \begin{cases} \alpha_k I_0\left(\beta_k \dfrac{\rho}{a_{sc}}\right) K_0\left(\beta_k \dfrac{R}{a_{sc}}\right), & \text{for} \quad \rho \leq R, \\ \alpha_k K_0\left(\beta_k \dfrac{\rho}{a_{sc}}\right) I_0\left(\beta_k \dfrac{R}{a_{sc}}\right), & \text{for} \quad \rho > R, \end{cases} \quad (21)$$

which is axially symmetric.

We also assume that distribution of the thermal vibrations $P_{th}(\rho)$ is isotropic, of standard deviation σ_{th}. Since σ_{th} is generally small, it can be shown that thermal averaged potential $U_{C_h}^{th}$ is given by the expression

$$U_{C_h}^{th}(\rho; R) \approx \frac{1}{\sqrt{2\pi}\sigma_{th}} \int_{R-6\sigma_{th}}^{R+6\sigma_{th}} U_{C_h}^{th}(\rho; R') \exp\left[-\frac{(R-R')^2}{2\sigma_{th}^2}\right] dR' \quad (22)$$

which represents an average of the potential U_{C_h} over the distribution of thermally induced changes of the SWCNT radius. The critical channeling angle for chiral SWCNT is $\Theta_c = (U_{C_h}^{th}(R - a_{sc}); R)/E_k)^{1/2}$

Axial symmetry of the potential considerably simplifies the finding of trajectories because particle motion in the polar direction is uncoupled from the motion in the radial directions. In the case of the motion of quantum particles, such separation is impossible. However, axial symmetry can be used to reduce the number of considered impact parameters, since wave functions for rotationally equivalent impact parameters can be generated by the application of the rotation operator.

In the first subsection, the manifestations of the classical rainbow effect will be explained on the example of the proton channeling in SWCNT. Subsequent subsections will be devoted to the analysis of the quantum rainbow channeling of positrons.

3.1. Interpretation of the Classical Rainbow Effect

Here we consider the transmission of the parallel, monochromatic, 1-GeV proton beam through SWCNT (11, 9), which is perfectly aligned with nanotube axis. The radius of the SWCNT is $R = 0.689$ nm, and the number of its atomic strings is $N = 1204$. Standard deviation of the carbon thermal motion σ_{th} can be estimated from the Debye theory, which for the room temperature (T = 300 K) gives $\sigma_{th} = 0.005$ nm. Screening length of carbon atom is $a_{sc} = 0.026$ nm.

The motion of the protons in the longitudinal direction is relativistic while its motion in the transverse direction is classical. The equations of motion (11) still hold. The only differences are that m should be replaced by the protons relativistic mass m_r ($m_r/m = 2.066$), and the relationship between initial kinetic energy E_k and longitudinal linear momentum p_z is $p_z^2 c^2 = E_k^2 + 2mc^2 E_k$, where c is the speed of light [9]. The relativistically corrected critical angle is $\Theta_c = \sqrt{\frac{2mc^2 + 2E_k}{2mc^2 + E_k}} \Theta_c = 0.268$ mrad.

We consider only protons whose impact parameters satisfy inequality $|b| \leq R - a_{sc}$. Since the proton beam is aligned with the SWCNT axes, each proton trajectory is confined to a plane defined by the impact parameter b and the nanotube axis. Figure 3 shows the obtained proton trajectories in the $x0z$ plane. Newton's equations of motion (11) were solved by Runge-Kutta method of the 4-th order [41]. Note that the amplitude of any proton trajectory is constant and corresponds to its impact parameter b_x. The corresponding trajectories in the angular space are shown in Figure 3b. The amplitude of any trajectory is also constant. Note that the maximal deflection angle of Θ_c corresponds to the trajectory of impact parameter $b_x = R - a_{sc}$. All these facts demonstrate that proton trajectories were accurately calculated.

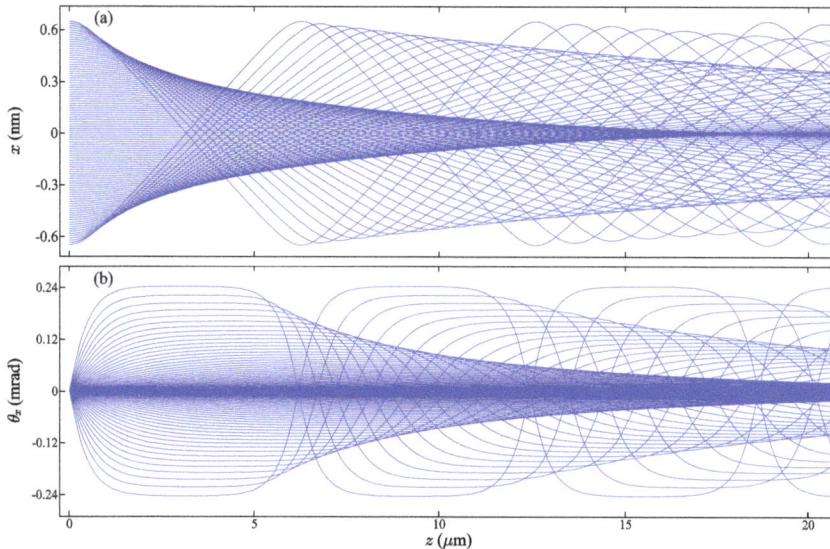

Figure 3. (a) Family of 1-GeV proton trajectories in the $x0z$ plane of the SWCNT (11,9). (b) Corresponding family of trajectories in the angular space.

   Proton trajectories shown in Figure 3a,b can be concisely represented as function $x(z; b_x)$, $\theta_x(z; b_x)$, depending on the parameter b_x. For SWCNT of length L functions $x(L; b_x)$, $\theta_x(L; b_x)$ define mappings of the impact parameter plane to the final transmitted position and final transmitted angle plane, called the spatial and angular transmission function, respectively, which will be denoted as $X(b_x)$ and $\Theta_x(b_x)$. The symmetry of the trajectory family requires that both transmission functions are odd functions $X(b_x) = -X(-b_x)$, $\Theta_x(b_x) = -\Theta_x(-b_x)$. Rainbow defining condition (13) now reduces to

$$\frac{dX(b_x)}{db_x} = 0, \quad \frac{d\Theta_x(b_x)}{db_x} = 0. \tag{23}$$

   Therefore, the critical points of transmission functions, which occur in symmetrical pairs, are rainbow points. Each critical point pair corresponds to the one circular rainbow line whose radius is equal to the absolute value of the critical point ordinate.

   For the 10-μm long SWCNT (11, 9) angular transmission function $\Theta_x(b)$, shown in Figure 4a, has only one critical point pair $(-b_x^{(1)}, \theta_x^{(1)}) = (-0.493 \text{ nm}, 0.856 \text{ mrad})$, and $(b_x^{(1)}, -\theta_x^{(1)}) = (0.493 \text{ nm}, -0.856 \text{ mrad})$, both labeled 1. This function shows that for any θ_x from the interval $(-\theta_x^{(1)}, \theta_x^{(1)})$ mrad, there are three corresponding impact parameters, while outside of this interval the correspondence is one-to-one. Therefore, there is only one circular angular rainbow line of radius 0.856 mrad. The interior of the line is the rainbow's bright side, while its exterior is the rainbow's dark side.

   The vertical slice through the corresponding angular distribution is shown in Figure 4b. The initial number of protons was 16,655,140, while the size of the bin in the θ_x space was 0.866 μrad. Note the small statistical fluctuation of the obtained distribution which reflects the randomness of the impact parameter selection process. Besides this the obtained distribution is perfectly axially symmetric. This distribution contains three prominent peaks. The central peak is the consequence of the fact that potential $U_{C_h}^{th}$ has its minimum at the coordinate origin. It represents the undeflected part of the proton beam, and it is not related to the rainbow effect. The two remaining peaks are located symmetrically around the central maximum. Angular positions of the critical points form Figure 4a are indicated by the number 1. It is obvious that their positions are in the perfect correspondence with the positions of the mentioned peaks. Note also high particle yield inside, and low yield outside the interval enclosed by the observed peak pair, which corresponds to the rainbow light and dark sides. Therefore, the angular distribution contains one circular rainbow line whose properties are determined by the critical points of the transmission function.

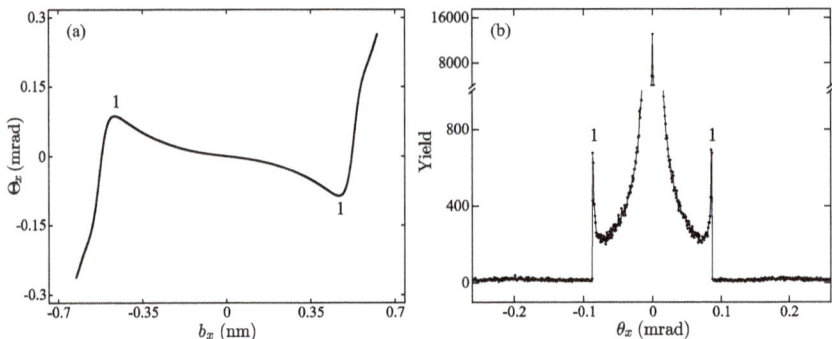

Figure 4. (a) Angular transmission function $\Theta_x(b)$ of 1-GeV protons transmitted through 10 μm long SWCNT (11,9). **(b)** The vertical slice through the corresponding angular distribution. All rainbow points are numbered, equivalent rainbow points are designated by the same number.

3.2. Semi-Classical Interpretation of Quantum Rainbow Effect

This subsection is devoted to the analysis of the channeling of 1-MeV positrons through 200-nm long chiral SWCNT (14, 4). The radius of this nanotube is $R = 0.650$ nm, while the number of atomic strings is $N = 536$. Longitudinal motion of 1-MeV positrons is free and relativistic ($m_r/m = 2.957$) while transverse motion is quantum and nonrelativistic. As in the previous example, Equation (16) still holds if bear mass m is replaced with relativistic mass m_r [9]. To observe channeling effect one need to use positron beam collimated better then critical channeling angle Θ_c. Here we assume that the incoming positron beam has angular standard deviation $\Delta_\theta = 0.1\Theta_c$, which gives $\sigma_\theta = 0.735$ mrad and $\sigma_\rho = 0.134$ nm. This value was selected because then transverse size of any wave packet is large enough that the self-interference effect becomes significant, while on the other hand it is small enough to allow explicit dependence of their dynamics on impact parameters to be analyzed. Let M represent the number of wave packets uniformly covering impact parameter plane. Expansion coefficients from Equation (19) are $w_b = 1/M$. In order that $Y_\rho(\rho, 0)$ be a uniform distribution in the entrance plane of the nanotube, the number of wave packets M must be very large (theoretically infinite). In order to minimize the number M an algorithm was devised which optimize the values of the coefficients w_b while keeping the difference between the distribution Y_θ and uniform distribution $1/(\pi(R - a_{sc})^2)$ in the region $\rho \leq R - a_{sc}$ below some prescribed tolerance. We have found that accurate representation of the initial distribution of the positron beam can be accomplished with only 142 Gaussian wave packets.

Let us examine the motion of the wave packet of impact parameter $b = (0.624, 0)$ nm. The corresponding Schrödinger Equation (16) is solved using the method of Chebyshev global propagation [42]. The obtained probability densities at the exit of the SWCNT are shown in Figure 5. In both representations, densities have a number of peaks, which can be attributed to self-interference of the incoming part of wave function with the part of wave function already reflected from SWCNT wall. However, since all peaks are the consequence of the wave packet self-interference looking only on the numerically obtained probability densities, it is very difficult to say which peak is connected with the rainbow effect, and which one is a simple manifestation of the positron wave nature.

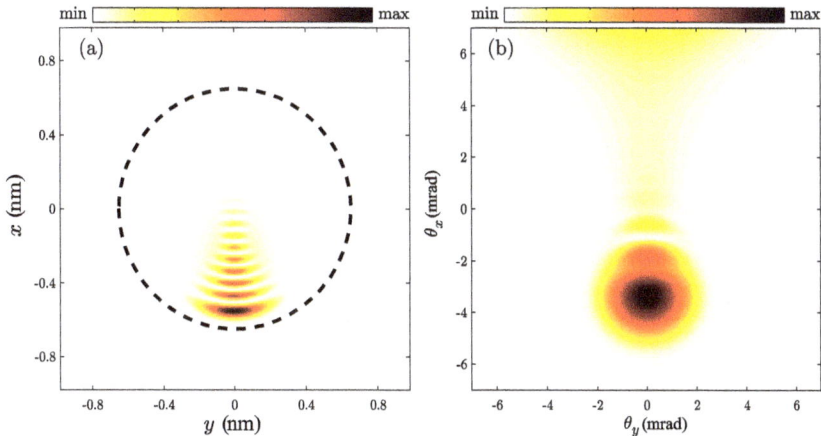

Figure 5. The probability density of the wave-packet of impact parameter $b = (0.624, 0)$ nm, in the logarithm scale, at the exit of 200-nm long chiral SWCNT (14,4) in (**a**) spatial, and (**b**) angular representation, respectively. The thick dashed line represents the SWCNT boundary.

In order to understand and classify self-interference peaks, a semiclassical approach can be applied. To avoid unnecessary complications here we will focus only on the vertical slices of the wave packets moving along the x axis. Transmission functions $X(b_x)$ and $\Theta_x(b_x)$ then fully characterize the classical motion of the particle. In the phase space, those two functions define a curve called the rainbow diagram which is shown in Figure 6a as a thin black line. The transmission function $X(b_x)$ has two pairs of critical points labeled 1_s and 2_s whose ordinates are ± 0.14 nm, and ± 0.59 nm, respectively, while transmission function $\Theta_x(b_x)$ has only one pair of critical points labeled 1_a whose ordinates are ± 4.97 mrad. The positions of the rainbow points in the rainbow diagram are indicated by the black arrows. Note that at rainbow points, the tangents of the rainbow diagram are vertical or horizontal. They represent points where different branches of the mappings $\Theta_x(X)$, and $X(\Theta_x)$, respectively, meet. Therefore, mapping $\Theta_x(X)$ has 5 branches, while mapping $X(\Theta_x)$ has only 3.

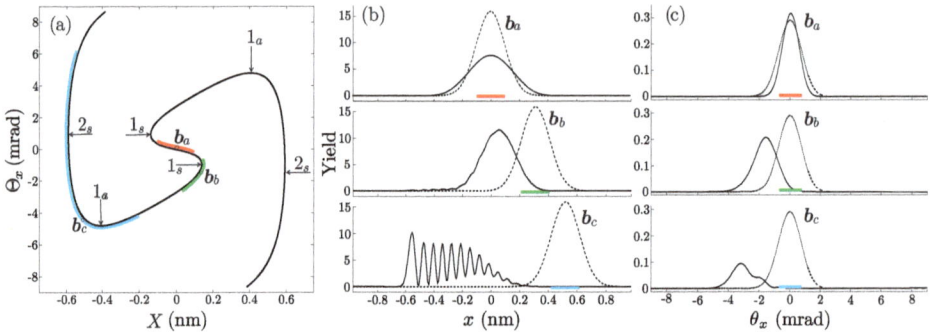

Figure 6. (**a**) Rainbow diagram of 1-MeV protons transmitted through 200-nm long SWCNT (14,4). Arrows show positions of the classical rainbow points. Motion of the wave packets having impact parameters $b_a = (0,0)$, $b_b = (0.31, 0)$, and $b_c = (0.52, 0)$ nm, respectively, in the: (**b**) spatial, and (**c**) angular representations. Initial wave packets are shown by the dashed lines while final wave functions are shown by the solid lines. Intervals containing trajectories giving dominant contribution to wave packets are denoted by the thick red, green, and blue lines, respectively.

In the semiclassical approach the quantum wave function at the time t can be constructed from classical rainbow diagram as a sum of the contributions of the wave trains coming from all branches [43]. The contribution of the branch μ is of the form $\rho_\mu^{1/2} \exp[iS_\mu/\hbar]$. Let density of the trajectories in the impact parameter plane be $K(b)$. The number of particles in the interval dx around x coming from the branch μ is equal to the number of particles in the interval db_x around the point b_x mapped to the corresponding interval. Therefore, $\rho_\mu dx = K_\mu(b_x)db_x$, where $K_\mu(b_x)$ is the density of the points in the branch μ. The phase S_μ is a type 2 canonical transformation $dS_\mu = \hbar k_z \theta_x dx$ in the spatial representation and type 1 canonical transformation $dS_\mu = \hbar k_z x d\theta_x$ in the angular representation, of the branch μ. Therefore, the total semiclassical wave functions in spatial and angular representations are given by the expressions

$$\psi_{sc}(x) = \sum_\mu \sqrt{K_\mu} \left| \frac{db_x}{dX_\mu} \right|^{\frac{1}{2}} \exp\left[ik_z \int_{X_\mu^{(0)}}^x \Theta_x^\mu(X')dX' \right],$$

$$\varphi_{sc}(\theta_x) = \sum_\nu \sqrt{K_\nu} \left| \frac{db_x}{d\Theta_x^\nu} \right|^{\frac{1}{2}} \exp\left[ik_z \int_{\Theta_\nu^{(0)}}^{\theta_x} X_\nu(\Theta_x')d\Theta_x' \right],$$

(24)

where indices μ, and ν count branches of the spatial and angular transmission functions $X_\mu(\Theta_x)$, and $\Theta_x^\nu(X)$, respectively, $X_\mu^{(0)}$, and $\Theta_\nu^{(0)}$ are referent points of the μ-th and ν-th branch respectively. Note that $\Theta_x^\mu(X)$ stands for the inverse function of the branch $X_\mu(\Theta_x)$ of spatial transmission function, while $X_\nu(\Theta_x)$ is inverse function of the branch $\Theta_x^\nu(X)$ of the angular transmission function. In the spatial representation the sum should be taken over all branches satisfying equation $X_\mu = x$, while in the angular representation the sum is over all branches satisfying equation $\Theta_x^\nu = \theta_x$.

Let us now apply semiclassical reasoning for interpretation of the quantum motion of wave packets labeled *a*, *b* and *c* of impact parameters $b_a = (0, 0)$, $b_b = (0.31, 0)$, and $b_c = (0.52, 0)$ nm, respectively. Their spatial, and angular representations are shown in Figsure 6b and c respectively. Wave packets in the impact parameter plane are shown by the dashed lines, wave packets in the exit plane of the SWCNT are shown by the solid line. Since the initial distributions are Gaussian, the dominant contribution comes from trajectories from the interval $[b_x - \sigma_\rho, b_x + \sigma_\rho]$. For reasons of simplicity, the contributions of all other trajectories will be neglected. The dominant intervals are in Figure 6b shown by the red, green, and blue lines, respectively. In the angular space, the dominant intervals of length of $2\sigma_\theta$ are in Figure 6c designated by the same colors. The corresponding exit positions of the trajectories from the dominant intervals are in Figure 6a indicated by thick red, green and blue lines, respectively. For wave packer *a*, the red curve in Figure 6a covers only one branch of the mappings $X(\Theta_x)$ and $\Theta_x(X)$, respectively. This is the reason why for wave packet *a* self-interference is not observable. For wave packet *b* the green line in Figure 6a covers one branch of mapping $X(\Theta_x)$ completely and slightly extends into the second branch, and covers only one branch of the mapping $\Theta_x(X)$. Therefore, the number of trajectories which interfere is small. This explains why in the spatial representation for wave packet *b* only weak self-interference can be observed, and there is no observable self-interference in the angular representation. For wave packet *c* the blue curve in Figure 6a covers two branches of mappings $X(\Theta_x)$ and $\Theta_x(X)$, respectively. In this case, the number of trajectories which interfere is large. This fact explains why self-interference is the strongest for wave packet *c*.

Let us now examine more closely shape of the classical, the semi-classical probability density of the wave packet *c* in the vicinity of the point 2_s, and compare it with the exact solution given on Figure 6b. The spatial transmission function $X(b_x)$ is shown in Figure 7a has a minimum labeled 2_s at the point at (b_{2_s}, x_{2_s}). It ends at the point *e* which correspond to maximal possible considered impact parameter $b_e = R - a_{sc}$. The inverse transmission mapping $b_x(x)$ have two branches labeled X_1 and X_2, respectively. The branch X_1 is formed by the end positions of positrons having impact parameters $b_x < b_{2_s}$, while positrons ending on the branch X_2 have impact parameters belong to the interval $[b_{2_s}, b_e]$. Therefore, for $x < x_{2_s}$ the mapping $b_x(x)$ is zero-valued for *x* in the interval $[x_{2_s}, x_e]$, the mapping is double-valued, while for $x > x_e$ it is single-valued. The classical probability density is defined by the equation

$$\rho(x) = \left\{ K_1(b_x(X_1)) \left| \frac{db_x}{dX_1} \right| \right\}_{X_1 = x} + \left\{ K_2(b_x(X_2)) \left| \frac{db_x}{dX_2} \right| \right\}_{X_2 = x}. \tag{25}$$

For an interval of impact parameters that is considered to be small from Figure 7a, function $K(b_x)$ can be approximated with a constant (this also means that $K_1(b_x) = K_2(b_x) = K$). The resulting normed distribution $\rho(x)$ is in Figure 7c shown by the black line. Since, $\frac{dX_1}{db_x} = \frac{dX_2}{db_x} = 0$ for $b_x = b_{2_s}$ both branches give singular contributions to the function ρ which is infinite at the rainbow point x_{2_s}. For $x < x_{2_s}$ density $\rho(x) = 0$, therefore this region is the dark side of the rainbow. For *x* in the interval $[x_{2_s}, x_e]$ the function $\rho(x)$ is monotonously decreasing. Note an abrupt jump of the function $\rho(x)$ at $x = x_e^+$ which is consequence of the change of the multiplicity of the mapping $b_x(x)$ from 2 to 1.

The exact normed probability density $|\psi_c(x)|^2$ is shown in the Figure 7c by the blue line. Comparison of these two solutions shows the following. The classical approximation correctly predicts overall order of the magnitude of the exact probability density $|\psi_c(x)|^2$. It predicts that the largest contribution to the density is at the rainbow point x_{2_s}, which is very close to the largest peak of the function $|\psi_c(x)|^2$, and that density is low for $x < x_{2_s}$ and high for $x \geq x_{2_s}$. Out of many peaks of the density $|\psi_c(x)|^2$, as is clearly visible on Figures 6b and 7c, the classical approximation explains the existence of only one, and clearly overestimates its amplitude.

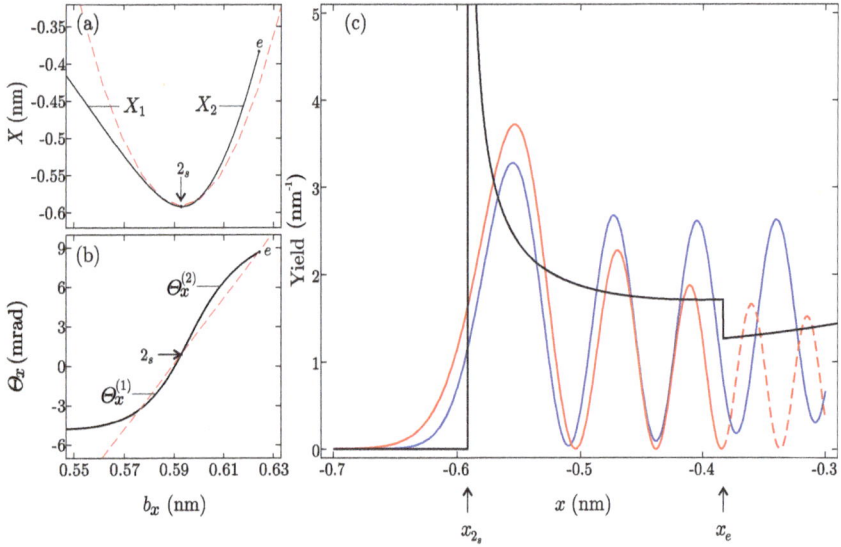

Figure 7. The spatial (**a**) and the angular (**b**) transmission function of the 1-MeV positrons transmitted through 200 nm long SWCNT (14, 4) in the vicinity of the spatial rainbow point 2_s. (**c**) The classical, the semiclassical, and the exact normed probability density shown by the blue, the red and black line respectively.

According to the Equation (24) to construct semiclassical solution $\psi_{sc}(x)$ one need to find all solution of the equation $X = x$, form semiclassical waves and sum their contributions. For example, if $x \in [x_{2_s}, x_e]$ amplitudes of the individual semiclassical waves are $(K\frac{db_x}{dX_1})^{1/2}$ and $(K\frac{db_x}{dX_2})^{1/2}$. To obtain phases of the semiclassical waves one need to consider also angular transmission function $\Theta_x(b_x)$ shown in Figure 7b. Since the mapping $b_x(X)$ is two-valued in the considered interval, the mapping $\Theta_x(b_x(X))$ has also two branches $\Theta_x^{(1)}(b_x(X_1))$, and $\Theta_x^{(2)}(b_x(X_2))$. Phases of the semiclassical waves are solutions of the equations $\frac{dS_1}{dX_1} = k_z\Theta_x^{(1)}(X_1)$, and $\frac{dS_2}{dX_2} = k_z\Theta_x^{(2)}(X_2)$.

Note that direct evaluation of the Equation (24) is not possible at $x = x_{2_s}$ since both wave amplitudes $(K\frac{db_x}{dX})^{1/2}$ diverge. To circumvent this limitation one can use the so-called transitional approximation [11,29]. Firstly, the spatial and the angular transmission functions are in the vicinity of the rainbow point 2_s approximated by the following polynomials

$$X(b_x) = \alpha_2 b^2 + \alpha_1 b + \alpha_0, \quad \Theta_x(b_x) = \beta_1 b + \beta_0. \tag{26}$$

Obtained approximations are in Figure 7a,b shown by the dashed red lines. By the analytical continuation validity of the equation $X = x$ is extended to the whole complex plane. Therefore, for $x \in [x_{2_s}, x_e]$ equation $X = x$ has two real solutions, for $x = x_{2_s}$, the equation has a double root, while for $x < x_{2_s}$ the equation has two conjugate complex solutions. Taking into account additional complex solutions it can be shown that transitional semiclassical density $|\psi_{sc}^{(t)}(x)|^2$ is given by the equation

$$|\psi_{sc}^{(t)}(x)|^2 = \frac{\beta_1^{1/3}k_z^{1/3}K}{2\pi\alpha_2^{2/3}} \left| \text{Ai} \left(\frac{\beta_1^{1/3}k_z^{1/3}}{\alpha_2^{1/3}}(x - x_{2_s}) \right) \right|^2, \tag{27}$$

where Ai is the Airy function [37]. The obtained semiclassical distribution $|\psi_{sc}^{(t)}|^2$ is in Figure 7c shown by the red line. The largest maxima which is now finite of the function $|\psi_{sc}^{(t)}|^2$ is the closest to the classical rainbow 2_s. For $x < x_{2_s}$ interference of the complex rays make density $|\psi_{sc}^{(t)}|^2$ exponentially decaying. For $x \geq x_{2_s}$ a number of smaller peaks can be observed which appear due to the constructive interference of the real rays.

Note that almost identical expression describe semiclassical intensity of the light in the vicinity of the optical rainbow [44–46]. Therefore, peaks of the function $|\psi_{sc}^{(t)}|^2$ can be classified in analogues way as interference peaks of the optical rainbow. The large maximum closest to the position of the classical rainbow is considered to be the primary rainbow maximum, while all other peaks are supernumeraries associated with the observed primary.

Comparison of the red and the blue curve from Figure 7c reveals that transitional semiclassical approximation almost perfectly predicts the position and size of the dominant peak of the exact distribution $|\psi_c|^2$. The constructive interference of real rays explains the existence of all other maxima, while the interference of the complex rays explains how probability density $|\psi_c|^2$ can have non-zero values in the region where there are no real rays at all. It could be said that the ray interference "assuages" the sharpness of the classical distribution $\rho(x)$. Therefore, the semiclassical approximation captures all qualitative features of the quantum rainbow scattering effect. However, its validity is limited. Note that accuracy of the semiclassical solution $|\psi_{sc}^{(t)}|^2$ actually decreases for $x > x_{2_s}$. Its range of validity is limited only to the region $x \leq x_e$. The reason for this is that in the vicinity of the point $x = x_e$ the multiplicity of the inverse transmission function $b_x(X)$ changes by one, while number of real roots of any polynomial approximation of the $X(b_x)$ can change only by an integer multiple of 2. This is the reason why semiclassical density $|\psi_{sc}^{(t)}|^2$ for $x > x_e$ in Figure 7c is shown by the dashed red line. For $x > x_e$ the correct semiclassical wave function is given by only one semiclassical wave. According to the Equation (24) semiclassical density is then $|\psi_{sc}(x)|^2 = K\frac{db_x}{dX_1}$, i.e., it is equal to the classical solution $\rho(x)$. Therefore, the semiclassical approximation is not capable of explaining the existence of the peaks of the exact density $|\psi_c|^2$ in the region where transmission function has only one branch.

Summing contributions of all wave packets according to Equation (19) gives distributions of the transmitted positron beam at the exit plane of the SWCNT. The vertical slices through the obtained distributions in spatial and angular space are shown in Figure 8a,b, respectively. Each pair of symmetric maxima visible in the Figure 8a,b corresponds to the circular maxima of the 2D distribution. Positions of the classical rainbow point pairs 1_s, 2_s, and 1_a are indicated by the arrows. Large maxima closest to the classical rainbow points, labeled 1_s^{qu}, and 2_s^{qu} in the Figure 8a at $x_{1s}^{qu} = \pm 0.55$ nm and $x_{2s}^{qu} = \pm 0.14$ nm are interpreted as the primary and the secondary rainbow point. All other peaks are considered to be supernumerary rainbows. In the case of angular distribution, the large maximum pair 1_a^{qu} at $\theta_{1a}^{qu} = \pm 3.27$ mrad, closest to the classical rainbow peaks 1_a, are interpreted as the primary rainbow points, all remaining peaks are considered to be supernumerary rainbows.

3.3. Morphological Interpretation of Quantum Rainbow Effect

The developed method was applied for classification of angular distributions of 1-MeV positrons transmitted through 200-nm long SWCNTs $(7, 3)$, $(8, 5)$, $(9, 7)$, $(14, 4)$, $(16, 5)$, and $(17, 7)$. The nanotubes considered here are the easiest to produce by the arch discharge method [1]. The radii of the considered nanotubes are in the range from 0.35 to 0.85 nm. The classical critical angles for the considered SWCNTs are very close to each other, ranging from 7.3 to 7.4 mrad. For all SWCNT we assume that initial angular standard deviation of the positron beam Δ_θ is always equal to the 10% of the corresponding classical critical angle Θ_c. This means that the used angular standard deviations of the positron beams are in the range [0.73, 0.74] mrad, the corresponding range of the standard deviations of the Gaussian wave packets are in range [0.094, 0.095] nm. Note that all wave packets have almost the same size.

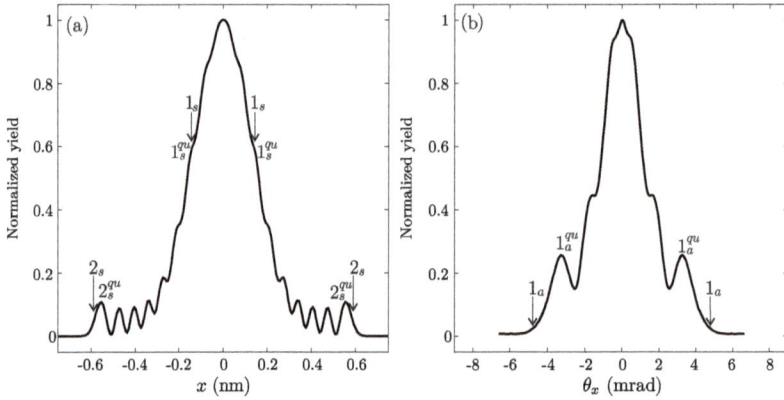

Figure 8. (**a**) The spatial distribution of the 1-MeV positron beam transmitted through SWCNT (14, 4). (**b**) The corresponding angular distribution. Positions of the classical rainbow lines are shown by the arrows.

Vertical slices through obtained distributions are shown in Figure 9. Initial spatial distributions were constructed using: 43, 71, 101, 141, 190, 241 Gaussians, respectively. All prominent peaks excluding central are labeled by numbers, with symmetrical maxima labeled by the same number. Numeration always starts from the outermost rainbow pair. In all analyzed cases there is only one classical rainbow point pair labeled 1'. Their positions in Figure 9 are shown by the arrows. It is clear that the quantum peak pairs labeled 1 in Figure 9a,d–f are quantum primary rainbow points. All other peaks are supernumerary rainbow.

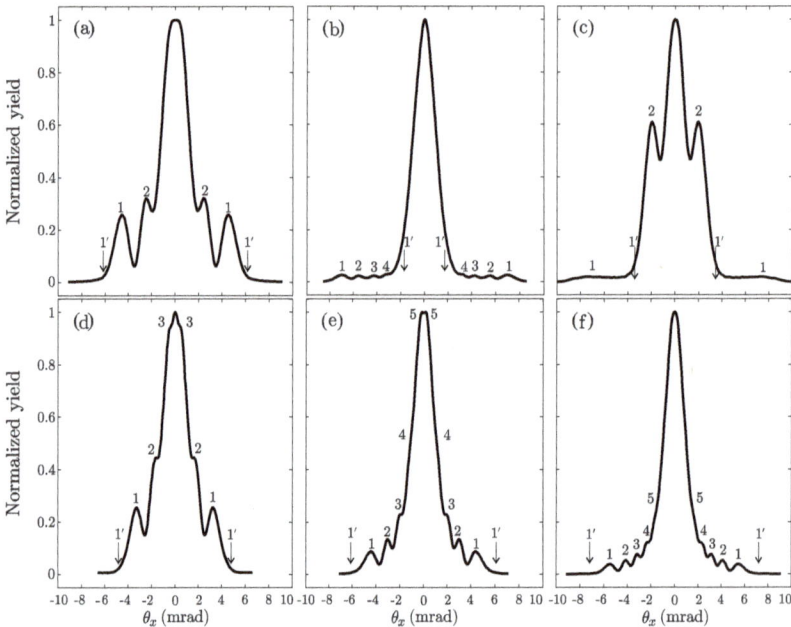

Figure 9. Angular distributions along θ_x axis for 1-MeV positrons transmitted through 200-nm long SWCNT: (**a**) (7, 3); (**b**) (8, 5); (**c**) (9, 7); (**d**) (14, 4); (**e**) (16, 5); and (**f**) (17, 7). Arrows show positions of maxima in the corresponding classical angular distributions. Symmetrical maxima are designated by the same number.

It should be noted that the semiclassical classification of the quantum peaks for the distributions from Figure 9b,c is ambiguous. This is not surprising since the applicability of the semiclassical approach is limited. Rather, it is surprising that the semiclassical interpretation works at all. It should be noted that accurate wave functions differ considerably from their semiclassical counterparts. If we take a closer look at the wave packet b_c shown in Figure 6b,c, the interference is clearly visible in the interval from −0.6 nm up to 0.2 nm in the spatial representation, and in interval from −5 mrad up to the −1 mrad. However, the corresponding relevant parts of the transmission functions are two-valued only in the intervals $[-0.6, -0.5]$ nm, and $[-6, -4]$ mrad, respectively. Outside the mentioned intervals the transmission functions are single-valued, and there should be no observable interference effects. Therefore, the semiclassical wave functions drastically underestimate real self-interference of the wave packets.

The problem with the semiclassical interpretation is that it intrinsically relay on classical concepts (such as exact position of the particle in the phase space) which do not have direct quantum analogue. A alternative approach would be to try to link the rainbow effect to certain morphological properties of the family of the classical trajectories, and quantum amplitude and phase function families. If morphological properties were found to be equal then both approaches are merely two descriptions of the same physical reality.

In this subsection it will be shown that it is possible unambiguously to classify quantum peaks relaying only on the information contained in the quantum amplitude and phase function families. In order to show that let us examine channeling of 1-MeV positrons through 400-nm long SWCNT (11, 9). The classical critical angle is $\Theta_c = 7.3$ mrad. The obtained trajectory family in xOz plane is shown in Figure 10a. The striking features of this family are three pairs of envelope lines, labeled c_1, c_2, and c_3, which are defined as a limiting line formed from intersections points of the neighboring family members [47]. The mathematical envelope is defined as a set of solutions of the equation

$$\frac{\partial}{\partial b_x} x(z; b_x) = 0. \tag{28}$$

Equation (28) shows that each envelope is a locus of one critical point of the transmission function, i.e., the envelope is caustic line of the trajectory family [48]. For example, for a nanotube of length $L = 150$ nm the spatial transmission function has one symmetrical pair of critical points whose ordinates $x_1^s = \pm 0.23$ nm are equal to the positions of envelope points $\pm 1^s$ in Figure 10a. Another important quantity is the Hamilton's principal function defined by the equation [49]

$$\frac{\partial}{\partial x} S(x) = \hbar k_z \theta_x, \tag{29}$$

which is directly related to the phase function of the quantum wave packet [50]. Hamilton's principal functions for nanotube whose length is $L = 150$ nm is shown in Figure 10b. It is multivalued singular curve composed of three branches. The caustic lines are also loci of singularities of the Hamilton's principal function [48]. Its cusped-like singular points, locally isomorphic to the cusp catastrophe [51], are also labeled $\pm 1^s$. Therefore, envelope lines and singularities of the Hamilton's principal function are inextricably linked to the manifestations of the rainbow effect.

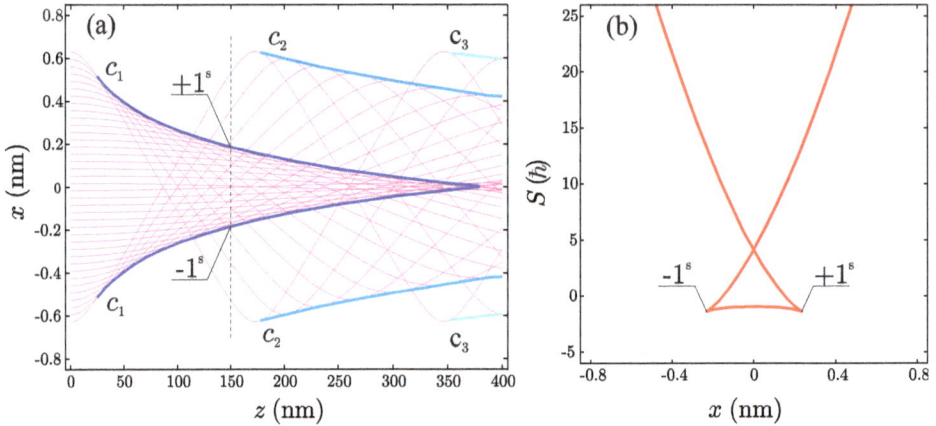

Figure 10. (**a**) Trajectories of the 1MeV positrons (magenta lines) in the xOz plane and associated envelope lines (blue hue lines). (**b**) Corresponding Hamilton's principal function at the $z = 150$ nm.

Now let us apply the same logic for the explanation of the quantum rainbow channeling. Figure 11a shows spatial distribution of the transmitted positron obtained assuming that initially its divergence was $\Delta_\theta = 0.1\Theta_c$. Spatial and angular standard deviations of the wave packets were $\sigma_\rho = 0.19$ nm and $\sigma_\theta = 0.73$ mrad. The vertical slice through the obtained spatial distribution of the positron beam, transmitted through 150-nm long SWCNT $(11, 9)$, is shown in Figure 11a. It consists of a large central peak which shows no signs of any internal structure, and six pairs of smaller peaks. The largest is the outermost peak pair, while remaining peaks are of approximately the same size. We need to provide a classification of the observed behaviour, and an explanation for its formation.

Since the obtained spatial distribution is axially symmetric we have focused only on the motion of wave packets with impact parameters belong to the nanotube vertical cross-section. We have followed evolution of 301 wave packets. Vertical cross-sections through obtained probability densities parameterized by the impact parameter are shown in Figure 11b. This distribution is dominated by two large maxima labeled $\pm 1^{m0}$ at $x^{m0} = \pm 0.17$ nm. Therefore, the dominant contribution to the large central peak in the Figure 11a actually comes from two smaller maxima. During their evolution, the wave packets become wrinkled (for example see Figure 5, or Figure 6b,c). This is the manifestation of the wave packet self-interference caused by the interaction with the nanotube walls. The wave packet ensemble shown in Figure 11b can be separated in two parts. The first one is formed by the wave packets showing no observable wrinkling (i.e., for $|b_x| < 0.33$ nm). The second subensemble is formed by the wave packets for which self-interference is considerable (i.e., for $|b_x| \geq 0.33$ nm), which is called the rainbow subensemble. Members of these two subensembles are separated by the magenta lines in Figure 11b. Note the formation of the vertical yellow stripes which occur for the wave packets of impact parameters approximately in the range $|b_x| \geq 0.50$ nm. This means that the wave packets wrinkle in the mutually coordinated way. The corresponding inverse classical spatial transmission function is shown by the thick black line. Positions of the classical rainbow points, labeled $\pm 1^s$, are also indicated. Quantum probability density is concentrated around the classical line. It behaves as if there is a virtual barrier preventing spreading of the wave packets in the areas beyond the line.

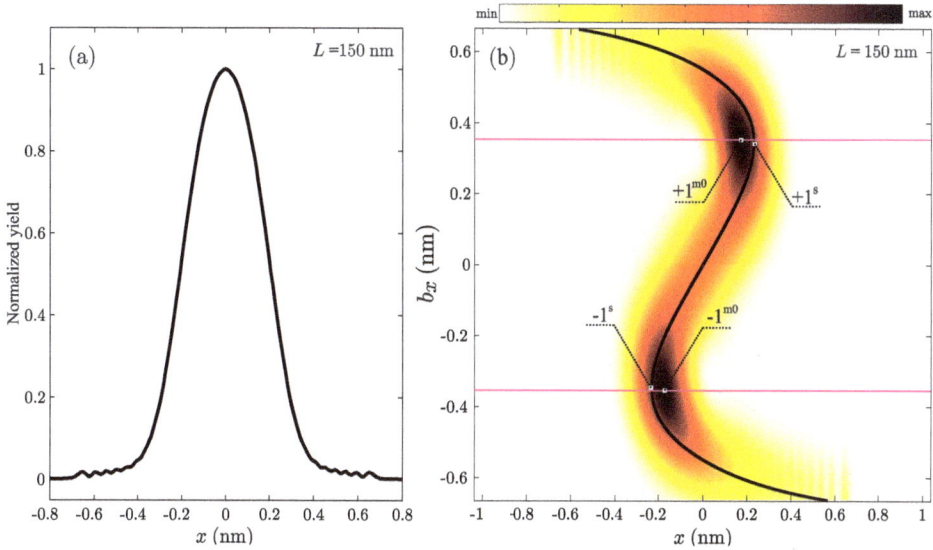

Figure 11. (**a**) Vertical slice through spatial probability density of 1-MeV positron beam in the exit plane of SWCNT (11, 9). (**b**) Corresponding slices through individual wave packets. Magenta lines separate wave packets belonging to the rainbow subensemble. The thick black line shows the corresponding inverse spatial transmission function.

It should be noted that the observed behavior is unexplainable by the semiclassical approach. For example, for $b_x > 0.33$ nm the wave packet wrinkling is the most noticeable in rainbow subensemble and in the region $x < 0.4$ nm, where the inverse transmission function is single valued. On the other hand the semiclassical approximation predicts that the most intense self-interference should be in the regions close to the classical rainbow points $\pm 1^s$, where no wave packet wrinkling can be observed.

Wrinkling, concentration and coordination of the wave packets are elementary processes clearly sufficient for description of the wave packet motion. Out of these three processes, the coordination is the most important for explanation of the rainbow effect. Next, we will show that coordinated evolution of wave packets generate the rainbow effect. Figure 12a shows the family of quantum probability densities. Members of the rainbow subensemble are designated by the magenta lines. Each member represents the motion of the wave packet reflected form the SWCNT boundary. The largest maxima of any member show the current position of the wave packet center, with a large number of self-interference maxima on its tail. The remaining probability densities are shown by the gray lines. Prominent peak pairs visible in Figures 11a and 12a labeled $\pm 1^{s0}$, $\pm 1^{s1}$, $\pm 1^{s2}$, $\pm 1^{s3}$, $\pm 1^{s4}$, $\pm 1^{s5}$, and $\pm 1^{s6}$, are at $x_1^{s1} = \pm 0.41$, $x_1^{s2} = \pm 0.45$, $x_1^{s3} = \pm 0.50$, $x_1^{s4} = \pm 0.55$, $x_1^{s5} = \pm 0.60$, and $x_1^{s6} = \pm 0.66$ nm, respectively. The numbering starts from the innermost peak pairs. The reason for such a convention will become apparent shortly. Note that due to the wave packet coordination, members of the rainbow subensemble have their respective maxima on the exactly same abscissas.

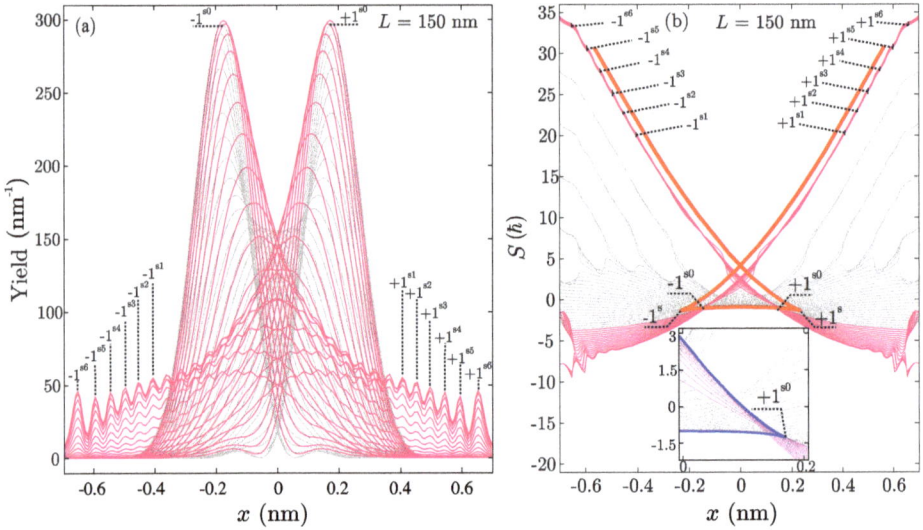

Figure 12. (**a**) The family of wave packet probability densities; (**b**) The corresponding family of wave packet phase functions. The red line shows the classical Hamilton's principal functions. The blue line shows the envelope function of the quantum phase function family. Inset show enlarged part of the phase function family in the vicinity of the classical singular point $+1^s$. The envelope function of the family is shown by the blue line. Members of the rainbow subensemble are designated by magenta lines, remaining wave packets are shown by the gray lines.

Figure 12b shows the obtained family of quantum phase functions expressed in the units of $\hbar$. The red line shows the corresponding classical Hamilton's principal function form Figure 10b. Phases of the members of the rainbow subensemble are designated by the magenta lines, remaining phases are shown by the gray lines. The family of phase functions can also be subdivided into subsets of lines which run in parallel, i.e., subsets of coordinated wave packets. Note that the subset of magenta lines is the largest, which also runs in parallel with the classical Hamilton's principal function. Therefore, wave packet coordination is the strongest in the rainbow subensemble. Detailed analysis of phases functions in the vicinity of classical singular points $\pm1^s$ have shown that envelope function of phase function family also have two cusp singular points $\pm1^{s0}$ at $x_1^{s0} = \pm0.16$ nm. This can be seen in inset in Figure 12b where the envelope line is shown by the blue line. The "vertical" branch of the envelope is defined by the members of the rainbow subensemble, while the "horizontal" is defined by the remaining phases of the ensemble. Therefore, both subensembles are important for the explanation of the formation of cusp singular points. The position of the points $\pm1^{s0}$ is also shown in Figure 12a and it proved to be very close to the peaks $\pm1^{m0}$ from Figure 11b. A careful examination revealed that phases of the rainbow subensemble have common inflection points labeled $\pm1^{s0}$, $\pm1^{s1}$, $\pm1^{s2}$, $\pm1^{s3}$, $\pm1^{s4}$, $\pm1^{s5}$, and $\pm1^{s6}$, respectively whose abscissas are equal with abscissas of the corresponding points in Figure 12a. Note that points $+1^{s1}$, $+1^{s2}$, $+1^{s3}$, $+1^{s4}$, $+1^{s5}$, and $+1^{s6}$, respectively belong to the branch generated by the point -1^{s0}.

Now the classification of the prominent peaks of the spatial distribution of the positron beam is straightforward. The central maximum consists of two primary quantum rainbow peaks $\pm1^{s0}$. Peaks $+1^{s1}$, $+1^{s2}$, $+1^{s3}$, $+1^{s4}$, $+1^{s5}$, and $+1^{s6}$, respectively, are supernumeraries of the primary rainbow -1^{s0}, while -1^{s1}, -1^{s2}, -1^{s3}, -1^{s4}, -1^{s5}, and -1^{s6} are supernumeraries of the primary rainbow $+1^{s0}$.

It should be stressed that although quantum mechanical description requires that amplitude and phase functions of any individual wave packet are smooth single valued functions, there is no such restriction regarding the behavior of the ensembles of amplitude or phase function families. The envelope function of the quantum phase functions can develop cusp singularities characteristic for the existence of the rainbow effect. Therefore, the morphological approach for the classification of the system behavior, based on the analysis of the singularities of the appropriate function family, is more general than other approaches considered here. The developed morphological method is geometrical in its nature, which makes it applicable for interpretation of the rainbow pattern for any physical system in which it can be observed. The true limitation of the method is the existence of the some random factors which can destroy the coordinated behavior of individual wave packets.

4. Conclusions

In this paper, the transmission of both quantum and classical particles was examined in detail. It has been proven that the quantum rainbow effect exists and that it can be explained in terms of: wave-packet wrinkling, concentration, and coordination. Both classical and quantum rainbows were found to be linked to the singularities of the Hamilton's principal function and quantum phase function family, respectively. The devised method for the classification of the prominent peaks in quantum distributions of transmitted particles was found to be more general than an alternative approach based on the semiclassical approximation.

More profoundly, we have found that the rainbow pattern is an inartistical property emerging out of a collective, i.e., its behavior is irreducible to the behaviour of any constitute member. This represents a very interesting example of the so-called deducible or computational emergence property [52]. It has been shown that nontrivial morphological properties of the trajectory family or family of quantum amplitude and phase functions are related to the nontrivial physical properties of channeled particles. The classical behavior of the particle beam seems to be embedded in the quantum ensemble and not in the behavior of individual wave packets. It emerges directly from the underlying quantum ensemble without the need for any additional approximations. It seems that, in the case of rainbows, scattering physical systems follow J. von Neumann's dictum that classical mechanics is merely a consequence of the law of large numbers.

Author Contributions: Conceptualization, M.Ć., S.P. and N.N.; methodology, M.Ć., S.P. and N.N.; software, M.Ć.; validation, S.P., and N.N.; investigation, M.Ć.; data curation, M.Ć.; writing—original draft preparation, M.Ć.; writing—review and editing, S.P.; visualization, M.Ć.

Funding: Authors M.Ć., M.H., and S.P. acknowledge the support to this work provided by the Ministry of Education, Science and Technological Development of Serbia through the project *Physics and Chemistry with Ion Beams*, No. III 45006.

Conflicts of Interest: The authors declare no conflict of interest.

Abbreviations

The following abbreviations are used in this manuscript:

SWCNT   Single Wall Carbon Nanotubes

References

1. Saito, R.; Dresselhaus, G.; Dresselhaus, M.S. *The Physical Properties of Carbon Nanotubes*; Imperial College Press: London, UK, 1998.
2. Baughman, R.H.; Zakhidov, A.A.; de Heer, W.A. Carbon nanotubes—The route toward applications. *Science* **2002**, *297*, 787–792. [CrossRef] [PubMed]
3. Bellucci, S. Carbon nanotubes: Physics and applications. *Phys. Status Solidi (c)* **2005**, *2*, 34–47. [CrossRef]
4. Bellucci, S. Nanotubes for particle channeling, radiation and electron sources. *Nucl. Instrum. Meth. Phys. Res. B* **2005**, *234*, 57–77. [CrossRef]
5. Lindhard, J. Influence of crystal lattice on motion of energetic charged particles. *Matematisk-Fysiske Meddelelser Det Kongelige Danske Videnskabernes Selskab* **1965**, *34*, 1–65.
6. Iijima, S. Helical microtubules of graphitic carbon. *Nature* **1991**, *354*, 56–58. [CrossRef]

7. Klimov, V.V.; Letokhov, V.S. Hard X-radiation emitted by a charged moving in a carbon nanotube. *Phys. Lett. A* **1996**, *222*, 424–428. [CrossRef]
8. Klimov, V.V.; Letokhov, V.S. Monochromatic γ-radiation emitted by a relativistic electron moving in a carbon nanotube. *Phys. Lett. A* **1997**, *226*, 244–252. [CrossRef]
9. Artru, X.; Fomin, S.P.; Shul'ga, N.F.; Ispirian, K.A.; Zhevago, N.K. Carbon nanotubes and fullerites in high-energy and X-ray physics. *Phys. Rep.* **2005**, *412*, 89–189. [CrossRef]
10. Adams, J.A. The mathematical physics of rainbows and glories. *Phys. Rep.* **2002**, *356*, 229–365. [CrossRef]
11. Ford, K.W.; Wheeler, J.A. Semiclassical description of scattering. *Ann. Phys.* **2002**, *281*, 608–635. (Reprinted) [CrossRef]
12. Connor, J.N.L.; Farrelly, D. Theory of cusped rainbows in elastic scattering: Uniform semiclassical calculations using Pearcey's integral. *J. Chem. Phys.* **1981**, *75*, 2831–2846. [CrossRef]
13. Ziegler, G.; Rädle, M.; Pütz, O.; Jung, K.; Ehrhardt, H.; Bergmann, K. Rotational rainbows in electron-molecule scattering. *Phys. Rev. Lett.* **1987**, *58*, 2642–2645. [CrossRef]
14. Kleyn, A.W.; Horn, T.C.M. Rainbow scattering from solid surfaces. *Phys. Rep.* **1991**, *199*, 191–230. [CrossRef]
15. Nešković, N. Rainbow effect in ion channeling. *Phys. Rev. B* **1986**, *33*, 6030–6035. [CrossRef]
16. Petrović, S.; Borka, D.; Nešković, N. Rainbows in transmission of high energy protons through carbon nanotubes. *Eur. Phys. J. B* **2005**, *44*, 41–45. [CrossRef]
17. Petrović, S.; Miletić, L.; Nešković, N. Theory of rainbows in thin crystals: The explanation of ion channeling applied to $Ne\ 10^+$ ions transmitted through a $\langle 100 \rangle$ Si thin crystal. *Phys. Rev. B* **2000**, *61*, 184–189. [CrossRef]
18. Borka, D.; Petrović, S.; Nešković, N. *Channeling of Protons through Carbon Nanotubes*; Nova Science Publishers: New York, NY, USA, 2011.
19. Petrović, S.; Nešković, N.; Ćosić, M.; Motapothula, M.; Breese, M.B.H. Proton-silicon interaction potential extracted from high-resolution measurements of crystal rainbows. *Nucl. Instrum. Meth. Phys. Res. B* **2015**, *360*, 23–29. [CrossRef]
20. Petrović, S.; Nešković, N.; Berec, V.; Ćosić, M. Superfocusing of channeled protons and subatomic measurement resolution. *Phys. Rev. A* **2012**, *85*, 032901. [CrossRef]
21. Petrović, S.; Borka, D.; Nešković, N. Rainbow effect in channeling of high energy protons through single-wall carbon nanotubes. *Nucl. Instrum. Methods Phys. Res. Sect. B* **2005**, *234*, 78–86. [CrossRef]
22. Borka, D.; Petrović, S.; Nešković, N. Channeling star effect with bundles of carbon nanotubes. *Phys. Lett. A* **2006**, *354*, 457–461. [CrossRef]
23. Petrović, S.; Ćosić, M.; Nešković, N. Quantum rainbow channeling of positrons in very short carbon nanotubes. *Phys. Rev. A* **2013**, *88*, 012902. [CrossRef]
24. Ćosić, M.; Petrović, S.; Nešković, N. Quantum rainbow characterization of short chiral carbon nanotubes. *Nucl. Instrum. Methods Phys. Res. Sect. B* **2014**, *323*, 30–35. [CrossRef]
25. Takabayashi, Y.; Pivovarov, Y.L.; Tukhfatullin, T.A. First observation of scattering of sub-GeV electrons in ultrathin Si crystal at planar alignment and its relevance to crystal-assisted 1D rainbow scattering. *Phys. Lett. B* **2018**, *785*, 347–353. [CrossRef]
26. Shul'ga, N.F.; Shul'ga, S.N. Scattering of ultrarelativistic electrons in ultrathin crystals. *Phys. Lett. B* **2017**, *769*, 141–145. [CrossRef]
27. Shul'ga, S.N.; Shul'ga, N.F.; Barsuk, S.; Chaikovska, I.; Chehab, R. On classical and quantum effects at scattering of ultrarelativistic electrons in ultrathin crystal. *Nucl. Instrum. Methods Phys. Res. Sect. B* **2017**, *402*, 16–20. [CrossRef]
28. Schüller, A.; Winter, H. Supernumerary rainbows in the angular distribution of scattered projectiles for grazing collisions of fast atoms with a LiF(001) surface. *Phys. Rev. Lett.* **2008**, *100*, 097602. [CrossRef] [PubMed]
29. Berry, M.V. Uniform approximation: A new concept in wave theory. *Sci. Prog. Oxf.* **1969**, *57*, 43–64.
30. Wong, R. *Asymptotic Approximations of Integrals*; SIAM: Philadelphia, PA, USA, 2001.
31. Howie, A.; Whelan, M.J. Diffraction contrast of electron microscope images of crystal lattice defects. II. The development of a dynamical theory. *Proc. R. Soc. A* **1961**, *263*, 217–237.
32. Howie, A. Diffraction channelling of fast electrons and positrons in crystals. *Philos. Mag.* **1966**, *14*, 223–237. [CrossRef]
33. Andersen, J.U.; Augustyniak, W.M.; Uggerhøj, E. Channeling of Positrons. *Phys. Rev. B* **1971**, *3*, 705–711. [CrossRef]

34. Haakenaasen, R.; Hau, L.V.; Golovchenko, J.A.; Palathingal, J.C.; Peng, J.P.; Asoka-Kumar, P.; Lynn, K.G. Quantum channeling effects for $1 MeV$ positrons. *Phys. Rev. Lett.* **1995**, *75*, 1650–1653. [CrossRef] [PubMed]

35. Berry, M.V.; Nye, J.F.; Wright, F.J. The elliptic umbilic diffraction catastrophe. *Phil. Trans. R. Soc. A* **1979**, *291*, 453–484. [CrossRef]

36. Molière, G. Therorie der Streuung schneller geladener Teilchen I. Einzelstreuung am abgeschirmten Coulomb-Feld. *Z. Naturforsch.* **1947**, *2*, 133–145. [CrossRef]

37. Abramowitz, M.; Stegun, I. *Handbook of Mathematical Functions*; National Bureau of Standards: Gaithersburg, MD, USA, 1972; p. 302.

38. Zhevhago, N.K.; Glebov, V.I. Diffraction and channeling in nanotubes. *ZhETF* **2000**, *91*, 504–514. [CrossRef]

39. Appleton, B.R.; Erginsoy, C.; Gibson, W.M. Channeling effects in the energy loss of 3-11-MeV protons in silicon and germanium single crystals. *Phys. Rev.* **1967**, *161*, 330–349. [CrossRef]

40. Watson, G.N. *Theory of Bessel Functions*; Cambridge University Press: Cambridge, UK, 1922.

41. Press, W.; Teukolsky, S.; Vetterling, W.; Flannery, B. *Numerical Recipes in FORTRAN*; Cambridge University Press: Cambridge, UK, 1993.

42. Kosloff, R. Time-dependent quantum-mechanical methods for molecular dynamics. *J. Chem. Phys.* **1988**, *92*, 2087–2100. [CrossRef]

43. Berry, M.V. The elliptic umbilic diffraction catastrophe. *Ann. N. Y. Acad. Sci.* **1980**, *357*, 183–202. [CrossRef]

44. Airy, G.B. On the intensity of light in the neighbourhood of a causti. *Trans. Camb. Phil. Soc.* **1838**, *6*, 379–402.

45. Nye, J.F. *Natural Focusing and Fine Structure of Light: Caustics and Wave Dislocations*; IOP Publishing: Bristol, UK, 1999; p. 132.

46. Nussenzveig, H.M. *Diffraction Effects in Semiclassical Scattering*; Cambridge University Press: Cambridge, UK, 1992; pp. 105–107.

47. Bruce, J.W.; Giblin, P.J. *Curves and Singularities*; Cambridge University Press: Cambridge, UK, 1984.

48. Arnold, V.I. *Singularities of Caustics and Wave Fronts*; Springer-Verlag: Dordrecht, The Netherlands, 1990.

49. Goldstein, H. *Classical Mechanics*; Addison-Wesley, Reading: Boston, MA, USA, 1982; p. 386.

50. Weyl, H. *The Group Theory and Quantum Mechanics*; Dover Publications: New York, NY, USA, 1950; p. 93.

51. Arnold, V.I. *Catastrophe Theory*; Springer-Verlag: Berlin/Heidelberg, Germany, 1986.

52. Baas, N.A.; Emmeche, C. On Emergence and explanation. *Intellectica* **1997**, *2*, 67–83. [CrossRef]

atoms

MDPI

Article

Quasars: From the Physics of Line Formation to Cosmology

Paola Marziani [1,*]**, Edi Bon** [2]**, Natasa Bon** [2]**, Ascension del Olmo** [3]**, Mary Loli Martínez-Aldama** [3]**,
Mauro D'Onofrio** [4]**, Deborah Dultzin** [5]**, C. Alenka Negrete** [5] **and Giovanna M. Stirpe** [6]

[1]   National Institute for Astrophysics (INAF), Astronomical Observatory of Padova, IT-35122 Padova, Italy
[2]   Astronomical Observatory, 11060 Belgrade, Serbia; ebon@aob.rs (E.B.); nbon@aob.rs (N.B.)
[3]   Instituto de Astrofisíca de Andalucía, IAA-CSIC, Glorieta de la Astronomia s/n, E-18008 Granada, Spain;
     chony@iaa.es (A.d.O.); mmary@cft.edu.pl (M.L.M.-A.)
[4]   Dipartimento di Fisica & Astronomia "Galileo Galilei", Università di Padova, IT-35122 Padova, Italy;
     Mauro.donofrio@unipd.it
[5]   Instituto de Astronomía, UNAM, Mexico D.F. 04510, Mexico; deborah@astro.unam.mx (D.D.);
     negrete@astro.unam.mx (C.A.N.)
[6]   INAF, Osservatorio di Astrofisica e Scienza dello Spazio, IT-40129 Bologna, Italy; giovanna.stirpe@inaf.it
*   Correspondence: paola.marziani@inaf.it; Tel.: +39-0498293415

Received: 26 November 2018; Accepted: 28 January 2019; Published: 4 February 2019

Abstract: Quasars accreting matter at very high rates (known as extreme Population A (xA) or
super-Eddington accreting massive black holes) provide a new class of distance indicators covering
cosmic epochs from the present-day Universe up to less than 1 Gyr from the Big Bang. The very
high accretion rate makes it possible that massive black holes hosted in xA quasars can radiate at a
stable, extreme luminosity-to-mass ratio. This in turn translates into stable physical and dynamical
conditions of the mildly ionized gas in the quasar low-ionization line emitting region. In this
contribution, we analyze the main optical and UV spectral properties of extreme Population A quasars
that make them easily identifiable in large spectroscopic surveys at low- ($z \lesssim 1$) and intermediate-z
($2 \lesssim z \lesssim 2.6$), and the physical conditions that are derived for the formation of their emission
lines. Ultimately, the analysis supports the possibility of identifying a virial broadening estimator
from low-ionization line widths, and the conceptual validity of the redshift-independent luminosity
estimates based on virial broadening for a known luminosity-to-mass ratio.

Keywords: black hole physics; cosmology; quasar spectroscopy; cosmological parameters;
ionized gas; broad line region

1. Introduction

1.1. Quasar Spectra: Emission from Mildly Ionized Gas

The spectra of quasars can be easily recognized by the presence of broad and narrow
optical and UV lines emitted by mildly-ionized species over a wide range of ionization potential.
The type-1 composite quasar spectrum from the Sloan Digital Sky Survey (SDSS) [1] reveals Broad
(FWHM $\gtrsim 1000$ km s^{-1}) and Narrow High Ionization lines (HILs, 50 eV) and Low Ionization
lines (LILs, <20 eV). Broad HILs encompass CIVλ1549, HeIIλ1640 and HeIIλ4686 as representative
specimens. Broad LILs include HI Balmer lines (Hβ, Hα), MgIIλ2800, the CaII IR Triplet, and FeII
features. The FeII emission deserves a particular mention, as it is extended over a broad range of
wavelengths (Figure 6 of [2]), and is especially prominent around MgIIλ2800 and Hβ. The FeII
emission is one of the dominant coolants in the broad line region (BLR) and therefore a main factor in
its energetic balance (the FeII emission extends from the UV to the far IR, and can reach the luminosity

of Lyα, [3,4]). Thus, it may not appear surprising that an estimator of its strength plays an important role in the systematic organization of quasar properties (Section 2).

This paper reviews results obtained in the course of two decades (Sections 2 and 3), attempting to explain how the spectral properties of a class of type-1 quasars and their physical interpretation can lead to the definition of "Eddington standard candles" (ESC, Section 4). In the following, we will restrict the presentation to type-1 quasars which are considered mainly "unobscured" sources with an unimpeded view of the BLR, and exclude type-2 active galactic nuclei (AGN) or quasars in which the broad lines are not detected in natural light (see [5] for an exhaustive review). We describe the physical basis of the method in Sections 3 and 4. We then introduce ESC selection criteria (Section 5) and preliminary cosmology results (Section 6).

1.2. Quasars for Cosmology: An Open Issue

The distribution of quasars in space and the intervening absorptions along the line of sight (i.e., the so called Lyα forest) has long been considered as a tracer of matter in the distant Universe (see [6] and references therein). However, a relevant question may be why intrinsic properties of quasars have never been successfully used as cosmological probes. On the one hand, (1) quasars are easily recognizable and plentiful ($\gtrsim$500,000 in the data release 14 of the SDSS, [7]). (2) They are very luminous and can reach bolometric luminosity $L \gtrsim 10^{48}$ erg s^{-1}; (3) they are observed in an extremely broad range of redshift $0 \lesssim z \lesssim 7$, and (4) they are stable compared to transients that are employed as distance indicators in cosmology, such as type Ia supernovæ (Section 2, Ref. [8] for a review). On the other hand, (1) quasars are anisotropic sources even if the degree of anisotropy is expected to be associated with the viewing angle of the accretion disk in radio-quiet quasars [9], and not large compared to radio-loud quasars whose optical continuum is in part beamed (see, for example [10]); (2) quasars have an open-ended luminosity function (i.e., without a clearly defined minimum, as the quasar highest spatial density occurs at the lowest luminosity); in other words, they are the "opposite" of a cosmological standard candle. In addition, (3) the long-term variability of radio-quiet quasars is poorly understood (see e.g., [11,12] and references therein) (4) and the internal structure of the active nucleus ($\lesssim$1000 r_g) is still a matter of debate (see, e.g., a summary of open issues [13] in [14]). Correlations with luminosity have been proved to be rather weak (see [15], for a synopsis up to mid-1999). The selection effect may even cancel out the "Baldwin effect" [16], a significant but weak anti-correlation between rest-frame equivalent width and continuum luminosity of CIVλ1549 that has been the most widely discussed luminosity correlation in the past several decades.

2. Definition of a Class of Type-1 Quasars with Properties of Eddington Standard Candles

Nonetheless, new developments in the past decades have paved the road to the possibility of exploiting quasars as cosmological distance indicators in a novel way that would make them literal "Eddington standard candles" (ESC) ([17–20]; see also [21] for a comprehensive review of secondary distance indicators including several techniques based on quasars). This possibility is based in the development of the concept of a quasar main sequence (MS), intended to provide a sort of H-R diagram for quasars [22]. The quasar MS can be traced in the plane defined by the prominence of optical FeII emission, $R_{FeII} = $ I(FeIIλ 4570)/I(Hβ) (see [15,23–26]). Figure 1 provides a sketch of the MS in the optical plane FWHM(Hβ) vs. R_{FeII}. It is possible to isolate spectral types in the optical plane of the MS as a function of R_{FeII} and FWHM Hβ and, at a coarser level, two populations: Population A (FWHM Hβ < 4000 km/s) and Population B of broader sources. Pop. A is rather heterogeneous, and encompasses a range of R_{FeII} from almost 0 to the highest values observed ($R_{FeII} \gtrsim 2$ are very rare, $\lesssim$1% in optically-selected samples, [25]). Along the quasar main sequence, the extreme Population A (xA) sources satisfying the condition $R_{FeII} > 1$ (about 10% of all quasars in optically-selected sample, green area in Figure 1) show remarkably low optical variability, so low that it is even difficult to estimate the BLR radius via reverberation mapping [27]. This is at variance with Pop. B sources that show more pronounced variability [28,29], the most extreme cases being observed among blazars which are low-accretors, at the opposite end in the quasar MS. Of the many multi-frequency trends along the main sequence (from the sources whose spectra show the broadest LILs (extreme Pop. B), and the weakest FeII emission, to sources with the narrowest LIL profiles and strongest FeII emission [extreme Pop. A]), we recall a systematic decrease of the CIV equivalent width, an increase in metallicity, and amplitude of HIL blueshifts (a more exhaustive list is provided by Table 1 of [30]). The Eddington ratio is believed to increase along with R_{FeII} [23,26,31,32]. The FWHM Hβ is strongly affected by the viewing angle (i.e., the angle between the line of sight and the accretion disk axis), so that at least the most narrow-line Seyfert 1s (NLSy1s) can be interpreted as Pop. A sources seen with the accretion disk oriented face-on or almost so [33]. At low-z ($\lesssim$0.7), Pop. A implies low black hole mass M_{BH}, and high Eddington ratio; on the converse, Pop. B is associated with high M_{BH} and low L/L_{Edd}. This trend follows from the "downsizing" of nuclear activity at low-z that helps give an elbow shape to the MS [34]: at low-z, very massive quasars ($M_{BH} \gtrsim 10^9$ M$_\odot$) do not radiate close to their Eddington limit but are, conversely, low-radiators ($L/L_{Edd} \lesssim 0.1$).

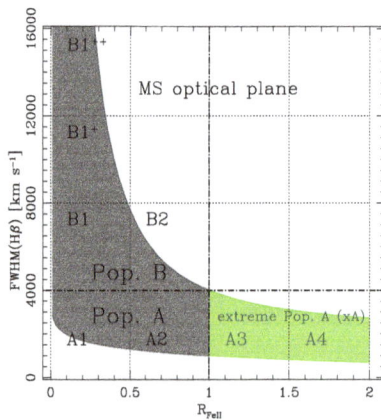

Figure 1. The plane FWHM(Hβ) vs. R_{FeII}. The MS is sketched as the grey strip, with the section occupied by xA sources colored pale green. The thick dot-dashed line separates Pop. A and B at 4000 km s^{-1}, while the vertical one at $R_{FeII} = 1$ traces the R_{FeII} lower value for xA identification. The spectral types with significant occupation at low-z are labeled.

The inter-comparison between CIVλ1549 and Hβ supports low-ionization lines virial broadening (in a system of dense clouds or in the accretion disk) + high-ionization lines (HILs) radial or vertical outflows, at least in Pop. A sources [35,36]. There is now a wide consensus on an accretion disk + wind system model [37], and therefore on the existence of a "virialized" low-ionization subregion + higher ionization, with the subregion outflowing up to the highest quasar luminosities [36,38,39].

The most extreme examples at high accretion rate are a population of sources with distinguishing properties. They have been called extreme Pop. A or extreme quasars (xA), and are also known as super-Eddington accreting massive black holes (SEAMBHs) [9,18,40,41]. Figure 2 shows a composite rest-frame UV spectrum of high-luminosity xA quasars. Observationally, xA quasars satisfy $R_{FeII} \geq 1$ and still show LIL Hβ profiles basically consistent with emission from a virialized system. xA quasars may well represent an early stage in the evolution of quasars and galaxies. In the hierarchical growth scenario for the evolution of galaxies, merging and strong interaction lead to accumulation of gas in the galaxy central regions, inducing enhanced star formation. Strong winds from massive stars and eventual Supernova explosions may ultimately provide enriched accretion fuel for the massive black hole at the galaxy [42–44]. The active nucleus radiation force and the mechanical thrust of the accretion disk wind can then sweep the dust surrounding the black hole, at least within a cone coaxial with the accretion disk axis (see Figure 7 of [45]). The fraction of mass that is accreted by the black hole and the fraction that is instead ejected in the wind are highly uncertain; the outflow kinetic power can become comparable to the radiative output [46,47], especially in sources accreting at very high rate [48]; interestingly, this seems to be true also for stellar-mass black holes [49]. Feedback effects on the host galaxies are maximized by the high kinetic power of the wind, presumably made of gas much enriched in metals [50].

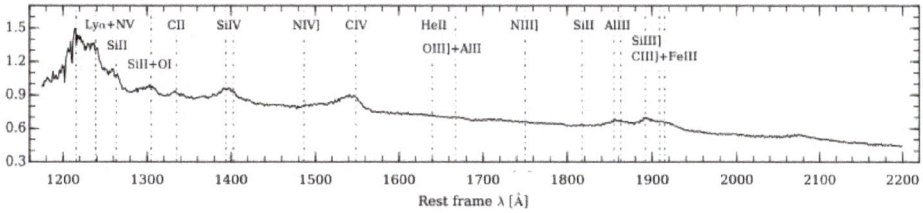

Figure 2. Composite UV spectrum of high-z xA sources. The abscissa is rest-frame wavelength, and the ordinate is normalized flux.

3. Diagnostics of Mildly-Ionized Gases

Diagnostics from the rest-frame UV spectrum takes advantage of the observations of strong resonance lines that are collisionally excited [51,52]. The point is that the rest-frame UV spectrum offers rich diagnostics that constrains at least gas density n_H, ionization parameter U, and chemical abundance Z. For instance, Si IIλ 1814/Si III]λ 1892 is sensitive to ionization CIVλ1549/Lyα, CIVλ1549/(Si IV + OIV])λ1400, CIVλ1549/HeIIλ1640, NVλ1240/HeIIλ1640 are sensitive to metallicity; and Al IIIλ1860/Si III]λ1892, Si III]λ1892/CIII]λ1909 are sensitive to density, since inter-combination lines have a well defined critical density [51].

The photoionization code *Cloudy* models the ionization, chemical, and thermal state of gas exposed to a radiation field, and predicts its emission spectra and physical parameters [53,54]. In *Cloudy*, collisional excitation and radiative processes typical of mildly ionized gases are included. *Cloudy* simulation requires inputs in terms of n_H, U, Z, quasar spectral energy distribution (SED), and column density N_c. The ionization parameter

$$U = \frac{\int_{\nu_0}^{\infty} \frac{L_\nu}{h\nu}}{4\pi r_{BLR}^2 c n_H} = \frac{Q(H)}{4\pi r_{BLR}^2 c n_H}, \tag{1}$$

where $Q(H)$ is the number of ionizing photons, provides the ratio between photon and hydrogen number density. More importantly, the inversion of equation provides a measure of the emitting region radius r_{BLR} once the ionizing photon flux i.e., the product $U n_H$ is known. As we will see, the photon flux can be estimated with good precision from diagnostic line intensity ratios.

Maps built on an array of 551 Cloudy 08.00–13.00 photoionization models for a given metallicity Z and N_c, constant n and U evaluated at steps of 0.25 dex covering the ranges $7 \leq \log n_H \leq 14$ [cm^{-3}], $-4.5 \leq \log U \leq 0$. Given the measured intensity ratios for xA quasars, *Cloudy* simulations show convergence toward a well-defined value of $\log (n_H U)$ [40,51]. UV diagnostic ratios in the plane ionization parameter versus density indicate extremely high n_H $10^{12.5-13}$ cm^{-3}, extremely low $\log U \sim -2.5$–3 (Figure 3). Note the orthogonal information provided by the AlIIIλ1860/SiIII]λ1892 that mainly depends on density. The left and right panels differ because of chemical abundances: the case with five times solar metallicity plus overabundance of Si and Al produces better agreement, displacing the solution toward lower density and higher ionization. Nonetheless, the product $U n_H$ remains fairly constant. Diagnostic ratios sensible to chemical composition suggest high metallicity. The metallicities in the quasar BLR gas are a function of the spectral type (ST) along the MS: relatively low (solar or slightly sub-solar in extreme Pop. B sources (as estimated recently for NGC 1275 [55]), and relatively high for typical Pop. A quasars with moderate FeII emission ($Z \sim 5$–$10 Z_\odot$, [56,57]). If the diagnostic ratios are interpreted in terms of scaled $Z_\odot$, they may reach $Z \gtrsim 20Z_\odot$, even $Z \sim 100Z_\odot$ for xA quasars [40]. Z values as high as $Z \sim 100Z_\odot$ are likely to be unphysical, and suggest relative abundances of elements deviating from solar values, as assumed in the previous example, or significant turbulence. The analysis of the gas chemical composition in the BLR of xA source has just begun. However, high or non-solar Z are in line with the idea of xA sources being high accretors surrounded by huge amount of gas and a circum-nuclear star forming system, possibly with a top-heavy initial mass function [51]. The high n_H is consistent with the low CIII]λ1909 emission that becomes undetectable in some cases. While in Pop. A and B we find evidence of ionization stratification within the low-ionization part of the BLR ([58–60] and references therein), xA sources show intensity ratios that are consistent with a very dense "remnant" of the BLR, perhaps after lower density gas has been ablated away by radiation forces.

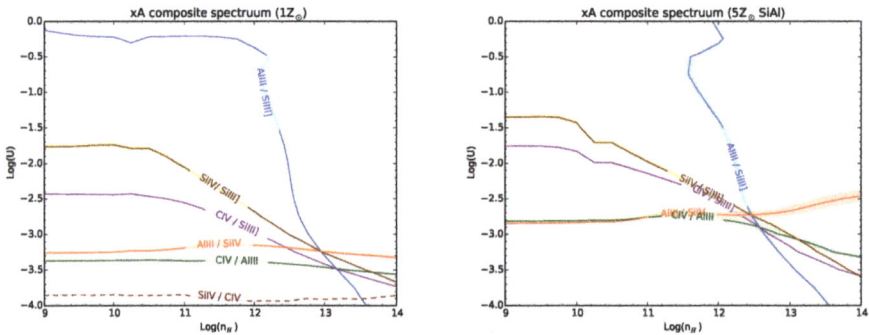

Figure 3. Intensity ratios in the plane ionization parameter vs. density, for the intensity ratios measured on the composite xA quasar spectrum shown in Figure 2 of [40]. left panel: solar chemical composition; right: 5× solar chemical composition with selective enrichment in Al and Si, following [51]. In this latter case, the SiIVλ1402/CIVλ1549 is degenerate.

4. xA Quasars as Eddington Standard Candles

There are several key elements that make it possible to exploit xA quasars as Eddington standard candles.

The first is the similarity of their spectra and hence of the physical condition in the mildly-ionized gas that is emitting the LILs. Line intensity ratios are similar (they scatter around a constant average with small dispersion). Since the line emitting gas is photoionized, intensity line ratios depend strongly on the ionizing continuum SED. Thus, the ionizing SED is also constrained within a small scatter. We remark that this is not true for the general population of quasars that show differences in line equivalent width and intensity ratios larger than an order of magnitude along the MS.

The mass reservoir in all xA sources is sufficient to ensure a very high accretion rate (possibly super-Eddington) that yields a radiative output close to the Eddington limit. The similarity of the SED and the presence of high rates of circumnuclear and galactic star formation as revealed by Spitzer [61] have led to the conjecture that xA sources may be in a particular stage of a quasar development, as mentioned above.

The second key element is the existence of a virialized low-ionization sub-region (possibly the accretion disk itself). This region coexists with outflowing gas even at extreme $L \gtrsim 10^{48}$ erg s^{-1} and highest Eddington ratios but is kinematically distinguishable on the basis of inter-line shifts between LILs and HILs—for example, Hβ and CIVλ1549.

In addition, xA quasars show extreme L/L_{Edd} along the MS with small dispersion. If the Eddington ratio is known, and constant, then $\natural = L/L_{\text{Edd}} \propto L/M_{\text{BH}}$. Accretion disk theory teaches low radiative efficiency at a high accretion rate, and that $\natural$ saturates toward a limiting value ([62–64] and references therein). Therefore, empirical evidence (the xA class of sources, easily identified by their self-similar properties, scatters around a well-defined, extremal L) and theoretical support (the saturation of the radiative output per unit M_{BH}) justified the consideration of xA sources potential ESCs.

Virial Luminosity

The use of xA sources as Eddington standard candles requires several steps which should considered carefully.

1. The first step is the actual estimate of the accretion luminosity via a virial broadening estimator (VBE). The luminosity can be written as

$$L \propto L M_{BH} \propto L r_{BLR}(\delta v)^2, \tag{2}$$

   assuming virial motions of the low-ionionization part of the broad-line region (BLR). The δv stands for a suitable VBE, usually the width of a convenient LIL (in practice, the FWHM of $H\beta$ or even $Pa\alpha$, [65]).

2. The r_{BLR} can be estimated from the inversion of Equation (1) [51,52], again taking advantage of the fact that the ionizing photon flux shows a small scatter around a well defined value. In addition, another key assumption is that

$$r_{BLR} \propto \left(\frac{L}{n_H U} \right)^{\frac{1}{2}}. \tag{3}$$

   Equation (3) implies that r_{BLR} scales with the square root of the luminosity. This is needed to preserve the U parameter. If U were going to change, then the spectrum would also change as a function of luminosity. This is not evident comparing spectra over a wide luminosity range (4.5 dex), although some second order effects are possible.[1]

3. We can therefore write the virial luminosity as

$$L \propto L \left(\frac{L}{n_H U} \right)^{\frac{1}{2}} (\delta v)^2. \tag{4}$$

Making explicit the dependence of the number of ionizing photons on the SED, the virial luminosity becomes:

$$L \approx 7.8 \cdot 10^{44} \frac{L_1^2 \kappa_{0.5} f_2^2}{\tilde{\nu}_{2.42 \cdot 10^{16}}^2} \frac{1}{(n_H U)_{9.6}} (\delta v_{1000})^4, \tag{5}$$

where κ is the fraction of ionizing luminosity scaled to 0.5, $\tilde{\nu}$ the average frequency of ionizing photons scaled to $2.42 \cdot 10^{16}$ Hz, and $(n_H U)$ to $10^{9.6}$.

Equation (5) is analogous to the Tully–Fisher [66] and the early formulation of the Faber–Jackson [67] laws for galaxies. Equation (5) is applicable to xA quasars with Eddington ratio $L \sim 1$ and dispersion $\delta L \ll 1$ but, in principle, could be used for every sample of quasars whose L is in a very restricted range.

5. Selection of Eddington Standard Candles

Selection criteria are based on emission line intensity ratios which are extreme along the quasar MS [20]:

---

[1] The maximum temperature of the accretion disk is $\propto M_{BH}^{-\frac{1}{4}}$; the SED is expected to become softer at high M_{BH}, but this effect has not been detected yet at a high confidence level.

1. $R_{FeII} > 1.0$,
2.   UV AlIIIλ1860/SiIII]λ1892 > 0.5,
3.   SiIII]λ1892/CIII]λ1909 > 1.

The first criterion can be easily applied to optical spectra of a large survey such as the SDSS for sources at $z \lesssim 1$. The second and third criterion can be applied to sources at $1 \lesssim z \lesssim 4.5$ for which the 1900 blend lines are shifted into the optical and near IR domains. UV and optical selection criterions are believed to be equivalent. Due to a small sample size at low z for which rest-frame optical and UV spectra are available, further testing is needed.

6. Tentative Applications to Cosmology and the Future Perspectives

Preliminary results were collected from three quasar samples (62 sources in total), unevenly covering the redshift range $0.4 \lesssim z \lesssim 2.6$. For redshift $z \gtrsim 2$, the UV AlIIIλ1860 FWHM was used as a VBE for the rest-frame UV range, save a few cases for which Hβ was available. This explorative application to cosmology yielded results consistent with concordance cosmology, and allowed the exclusion of some extreme cosmologies [20]. A more recent application involved the [20] sample, along with the Hβ sample of [9] and preliminary measurements from [40]. The resulting Hubble diagram is shown in Figure 4. The plots in Figure 4 involve ≈ 220 sources and indicate a scatter $\delta\mu \approx 1.2$ mag. The slope of the residuals ($b \approx -0.002 \pm 0.104$) is not significantly different from 0, indicating good statistical agreement between luminosities derived from concordance cosmological parameters and from the virial equation. The Hubble diagram of Figure 4 confirms the conceptual validity of the virial luminosity relation, Equation (5).

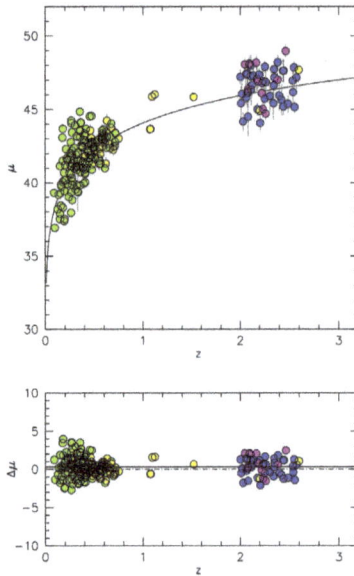

Figure 4. Hubble diagram distance modulus μ vs. z obtained from the analysis of the [20] data (yellow: Hβ, navy blue: AlIIIλ1860 and SiIII]λ1892) supplemented by new Hβ measurements from the SDSS obtained in this work (green) and from Gran Telescopio Canarias (GTC) observations of Martinez-Aldama et al. [68] (magenta). The lower panel shows the distance modulus residuals with respect to concordance cosmology. The filled line in the upper panel is the $\mu(z)$ expected from Λ cold dark matter (CDM) cosmology. The filled line in the lower panel represents a least-square fit to the residuals as a function of z. The figure is an updated version of Figure 1 of Marziani et al. [69].

Mock samples of several hundreds of objects, even with significant dispersion in luminosity with rms(log L) = 0.2–0.3, indicate that quasars covering the redshift range between 0 and 3 (i.e., a range of cosmic epochs from now to 2 Gyr since the Big Bang) could yield significant constraints on the cosmological parameters. A synthetic sample of 200 sources uniformly distributed in the redshift range 0–3 with a scatter of 0.2 dex yields $\Omega_M \approx 0.28 \pm 0.02$ at 1σ confidence level, assuming $H_0 = 70$ km s^{-1} Mpc^{-1}, and flatness ($\Omega_M + \Omega_\Lambda$=1). If $\Omega_M + \Omega_\Lambda$ is unconstrained, $\Omega_M \approx 0.30^{+0.12}_{-0.09}$ at 1σ confidence level [20]. The comparison between the constraints set by supernova surveys and by a mock sample of 400 quasars with rms = 0.3 dex in log L shows the potential ability of the quasar sample to better constrain Ω_M [70]. The scheme of Figure 5 illustrates the difference in sensitivity to cosmological parameters over the redshift range 0–4: supernovæ are sensitive to Ω_Λ since the effect of Ω_Λ, in a concordance cosmology scenario, became appreciable only at relatively recent cosmic epochs. High redshift quasars provide information on a redshift range where the expansion of the Universe was still being decelerated by the effect of Ω_M, a range that is not yet covered by any standard ruler or candle.

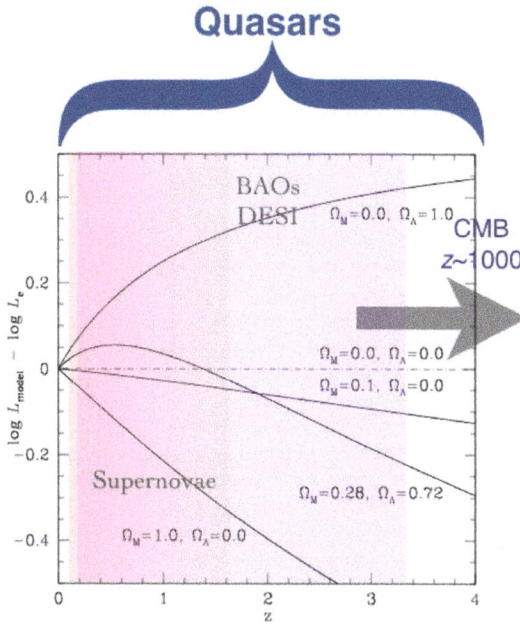

Figure 5. Luminosity difference with respect to an empty Universe for several cosmological models, identified by their values of Ω_M and Ω_Λ. The domain of supernovæ and of the baryonic acoustic oscillations (within the expectation of the future Dark Energy Spectroscopic Instrument (DESI) survey, [71]) are shown.

Error Budget

The large scatter in the luminosity estimates is apparently daunting in the epoch of precision cosmology. Statistical errors could be reduced to rms $\approx$ 0.2 dex in L by increasing numbers, collecting large samples ($\sim$500 quasars), but they still would remain high.

The xA quasar SED cannot vary much since spectra are almost identical in terms of line ratios (a second order effect [72] not yet detected as significant in the data considered by [20,40] may become significant with larger samples). The scaling $r_{BLR} \propto L^{0.5}$ should hold strictly: a small deviation would imply a systematic change in the ionization parameter and hence of ST with luminosity.

A simplified error budget for statistical errors [20] indicates that virial luminosity estimates are mostly affected by VBE uncertainties which enter with the fourth power in Equation (5). In addition to measurement uncertainties, orientation effects are expected to be determinant in the FWHM uncertainties, as they can contribute 0.3 dex of scatter in luminosity if Hβ or any other line used as a VBE is emitted in a highly-flattened configuration.[2] Modeling the effect of orientation by computing the difference between L from concordance cosmology and virial luminosity indeed reduces the sample standard deviation in the Hubble diagram by a factor $\approx$5 to $\approx$0.2 mag, and accounts for most of the rms $\approx$ 0.4 dex in the virial luminosity estimates of the sample shown in Figure 4 [9]. The rms $\approx$ 0.2 mag value is comparable to the uncertainty in supernova magnitude measurements. Work is in progress in order to make viewing angle estimates of xA quasars usable for cosmology.

7. Conclusions

This paper provided an overview of the physical conditions in the broad line emitting region of extreme spectral types of type-1 quasars (the extreme Pop. A). There is strong evidence that xA sources are radiating close to their Eddington limit (i.e., with Eddington ratio scattering around a well-defined value), at high accretion rates. Their physical properties appear to be very stable across a very wide range of luminosity, 4–5 dex. The assumption of a constant $\natural$ makes it possible to write a relation between luminosity and virial broadening, analogous to the one expressed by the Tully–Fisher and the early formulation of Faber–Jackson laws.

The scatter in the Hubble diagram obtained from virial luminosity estimates is still very high, about 1 mag (although comparable to the scatter from a method based on the nonlinear relation between the X-ray and the UV emission of quasars [75]). Very large samples are needed for reduction of scatter (and statistical error). In addition, the inter-calibration of rest-frame visual and UV properties and their dependence on L (expect systematic errors!) needs to be extended by dedicated observations of xA sources covering the rest frame UV and visual range. Simulations of statistical and systematic effects which influence the estimates of the cosmic parameters are also needed.

In principle, Eddington standard candles can cover a range of distances where the metric of the Universe has not been "charted" as yet to retrieve an independent estimate of Ω_M. If samples with uniform coverage over a wide range of redshift would become available, xA sources could also address the physics of accelerated expansion (i.e., provide measurements of the dark energy equation of state).

Author Contributions: E.B., N.B., A.d.O. and D.D. contributed with funding acquisition and resources. P.M., E.B., N.B., A.d.O., M.L.M.-A., C.A.N., M.D., G.M.S. significantly contributed to the papers on which this review is based. P.M. wrote the review paper.

Funding: P.M. wishes to thank the Scientific Organizing Committee of the Symposium on the Physics of Ionized Gases (SPIG 2018) meeting for inviting the topical lecture on which this paper is based, and acknowledges the Programa de Estancias de Investigación (PREI) No. DGAP/DFA/2192/2018 of Universidad Nacional Autónoma de México (UNAM), where this paper was written. The relevant research is part of the project 176001 "Astrophysical spectroscopy of extragalactic objects" and 176003 "Gravitation and the large scale structure of the Universe" supported by the Ministry of Education, Science and Technological Development of the Republic of Serbia. M.L.M.-A. acknowledges a CONACyT postdoctoral fellowship. A.d.O. and M.L.M.-A. acknowledge financial support from the Spanish Ministry for Economy and Competitiveness through Grant Nos. AYA2013-42227-P and AYA2016-76682-C3-1-P. M.L.M.-A, P.M. and M.D. acknowledge funding from the INAF PRIN-SKA 2017 program 1.05.01.88.04. D.D. and A.N. acknowledge support from CONACyT through Grant No. CB221398. D.D. and C.A.N. are also thankful for the support from Grant No. IN108716 53 PAPIIT, UNAM.

Conflicts of Interest: The authors declare no conflict of interest.

---

2   The M_{BH} is not computed explicitly for the estimate of L following Equation (5). However, the VBE uncertainty associated with orientation is the main source of uncertainty on M_{BH} for the xA sample. This is most likely the case also for the general quasar population [73,74].

Abbreviations

The following abbreviations are used in this manuscript:

AGN	Active Galactic Nucleus
BLR	Broad Line Region
DESI	Dark Energy Spectroscopic Instrument
ESC	Eddington Standard Candles
FWHM	Full Width Half-Maximum
HIL	High-Ionization Line
LIL	Low-Ionization Line
MDPI	Multidisciplinary Digital Publishing Institute
MS	Main Sequence
NLSy1	Narrow-Line Seyfert 1
SDSS	Sloan Digital Sly Survey
VBE	Virial Broadening Estimator

References

1. Vanden Berk, D.E.; Richards, G.T.; Bauer, A.; Strauss, M.A.; Schneider, D.P.; Heckman, T.M.; York, D.G.; Hall, P.B.; Fan, X.; Knapp, G.R.; et al. Composite Quasar Spectra from the Sloan Digital Sky Survey. *Astron. J.* **2001**, *122*, 549–564. [CrossRef]
2. Marziani, P.; Dultzin-Hacyan, D.; Sulentic, J.W. Accretion onto Supermassive Black Holes in Quasars: Learning from Optical/UV Observations. In *New Developments in Black Hole Research*; Kreitler, P.V., Ed.; Nova Press: New York, NY, USA, 2006; p. 123.
3. Netzer, H. AGN emission lines. In *Active Galactic Nuclei*; Blandford, R.D., Netzer, H., Woltjer, L., Courvoisier, T.J.-L., Mayor, M., Eds.; Springer: Berlin/Heidelberg, Germany, 1990; pp. 57–160.
4. Marinello, A.O.M.; Rodriguez-Ardila, A.; Garcia-Rissmann, A.; Sigut, T.A.A.; Pradhan, A.K. The FeII emission in active galactic nuclei: Excitation mechanisms and location of the emitting region. *arXiv* **2016**, arXiv:1602.05159.
5. Antonucci, R. Unified models for active galactic nuclei and quasars. *Annu. R. Astron. Astrophys.* **1993**, *31*, 473–521. [CrossRef]
6. D'Onofrio, M.; Burigana, C. *Questions of Modern Cosmology: Galileo's Legacy*; Springer: Berlin/Heidelberg, Germany, 2009. [CrossRef]
7. Pâris, I.; Petitjean, P.; Aubourg, É.; Myers, A.D.; Streblyanska, A.; Lyke, B.W.; Anderson, S.F.; Armengaud, É.; Bautista, J.; Blanton, M.R.; et al. The Sloan Digital Sky Survey Quasar Catalog: Fourteenth data release. *Astron. Astrophys.* **2018**, *613*, A51. [CrossRef]
8. Marziani, P.; Dultzin, D.; Sulentic, J.W.; Del Olmo, A.; Negrete, C.A.; Martinez-Aldama, M.L.; D'Onofrio, M.; Bon, E.; Bon, N.; Stirpe, G.M. A main sequence for quasars. *Front. Astron. Space Sci.* **2018**, *5*, 6. [CrossRef]
9. Negrete, C.A.; Dultzin, D.; Marziani, P.; Esparza, D.; Sulentic, J.W.; del Olmo, A.; Martínez-Aldama, M.L.; García López, A.; D'Onofrio, M.; Bon, N.; et al. Highly accreting quasars: The SDSS low-redshift catalog. *Astron. Astrophys.* **2018**, *620*, A118. [CrossRef]
10. Liu, Y.; Jiang, D.R.; Gu, M.F. The Jet Power, Radio Loudness, and Black Hole Mass in Radio-loud Active Galactic Nuclei. *Astrophys. J.* **2006**, *637*, 669–681. [CrossRef]
11. Bon, E.; Zucker, S.; Netzer, H.; Marziani, P.; Bon, N.; Jovanović, P.; Shapovalova, A.I.; Komossa, S.; Gaskell, C.M.; Popović, L.Č.; et al. Evidence for Periodicity in 43 year-long Monitoring of NGC 5548. *Astrophys. J. Suppl. Ser.* **2016**, *225*, 29. [CrossRef]
12. Bon, E.; Marziani, P.; Bon, N. Periodic optical variability of AGN. *IAU Symp.* **2017**, *324*, 176–179. [CrossRef]
13. Netzer, H. Meeting Summary: A 2017 View of Active Galactic Nuclei. *Front. Astron. Space Sci.* **2018**, *5*, 10. [CrossRef]
14. Marziani, P.; D'Onofrio, M.; del Olmo, A.; Dultzin, D. (Eds.) *Quasars at All Cosmic Epochs*; Frontiers Media: Lausanne, Switzerland, 2018. [CrossRef]
15. Sulentic, J.W.; Marziani, P.; Dultzin-Hacyan, D. Phenomenology of Broad Emission Lines in Active Galactic Nuclei. *Annu. Rev. Astron. Astrophys.* **2000**, *38*, 521–571. [CrossRef]

16. Baldwin, J.A.; Burke, W.L.; Gaskell, C.M.; Wampler, E.J. Relative quasar luminosities determined from emission line strengths. *Nature* **1978**, *273*, 431–435. [CrossRef]

17. Teerikorpi, P. On Öpik's distance evaluation method in a cosmological context. *Astron. Astrophys.* **2011**, *531*, A10. [CrossRef]

18. Wang, J.M.; Du, P.; Valls-Gabaud, D.; Hu, C.; Netzer, H. Super-Eddington Accreting Massive Black Holes as Long-Lived Cosmological Standards. *Phys. Rev. Lett.* **2013**, *110*, 081301. [CrossRef] [PubMed]

19. Wang, J.M.; Du, P.; Li, Y.R.; Ho, L.C.; Hu, C.; Bai, J.M. A New Approach to Constrain Black Hole Spins in Active Galaxies Using Optical Reverberation Mapping. *Astrophys. J. Lett.* **2014**, *792*, L13. [CrossRef]

20. Marziani, P.; Sulentic, J.W. Highly accreting quasars: Sample definition and possible cosmological implications. *Mon. Not. R. Astron. Soc.* **2014**, *442*, 1211–1229. [CrossRef]

21. Czerny, B.; Beaton, R.; Bejger, M.; Cackett, E.; Dall'Ora, M.; Holanda, R.F.L.; Jensen, J.B.; Jha, S.W.; Lusso, E.; Minezaki, T.; et al. Astronomical Distance Determination in the Space Age. Secondary Distance Indicators. *Space Sci. Rev.* **2018**, *214*, 32. [CrossRef]

22. Sulentic, J.W.; Zamfir, S.; Marziani, P.; Dultzin, D. Our Search for an H-R Diagram of Quasars. *Revista Mexicana de Astronomia y Astrofisica Conference Series* **2008**, *32*, 51–58.

23. Boroson, T.A.; Green, R.F. The emission-line properties of low-redshift quasi-stellar objects. *Astrophys. J. Suppl. Ser.* **1992**, *80*, 109–135. [CrossRef]

24. Sulentic, J.W.; Marziani, P.; Zamanov, R.; Bachev, R.; Calvani, M.; Dultzin-Hacyan, D. Average Quasar Spectra in the Context of Eigenvector 1. *Astrophys. J. Lett.* **2002**, *566*, L71–L75. [CrossRef]

25. Zamfir, S.; Sulentic, J.W.; Marziani, P.; Dultzin, D. Detailed characterization of Hβ emission line profile in low-z SDSS quasars. *Mon. Not. R. Astron. Soc.* **2010**, *403*, 1759. [CrossRef]

26. Shen, Y.; Ho, L.C. The diversity of quasars unified by accretion and orientation. *Nature* **2014**, *513*, 210–213. [CrossRef] [PubMed]

27. Du, P.; Zhang, Z.X.; Wang, K.; Huang, Y.K.; Zhang, Y.; Lu, K.X.; Hu, C.; Li, Y.R.; Bai, J.M.; Bian, W.H.; et al. Supermassive Black Holes with High Accretion Rates in Active Galactic Nuclei. IX. 10 New Observations of Reverberation Mapping and Shortened Hβ Lags. *Astrophys. J.* **2018**, *856*, 6. [CrossRef]

28. Dultzin-Hacyan, D.; Schuster, W.J.; Parrao, L.; Pena, J.H.; Peniche, R.; Benitez, E.; Costero, R. Optical variability of the Seyfert nucleus NGC 7469 in timescales from days to minutes. *Astron. J.* **1992**, *103*, 1769–1787. [CrossRef]

29. Giveon, U.; Maoz, D.; Kaspi, S.; Netzer, H.; Smith, P.S. Long-term optical variability properties of the Palomar-Green quasars. *Mon. Not. R. Astron. Soc.* **1999**, *306*, 637–654. [CrossRef]

30. Sulentic, J.; Marziani, P.; Zamfir, S. The Case for Two Quasar Populations. *Balt. Astron.* **2011**, *20*, 427–434. [CrossRef]

31. Kuraszkiewicz, J.K.; Green, P.J.; Crenshaw, D.M.; Dunn, J.; Forster, K.; Vestergaard, M.; Aldcroft, T.L. Emission Line Properties of Active Galactic Nuclei from a Post-COSTAR Hubble Space Telescope Faint Object Spectrograph Spectral Atlas. *Astrophys. J. Suppl. Ser.* **2004**, *150*, 165–180. [CrossRef]

32. Sun, J.; Shen, Y. Dissecting the Quasar Main Sequence: Insight from Host Galaxy Properties. *Astrophys. J. Lett.* **2015**, *804*, L15. [CrossRef]

33. Dultzin, D.; Martinez, M.L.; Marziani, P.; Sulentic, J.W.; Negrete, A. Narrow-Line Seyfert 1s: A luminosity dependent definition. In Proceedings of the Conference "Narrow-Line Seyfert 1 Galaxies and Their Place in the Universe", Milano, Italy, 4–6 April 2011.

34. Fraix-Burnet, D.; Marziani, P.; D'Onofrio, M.; Dultzin, D. The Phylogeny of Quasars and the Ontogeny of Their Central Black Holes. *Front. Astron. Space Sci.* **2017**, *4*, 1. [CrossRef]

35. Leighly, K.M. Hubble Space Telescope STIS Ultraviolet Spectral Evidence of Outflow in Extreme Narrow-Line Seyfert 1 Galaxies. II. Modeling and Interpretation. *Astrophys. J.* **2004**, *611*, 125–152. [CrossRef]

36. Sulentic, J.W.; del Olmo, A.; Marziani, P.; Martínez-Carballo, M.A.; D'Onofrio, M.; Dultzin, D.; Perea, J.; Martínez-Aldama, M.L.; Negrete, C.A.; Stirpe, G.M.; et al. What does CivλA1549 tell us about the physical driver of the Eigenvector Quasar Sequence? *arXiv* **2017**, arXiv:1708.03187.

37. Elvis, M. A Structure for Quasars. *Astrophys. J.* **2000**, *545*, 63–76. [CrossRef]

38. Bisogni, S.; di Serego Alighieri, S.; Goldoni, P.; Ho, L.C.; Marconi, A.; Ponti, G.; Risaliti, G. Simultaneous detection and analysis of optical and ultraviolet broad emission lines in quasars at z 2.2. *arXiv* **2017**, arXiv:1702.08046.

39. Vietri, G.; Piconcelli, E.; Bischetti, M.; Duras, F.; Martocchia, S.; Bongiorno, A.; Marconi, A.; Zappacosta, L.; Bisogni, S.; Bruni, G.; et al. The WISSH Quasars Project IV. BLR versus kpc-scale winds. *arXiv* **2018**, arXiv:1802.03423.

40. Martínez-Aldama, M.L.; Del Olmo, A.; Marziani, P.; Sulentic, J.W.; Negrete, C.A.; Dultzin, D.; D'Onofrio, M.; Perea, J. Extreme quasars at high redshift. *Astron. Astroph.* **2018**, *618*, A179. [CrossRef]

41. Du, P.; Lu, K.X.; Hu, C.; Qiu, J.; Li, Y.R.; Huang, Y.K.; Wang, F.; Bai, J.M.; Bian, W.H.; Yuan, Y.F.; et al. Supermassive Black Holes with High Accretion Rates in Active Galactic Nuclei. VI. Velocity-resolved Reverberation Mapping of the Hβ Line. *Astrophys. J.* **2016**, *820*, 27. [CrossRef]

42. Heller, C.H.; Shlosman, I. Fueling nuclear activity in disk galaxies: Starbursts and monsters. *Astrophys. J.* **1994**, *424*, 84–105. [CrossRef]

43. Collin, S.; Zahn, J.P. Star formation and evolution in accretion disks around massive black holes. *Annu. Rev. Astron. Astrophys.* **1999**, *344*, 433–449.

44. Williams, R.J.R.; Baker, A.C.; Perry, J.J. Symbiotic starburst-black hole active galactic nuclei—I. Isothermal hydrodynamics of the mass-loaded interstellar medium. *Mon. Not. R. Astron. Soc.* **1999**, *310*, 913–962. [CrossRef]

45. D'Onofrio, M.; Marziani, P. A Multimessenger View of Galaxies and Quasars From Now to Mid-century. *Front. Astron. Space Sci.* **2018**, *5*, 31. [CrossRef]

46. King, A.; Pounds, K. Powerful Outflows and Feedback from Active Galactic Nuclei. *Annu. R. Astron. Astrophys.* **2015**, *53*, 115–154. [CrossRef]

47. Marziani, P.; Negrete, C.A.; Dultzin, D.; Martínez-Aldama, M.L.; Del Olmo, A.; D'Onofrio, M.; Stirpe, G.M. Quasar massive ionized outflows traced by CIV λ1549 and [OIII]λλ4959,5007. *Front. Astron. Space Sci.* **2017**, *4*, 16. [CrossRef]

48. Nardini, E.; Reeves, J.N.; Gofford, J.; Harrison, F.A.; Risaliti, G.; Braito, V.; Costa, M.T.; Matzeu, G.A.; Walton, D.J.; Behar, E.; et al. Black hole feedback in the luminous quasar PDS 456. *Science* **2015**, *347*, 860–863. [CrossRef] [PubMed]

49. Tetarenko, B.E.; Lasota, J.P.; Heinke, C.O.; Dubus, G.; Sivakoff, G.R. Strong disk winds traced throughout outbursts in black-hole X-ray binaries. *Nature* **2018**, *554*, 69. [CrossRef] [PubMed]

50. Baskin, A.; Laor, A. Metal enrichment by radiation pressure in active galactic nucleus outflows - theory and observations. *Mon. Not. R. Astron. Soc.* **2012**, *426*, 1144–1158. [CrossRef]

51. Negrete, A.; Dultzin, D.; Marziani, P.; Sulentic, J. BLR Physical Conditions in Extreme Population A Quasars: A Method to Estimate Central Black Hole Mass at High Redshift. *Astrophys. J.* **2012**, *757*, 62. [CrossRef]

52. Negrete, C.A.; Dultzin, D.; Marziani, P.; Sulentic, J.W. Reverberation and Photoionization Estimates of the Broad-line Region Radius in Low-z Quasars. *Astrophys. J.* **2013**, *771*, 31. [CrossRef]

53. Ferland, G.J.; Porter, R.L.; van Hoof, P.A.M.; Williams, R.J.R.; Abel, N.P.; Lykins, M.L.; Shaw, G.; Henney, W.J.; Stancil, P.C. The 2013 Release of Cloudy. *Revista Mexicana de Astronomía y Astrofísica* **2013**, *49*, 137–163.

54. Ferland, G.J.; Chatzikos, M.; Guzmán, F.; Lykins, M.L.; van Hoof, P.A.M.; Williams, R.J.R.; Abel, N.P.; Badnell, N.R.; Keenan, F.P.; Porter, R.L.; et al. The 2017 Release Cloudy. *Revista Mexicana de Astronomía y Astrofísica* **2017**, *53*, 385–438.

55. Punsly, B.; Marziani, P.; Bennert, V.N.; Nagai, H.; Gurwell, M.A. Revealing the Broad Line Region of NGC 1275: The Relationship to Jet Power. *Astrophys. J.* **2018**, *869*, 143. [CrossRef]

56. Shin, J.; Woo, J.H.; Nagao, T.; Kim, S.C. The Chemical Properties of Low-redshift QSOs. *Astrophys. J.* **2013**, *763*, 58. [CrossRef]

57. Sulentic, J.W.; Marziani, P.; del Olmo, A.; Dultzin, D.; Perea, J.; Alenka Negrete, C. GTC spectra of z ≈ 2.3 quasars: Comparison with local luminosity analogs. *Astron. Astrophys.* **2014**, *570*, A96. [CrossRef]

58. Peterson, B.M.; Wandel, A. Evidence for Supermassive Black Holes in Active Galactic Nuclei from Emission-Line Reverberation. *Astrophys. J. Lett.* **2000**, *540*, L13–L16. [CrossRef]

59. Peterson, B.M.; Wandel, A. Keplerian Motion of Broad-Line Region Gas as Evidence for Supermassive Black Holes in Active Galactic Nuclei. *Astrophys. J. Lett.* **1999**, *521*, L95–L98. [CrossRef]

60. Gaskell, C.M. What broad emission lines tell us about how active galactic nuclei work. *New Astron. Rev.* **2009**, *53*, 140–148. [CrossRef]

61. Sani, E.; Lutz, D.; Risaliti, G.; Netzer, H.; Gallo, L.C.; Trakhtenbrot, B.; Sturm, E.; Boller, T. Enhanced star formation in narrow-line Seyfert 1 active galactic nuclei revealed by Spitzer. *Mon. Not. R. Astron. Soc.* **2010**, *403*, 1246–1260. [CrossRef]

62. Abramowicz, M.A.; Czerny, B.; Lasota, J.P.; Szuszkiewicz, E. Slim accretion disks. *Astrophys. J.* **1988**, *332*, 646–658. [CrossRef]

63. Mineshige, S.; Kawaguchi, T.; Takeuchi, M.; Hayashida, K. Slim-Disk Model for Soft X-Ray Excess and Variability of Narrow-Line Seyfert 1 Galaxies. *Publ. Astron. Soc. Jpn.* **2000**, *52*, 499–508. [CrossRef]

64. Abramowicz, M.A.; Straub, O. Accretion discs. *Scholarpedia* **2014**, *9*, 2408. [CrossRef]

65. La Franca, F.; Bianchi, S.; Ponti, G.; Branchini, E.; Matt, G. A New Cosmological Distance Measure Using Active Galactic Nucleus X-Ray Variability. *Astrophys. J. Lett.* **2014**, *787*, L12. [CrossRef]

66. Tully, R.B.; Fisher, J.R. A new method of determining distances to galaxies. *Astron. Astrophys.* **1977**, *54*, 661–673.

67. Faber, S.M.; Jackson, R.E. Velocity dispersions and mass-to-light ratios for elliptical galaxies. *Astrophys. J.* **1976**, *204*, 668–683. [CrossRef]

68. Martínez-Aldama, M.L.; Del Olmo, A.; Marziani, P.; Sulentic, J.W.; Negrete, C.A.; Dultzin, D.; Perea, J.; D'Onofrio, M. Highly Accreting Quasars at High Redshift. *Front. Astron. Space Sci.* **2018**, *4*, 65. [CrossRef]

69. Marziani, P.; Negrete, C.A.; Dultzin, D.; Martinez-Aldama, M.L.; Del Olmo, A.; Esparza, D.; Sulentic, J.W.; D'Onofrio, M.; Stirpe, G.M.; Bon, E.; et al. Highly accreting quasars: A tool for cosmology? *IAU Symp.* **2017**, *324*, 245–246. [CrossRef]

70. Marziani, P.; Sulentic, J.W. Quasars and their emission lines as cosmological probes. *Adv. Space Res.* **2014**, *54*, 1331–1340. [CrossRef]

71. Levi, M.; Bebek, C.; Beers, T.; Blum, R.; Cahn, R.; Eisenstein, D.; Flaugher, B.; Honscheid, K.; Kron, R.; Lahav, O.; et al. The DESI Experiment, a whitepaper for Snowmass 2013. *arXiv* **2013**, arXiv:1308.0847.

72. Shemmer, O.; Lieber, S. Weak Emission-line Quasars in the Context of a Modified Baldwin Effect. *Astrophys. J.* **2015**, *805*, 124. [CrossRef]

73. Jarvis, M.J.; McLure, R.J. Orientation dependency of broad-line widths in quasars and consequences for black hole mass estimation. *Mon. Not. R. Astron. Soc.* **2006**, *369*, 182–188. [CrossRef]

74. Marziani, P.; Olmo, A.; Martínez-Aldama, M.; Dultzin, D.; Negrete, A.; Bon, E.; Bon, N.; D'Onofrio, M. Quasar Black Hole Mass Estimates from High-Ionization Lines: Breaking a Taboo? *Atoms* **2017**, *5*, 33. [CrossRef]

75. Risaliti, G.; Lusso, E. A Hubble Diagram for Quasars. *Astrophys. J.* **2015**, *815*, 33. [CrossRef]

Review

Dynamic Instability of Rydberg Atomic Complexes

Milan S. Dimitrijević [1,2,*], Vladimir A. Srećković [3], Alaa Abo Zalam [4], Nikolai N. Bezuglov [4,5] and Andrey N. Klyucharev [4]

[1]   Astronomical Observatory, Volgina 7, 11060 Belgrade, Serbia
[2]   Sorbonne Université, Observatoire de Paris, Université PSL, CNRS, LERMA, F-92190 Meudon, France
[3]   Institute of physics, University of Belgrade, P.O. Box 57, 11001 Belgrade, Serbia; vlada@ipb.ac.rs
[4]   Department of Physics, Saint Petersburg State University, 7/9 Universitetskaya nab., 199034 St. Petersburg, Russia; zalam@mail.ru (A.A.Z.); bezuglov50@mail.ru (N.N.B.); anklyuch@gmail.com (A.N.K.)
[5]   Rzhanov Institute of Semiconductor Physics SB RAS, 630090 Novosibirsk, Russia
*   Correspondence: mdimitrijevic@aob.rs; Tel.: +381-642-978-021

Received: 23 November 2018; Accepted: 31 January 2019; Published: 8 February 2019

Abstract: Atoms and molecules in highly excited (Rydberg) states have a number of unique characteristics due to the strong dependence of their properties on the values of principal quantum numbers. The paper discusses the results of an investigation of collisional Rydberg complexes specific features, resulting in the development of dynamic chaos and the accompanying diffusion autoionization processes. It is shown (experiment and theory) that, in subthermal low energies, the global chaotic regime that evolved in quasimolecular systems leads to significant changes in the Rydberg gases radiation/ionization kinetics. The effect of Förster resonance on the width of the fluorescence spectra and stochastic ionization processes in Rydberg systems is also discussed.

Keywords: Rydberg atoms; dynamic instability; control of atomic states; Förster resonance

1. Introduction

The interest in the research of physical processes involving highly excited (Rydberg) atomic systems is caused by their significance in the fundamental issues of science (due to the combination of quantum and classical properties) [1,2] and the prospects of their wide implementations in modern applied knowledge-intensive technologies (see, e.g., [3]). The main feature of Rydberg particles is their extremely big size $\sim n^2$ (where n is the principal quantum number), which results in huge dipole moments. This opens up unique opportunities for both the controlled and addressed management of quantum states by external electromagnetic fields [4], and for the creation of long-lived coherent (entangled) states in cold Rydberg media due to the long range dipole-dipole interaction between the medium particles [5,6]. Therefore, the cold Rydberg atoms are considered to be promising objects for solving the problems of quantum information. With their help, the physical carriers of quantum bits [7] can be realized with the simultaneous execution of the basic quantum operations [8].

Another class of interesting phenomena in dense gaseous media is associated with the collision and radiation kinetics of Rydberg electrons when they are scattered on cold atoms in the ground states, which can lead to the formation of exotic molecules [9,10] and specific chemical reactions [11]. Under certain conditions, cold Rydberg media quickly evolve into cold neutral plasma, which is accompanied by the formation of free electrons and the development of various phenomena, such as spontaneous plasma expansion, recombination of Rydberg atoms, plasma instabilities, and the propagation of collective waves [12]. The physics of ultra-low-temperature plasma should take into account not only traditional ionization and recombination processes involving electrons and ions [13,14], but also the formation of charged particles, which are both due to the ionization of

Rydberg atoms by thermal radiation [12,15] and as a result of the Penning autoionization of Rydberg atomic pairs [16,17].

Recombination processes leading to populating highly excited states are the sources of Rydberg atoms/molecules formation in stellar atmospheres of late spectral types, interstellar nebulae, and other space objects, including our solar system (see "Rydberg atoms in astrophysics" by Dalgarno in [1,18]). Rydberg particles play a fundamental role in the Earth's lower ionosphere, primarily affecting the propagation of satellite radio signals of the global positioning system (GPS) or radar stations [19]. Besides, they are a source of super background incoherent radiation in the decimeter and infrared ranges.

An important feature of the Rydberg systems is related to the "Coulomb" condensation of their quantum states near the energy continuum ($n \to \infty$). Due to strong Stark/Zeeman effects, interaction with electromagnetic fields can lead to multiple quasi-crossings and the mixing of Rydberg electron sublevels with different orbital (l), azimuthal (m), and principal (n) quantum numbers. This allows for the selectively excited initial state to begin a chaotic motion across a dense grid of Rydberg levels with the subsequent transition to the energy continuum. Such uncontrolled drift of a highly excited electron, leading to diffusion ionization, can be observed for single Rydberg atoms under the influence of external fields [20–22], as well as for an ensemble of Rydberg atoms with strong long-range dipole-dipole interaction [23]. In molecular or quasi-molecular collisional Rydberg complexes, the diffusion migration of the initial excitation causes both the dissociation of molecules [24] and the formation of molecular/atomic ions [25]. These diffusion processes, as induced by either external controlling or internal molecular fields, lead to the development of instability with the loss of initial coherence in the ensemble of Rydberg particles. The analysis of the dynamics of quantum complexes with a complex branched quasi-crossing structure of energy surfaces is extremely challenging within the framework of traditional quantum mechanical methods of calculation. The purpose of our work is to describe an alternative, semiclassical approach that is based on the concept of dynamic chaos evolution in Hamiltonian systems [26–28]. Another set of questions under consideration concerns the features of the radiation kinetics of Rydberg atoms in the vicinity of the Förster resonance. The latter is used as a controlling mechanism for varying the long-range interatomic interaction [29] and it has numerous applications in applied problems of Rydberg media. Noteworthy, the peculiarities of the radiative rate constants, as discussed below, are of potential interest for the interpretation of spectroscopic data, as obtained from fluorescence spectra of cold media of astrophysical relevance, such as different modifications of cold white dwarfs [30] or neutral sodium clouds near Jovian moon Io.

2. Kinetics of Radiative Transitions for Highly Excited Atoms

2.1. Spectral Parameters of an Excited Atom

We note that the atomic system of units is used, unless otherwise stated.

Knowledge of the probabilities of optical transitions lies at the basis of any analysis of processes involving excited atoms. Despite the fact that literature on this subject today has many dozens of works, the topic has not lost its actuality. As a rule, the existing theoretical methods for calculating the lifetime τ, or the radiation width $A = 1/\tau$ of the quantum state of an excited atom, leads to better agreement with the experiment in comparison with the probability of an individual optical transition. According to the review work from 1991 [31], even then, the discrepancy between the experimental and theoretical values of the radiation width for excited states of alkali metal atoms did not exceed 10%. Such accuracy was sufficient to use these data in an analysis of collisional processes in optically thin gas media and low-temperature plasmas. In the framework of the semiclassical approximation of quantum mechanics [32,33], the principal quantum number n determines in the first approximation the parameters of the orbit of the valence electron of the hydrogen atom and its energy. The orbital quantum number l determines the magnitude of the orbital angular momentum $L = l + 0.5$ (Langer's correction [33]), and also the degree of perturbation of the valence electron in the field of the atomic residue; the field of which may differ from the Coulomb. The corresponding "perturbed" value of energy ε is found, introducing the concepts of a quantum defect $\Delta\mu_l$ and an effective quantum number $n^* = n - \Delta\mu_l$: $\varepsilon = 1/(2n^{*2})$.

The experimental data on the lifetimes τ, as known today, can be described by a power law [31]:

$$\tau = \pi\sqrt{3}c^3/4 \cdot (l+0.5)^2 \cdot \alpha_l(n^*)^{\beta_l} \tag{1}$$

where α_l and β_l are dimensionless constants for a given l-series [31] and c is the speed of light. Accordingly, for example, for hydrogen and alkali atoms $\beta_l = 3$ [31,34]. At the primary selective excitation of Rydberg atoms (with a fixed value $n > 5$) and an increase of the concentration of normal atoms, the formation of a block of states with values of the quantum number l from 0 to $n-1$ and with an average lifetime $\sim n^5$, becomes possible due to intense atom-atom interactions. In this case, the principal fraction of the population of excited states is from those with large l values. In the case of a complete l mixing of the Hydrogen atom Rydberg states, the expression for the probabilities $\langle A \rangle_n$ of radiative decay of the block of n-states is written in the form [31,34]:

$$\langle A \rangle_n = \frac{8}{\sqrt{3}\pi c^3 n^5} \ln(1.414 \cdot n) \tag{2}$$

The current level of fundamental research and their engineering applications requires reliable data on the lifetimes of excited atoms. Until recently, such information for highly excited atoms in modern databases, as a rule, was absent. Importantly, Rydberg's lifetimes are highly dependent on ambient temperatures [15], so the correct measurement of their natural widths remains a challenge [35]. Concerning this, an extrapolation scheme for estimating the lifetimes of Rydberg states, tested while using the example of alkali metal atoms, has been developed [31].

2.2. Blocking of the Spectral Transitions. Double Stark (Förster) Resonance

One should have in mind that the term "hydrogen-like" atom, which is often used in the literature for Rydberg states, does not guarantee the complete coincidence of the structure of the energy levels of the excited atom with the energy structure of the hydrogen atom. Thus, for a "hydrogen-like" alkali metal atom with a standard [L-S] bond between the moments [33], the level $\{n, l+1\}$ may be located between the $\{n, l\}$ and $\{n+1, l\}$ states of the excited atom (see Figure 1), which does not correspond to the case of the hydrogen atom [33]. In the case when the $\{n, l+1\}$ level is located exactly in the middle between the $\{n, l\}$ and $\{n+1, l\}$ states, the Stark two-photon resonance condition is realized, otherwise the Förster resonance (hereinafter FR) [29] achieved in practice by an electric field of the order of 5 Vcm^{-1} [36]. The criterion for the emergence of the FR is simply formulated in terms of a quantum defect of atomic series:

$$\Delta\mu_l \equiv \mu_l - \mu_{l+1} = 0.5 \tag{3}$$

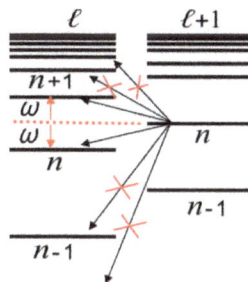

Figure 1. Scheme of Rydberg atomic levels, illustrating the effect of the double Stark (Förster) resonance in $\{l+1, l\}$ atomic series on blocking of the "long" transitions between remote states.

The simplest example is a three-dimensional oscillator with a frequency ω, for which the energies ε of the levels [33]

$$\varepsilon = \omega \cdot (2n - l + 3/2) \tag{4}$$

for all l-series satisfies FR. The unique frequency of oscillator emission is its natural frequency ω. Therefore, according to the Bohr-Heisenberg correspondence principle [32,37,38], the dipole matrix elements of the spectral transitions are nonzero only for "short" transitions between adjacent quantum states (see, for instance, Equations (2) and (9) in review work [39]). It means that, under conditions of FR, one should expect a significant weakening of the optical oscillator strengths for the "long" transitions (see Figure 1). This situation is most closely realized for the sodium atom for the s- and p-series (see Table 1), which causes the anomalously small radiation widths of the Rydberg states of sodium p-series, whose value is ~ 5 times smaller than the widths of the other alkali atoms (see Table 1, last row where $1/\tau_P$ are presented in) at the same energies of the corresponding states [31].

Table 1. Qantum defect μ_l for s-, p-series [32] along with $\Delta\mu_P$ and the factor $1/\tau_P$ from Equation (1).

	Li	Na	K	Rb	Cs	H
s	0.40	1.35	2.19	3.13	4.06	0
p	0.04	0.85	1.71	2.66	3.59	0
$\Delta\mu_P$	0.36	0.50	0.48	0.47	0.47	0
$10/\tau_P$	0.69	0.14	0.51	0.75	0.61	10

Thus, the Förster resonance is a unique phenomenon, in which anomalies in the spectral characteristics of excited atomic systems should be expected. Accordingly, in Ref. [36], under the conditions of the FR, the processes of blocking microwave transitions between Rydberg levels of rubidium in (A** + A) quasimolecular complex were investigated (experiment). From the academic point of view, the RF influence on the spectra of the excited atoms is considered in [40] within the framework of a model one-electron atom with the Sommerfeld potential [41]

$$U_Z(r) = -1/r + \alpha/(2r^2), \tag{5}$$

where r is the distance of the valence electron to the center of the atomic core and α is a parameter of the model. Sommerfeld introduced a potential of such a type in describing relativistic corrections to the theory of a hydrogen atom. An important particularity of Sommerfeld's potential is the possibility of an accurate analytical description of atomic parameters. For example, the energy of atomic states is given by the following expression

$$\varepsilon = -\frac{1}{2(n + l_{ef} - 1)^2}; \quad l_{ef} = \sqrt{(l + 0.5)^2 + \alpha} - 0.5 \tag{6}$$

The quantity l_{ef} is called the effective orbital number and is directly connected with the quantity of the quantum defect: $\Delta\mu_l = l - l_{ef}$. The concept of an effective orbital number l_{ef} is widely used in calculating the probability of radiative transitions in alkali atoms [42]. It allows, in particular, for describing the FR between l and $l - 1$ atomic series [43] by selecting the parameter $\alpha = \alpha_{l,l-1}$:

$$\alpha_{l,l-1} = 3 \cdot (l^2 - 0.25^2), \tag{7}$$

Accordingly, in the case of {p, s} series, $\alpha_{p,s}$ = 2.81. For {d, p} series, $\alpha_{d,p}$ = 11.8.

In Figure 2, the dependence of the radiative decay probability $A_{nl} = 1/\tau_{nl}$ (Einstein's coefficients) on the parameter α for fixed 30s and 25p states of the model Sommerfeld atom is shown. It can be seen that, in the vicinity of the FR, the radiative lifetimes of the Rydberg states can vary significantly (by orders of magnitude). Similar effects manifest themselves, and in collision Rydberg complexes [25,43], which will be discussed in the following sections.

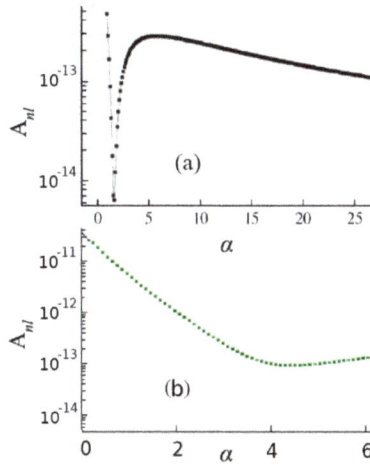

Figure 2. (a) Natural width A_{nl} of Rydberg state of s-series ($l = 0$) with $n = 30$ as a function of parameter α of Sommerfeld's atom. **(b)** The same for the state of p-series ($l = 1$) with $n = 25$. The Förster resonance corresponds to the value $\alpha_{p,s} = 2.81$.

3. Rydberg Quasimolecular Complex in the Framework of the Dipole Resonance Mechanism Model

A wide range of collisional processes

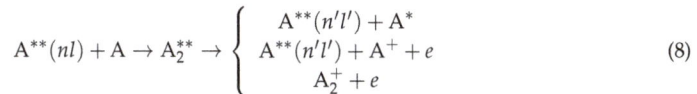

$$A^{**}(nl) + A \rightarrow A_2^{**} \rightarrow \begin{cases} A^{**}(n'l') + A^* \\ A^{**}(n'l') + A^+ + e \\ A_2^+ + e \end{cases} \tag{8}$$

with the participation of Rydberg states pass through the phase of formation of the intermediate Rydberg complex A_2^{**}. A fundamental contribution to "Rydberg Physics" was made by Fermi [44], who proposed considering the quasimolecular formation of A_2^{**} as a structure that consists of two positively charged atomic cores A^+, a quasi-free Rydberg electron (RE) e_{nl}^- in the Keplerian orbit, and the generalized valence electron e^- (see Figure 3). The further development of the Fermi approach [45–47], the so-called dipole resonance ionization (DRI), has found wide application in the solution of a diverse range of problems (see, for example, the review [13]) from the broadening and shift of spectral lines to the balance of ionization processes in the solar photosphere (see, for example, [25,48]).

Within the framework of the DRI model, the probability of realization of various final channels for the collision (8) is determined by the internal dipole moment **D** of the quasimolecule A_2^{**}. The moment **D** arises in the process of the charge exchange in the system ($A + A^+$) and it induces an alternating electric field $E(t)$, perturbing the motion of the RE e_{nl}^- on the Coulomb orbit (see Figure 3). With respect to the ionization channels for the DRI model, the following simplifying assumptions are accepted: (i) both the trajectory of the external electron e_{nl}^- and relative motion of the ion A^+ and unexcited atom A are semiclassical (ii) with the initial impact parameter ρ; and, (iii) ionization proceeds within a certain region with a given limiting distance R_{ion} between colliding atoms, which depends on the type of ion-atom residue and it is a parameter of the theory. The system is traditionally described in the adiabatic approximation using the appropriate potential curves of the Rydberg complex and the molecular ion A_2^+.

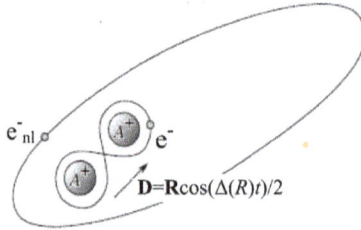

Figure 3. Scheme of highly excited collisional complex A_2^{**} where **D**—vector of the quasi-molecular ion dipole moment, **R**—vector of internuclear distance, e_{nl}^-—Rydberg electron, which is shared by the atomic cores of the quasi-molecular ion A_2^+.

A graphical illustration of the ionization process is given in Figure 4. Under the assumptions made, the charge exchange between atomic residues A^+ within the quasimolecule ion A_2^+ leads to the splitting $\Delta(R)$ (known as "exchange interaction" [45]) of its energy levels and it creates a time-dependent dipole moment $\mathbf{D} = \mathbf{R}^{\cos(\Delta(R)t)/2}$, which oscillates with frequency $\omega = \Delta(R)$, i.e., outside the complex A_2^+ an alternating quasimonochromatic microwave electric field $E(t)$ is induced (see Figure 3). We note that, as shown in [49], the effect of the field $E(t)$ on the Rydberg electron e_{nl}^- is equivalent to its perturbation by an external, spatially uniform field with the frequency $\omega_L = \Delta(R)$, and with polarization along the interatomic axis **R**. Ionization occurs inside the range of distances ($R < R_{ion}$), where the exchange interaction $\omega = \Delta(R)$), starting from the threshold value $\Delta(R_{ion})$, exceeds the binding energy $|\varepsilon_{nl}| = 1/(2n^{*2})$ of the e_{nl}^- electron and, thus, opens the autoionization channel of the quasimolecule complex A_2^{**}. The probability of ionization per unit time, or the autoionization width of the process, is expressed in terms of the photoionization cross section $\sigma_{ph}(nl, \omega)$.

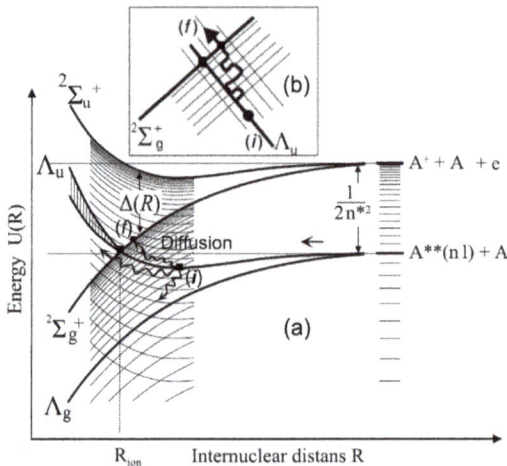

Figure 4. Mechanism of collisional ionization involving a Rydberg atom and an atom in normal state.

The considered mechanism of dipole resonance, which played an important role in the development of the physics of thermal collisions of heavy particles, has a strictly deterministic character. However, as the main quantum number n of the Rydberg atom $A^{**}(nl)$ increases, when the initial Λ_u energy curve (as shown in Figure 4) passes through a set of quasi-intersections with the neighboring Λ_g energy curves during the collision, the trajectory instability of the Rydberg electron e_{nl}^- begins to appear in the simiclassical approximation. As a result, the random migration of excitation occurs along the grid of energy terms, which have the energy separation $\Delta\varepsilon = 1/(n^*)^3$ between them. In the framework of modern concepts of nonlinear mechanics [26], this makes it possible to introduce the idea of dynamic chaos [27], in chemi-ionization processes in thermal and subthermal collisions that involve RA.

4. Rydberg Collisional Complex A_2^{**} in Approximation of Dynamic Chaos

At the end of the 19th century, in mathematics, Poincaré [27] introduced the notion of integrable and nonintegrable systems. In the first case, we meant a system with a "smooth" response to a small external perturbation and the conditions for the motion of individual particles are amenable to a direct description. However, in the general case, dynamical systems are not integrable, since perturbations that violate the total symmetry, as a rule, cannot be eliminated.

4.1. Nonlinear Dynamic Resonances and the Emergence of Deterministic Chaos

The main reason for the nontrivial influence of external periodic perturbations on the dynamic properties of integrable systems is associated with the appearance of nonlinear dynamic resonances [26], which, for large time intervals, leads to strong instability of the solutions obtained. In the framework of the KAM theory (Kolmogorov, Arnold, Moser) [27], it was shown that the presence of multiple resonances leads to the appearance of layers in the phase space of the considered ensembles of particles with diffusion motion that form a branched "stochastic web" [50]. To get into the element of the "stochastic web", the excitation energy of the initial state should be as close as possible to one of the "resonant" values. For given perturbation parameters, a set of quantum states associated with initial dynamic nonlinear resonances can be indicated.

The quantitative consideration of the instability of the RE dynamics in a periodic external microwave electromagnetic field $E(t) = \sum\limits_{m=0}^{\infty} \overline{E}_m \cos(m\,\omega_L)$ (where ω_L is the frequency) is based in the frame of the semiclassical approach on the Bohr–Heisenberg correspondence principles [32,37,38]. The largest perturbation by the field $E(t)$ (i.e., strong mixing between populations of two levels with fixed n_0 and $n_0 + k_0$ values of principal quantum numbers) should be expected in the case of m_0-photon resonance ($m_0 \geq 1$). For this, a "dynamic" coincidence of the $m_0\omega_L$ "photon" energy with the energy distance $k_0\omega_\varepsilon$ between the levels is necessary (see Figure 5):

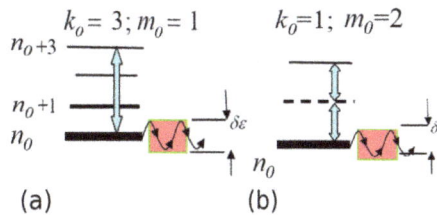

Figure 5. Schematic of nonlinear dynamic resonances of different $\{k_0 m_0\}$ orders for the initial quantum state n_0: (**a**) single-photon ($m_0 = 1$) and (**b**) double-photon ($m_0 = 2$) resonances.

$$m_0\,\omega_L = \varepsilon\,(n_0 + k_0) - \varepsilon\,(n_0) \approx k_0\,d\varepsilon/dn = k_0\,\omega_\varepsilon. \tag{9}$$

Note that the energy density $d\varepsilon/dn$ of the levels can be expressed through the frequency $\omega_\varepsilon = 1/n^{*3}$ of the classical revolution of the electron along the Keplerian orbit [32]. The realization of relation (9) is termed in the literature (see, for example, [27]) as the manifestation of a dynamic nonlinear resonance of the order $\{k_0 m_0\}$ in the vicinity of the energy $\varepsilon_{k_0,m_0} = \varepsilon(n_0)$ (see Figure 5). In this case, the electron energy ε begins to oscillate with respect to its unperturbed value with the amplitude $\delta\varepsilon$ [26,27]:

$$\delta\varepsilon = 2\sqrt{\frac{\omega_\varepsilon |\overline{E}_{m_0}|}{d\omega_\varepsilon/d\varepsilon}}; \qquad \varepsilon = \varepsilon_{k_0,m_0} \tag{10}$$

where $\overline{E}_{m_0}$ is an amplitude of the m_0-term at the frequency $m_0\omega_L$ in the Fourier series of the field $E(t)$.

The width $\delta\varepsilon$ of the nonlinear resonance (the amplitude of the energy oscillations) is characterized by a root dependence on the amplitude $\overline{E}_{m_0}$ of the perturbation. Out of resonance, the amplitude of the energy variations corresponds to a linear dependence on $E(t)$.

For linear systems $d\omega_\varepsilon/d\varepsilon = 0$, which formally lead to an arbitrarily large excitation of the electron ($\delta\varepsilon \to \infty$) without transition into the stochastic regime. For a nonlinear system, the finiteness of the widths of nonlinear resonances causes the emergence of a global chaos regime according to the Chirikov criterion [27]

$$K = \delta\varepsilon/\Delta\varepsilon > 1, \tag{11}$$

is satisfied, where $\Delta\varepsilon = \omega_\varepsilon = 1/n^{*3}$ is energy distance between neighboring levels. In the general case, for a particular system of quantum states of an excited atom, a whole set of quantum number values n_0 corresponding to different orders of dynamic resonances can be realized (see Figure 5). In the absence of the overlapping effect ($\delta\varepsilon < \Delta\varepsilon$) between the widths $\delta\varepsilon$ of neighboring resonances, separate islands of instability can arise in the energy space. Figure 6a shows the dynamics of the change in the energy of the Rydberg electron for such a case. It can be seen that the "overlap" of resonances is still not enough for the onset of global chaos, and the energy of the initial level n_0 undergoes finite oscillations within the width $\delta\varepsilon$ (10). In the case of the overlapping of resonances (see Figure 6b), an electron can go far from its initial state up to the states of a continuous spectrum (ionization) during diffusion. This corresponds to the onset of a global chaos regime. Here, the essential point is the threshold of the intensity of external perturbation, leading to a global chaos regime, the criterion of which appears in the inequality (11).

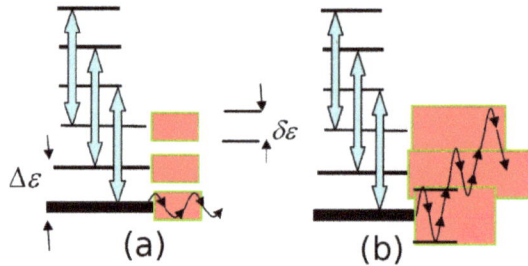

Figure 6. (**a**) Occurrence of dynamic nonlinear resonances without their widths $\delta\varepsilon$ overlapping. (**b**) Complete overlapping of widths $\delta\varepsilon$, corresponding to the formation of the global dynamic chaos. The initial states correspond to the lowest levels n_0 (bold lines).

4.2. The Standard Map (SM)

A nice illustration of the mechanisms that are involved in the transition to the global chaos is an example of the so-called standard mapping [26]. The standard map (SM), also known as the Chirikov–Taylor map [26], is one of the simplest models of chaos, in which the most characteristic and complex features of this problem are preserved. The corresponding model Hamiltonian, resulting in SM, has the form

$$H = H_0(I) + \frac{\overline{K}}{2\pi}\cos(\theta)\sum_{n=-\infty}^{\infty}\delta\left(\frac{t}{2\pi} - n\right); \qquad H_0(I) = \frac{I^2}{2} \tag{12}$$

and describes a rotor (an initial free atomic system with Hamiltonian $H_0(I)$) periodically kicked by delta-impulses of some external force (a perturbation). Here, $T = 2\pi$ is the corresponding period, while $\overline{K}$ is a dimensionless parameter that characterizes the force amplitude. The rotor phase space variables "angle-action" (θ, I) are defined on a cylinder ($-\infty < I < \infty; 0 < \theta < 2\pi$).

When considering the identity

$$\sum_{n=-\infty}^{\infty} \delta(t/T - n) = \sum_{m=-\infty}^{\infty} \exp(i2\pi mt/T), \tag{13}$$

we can rewrite the Hamiltonian (12) in the form

$$H = \frac{1}{2}I^2 + \frac{\overline{K}}{4\pi} \sum_{m=-\infty}^{\infty} [\exp(im\omega_L t + \theta) + \exp(im\omega_L t - \theta)] \tag{14}$$

with $\omega_L = 2\pi/T = 1$. Expression (14) corresponds to the motion of the rotor in a periodic wave packet with an infinite number of harmonics. The amplitudes of the harmonics have the same magnitude, while their phases depend on the rotor angle θ.

To identify the range of applicability of the Chirikov criterion (11), consider the Hamiltonian equations of the perturbed rotor (14)

$$\dot{I} = i\frac{\overline{K}}{4\pi} \sum_{m=-\infty}^{\infty} [\exp(im\omega_L t + i\theta) - \exp(im\omega_L t - i\theta)]; \tag{15}$$

$$\dot{\theta} = \omega(I); \qquad \omega(I) = dH_0/dI = I, \tag{16}$$

where $\omega(I)$ is an angular frequency of the rotor having the angular momentum I. The nonlinear resonances (9) arise in the frame of the zero order of the perturbation theory when we use in (15) unperturbed temporal dependence for the angle variable: $\theta(t) = I \cdot t$. The external force provides the strongest influence at the rotor dynamics if the series (15) contain stationary terms, i.e., the action variable I should satisfy $m\omega_L = \pm\omega(I) = \pm I$. This relation corresponds to Equation (11) provided $k_0 = \pm 1$, which means that the nonlinear resonances under consideration have orders $\{k = \pm 1, m\}$ with $I_{k,m} = m/k \cdot \omega_L = m/k$ and $\varepsilon_{k,m} = m^2/2$. Since $I_{-1,-m} = I_{1,m}$, we may restrict ourselves to the case of the resonance values $I_{1,m}$. In the vicinity of one of these resonances ($I \approx I_{1,m_0} = m_0$), one may simplify series (15) by dropping all the oscillating terms. This transforms the system (15), (16) into

$$\dot{\delta I} \approx \overline{K}/(2\pi) \sin(\theta - m_0 t); \qquad \dot{\Psi} = \delta I \tag{17}$$

with $\delta I = I - I_{1,m_0} = I - m_0$ and $\Psi = \theta - m_0 t$. System (17) is reduced to the nonlinear equation of the phase oscillation

$$\ddot{\Psi} \approx \overline{K}/(2\pi) \sin(\Psi) \tag{18}$$

Equations (17), (18) result in the oscillation of the action δI with the maximum amplitude $\max\delta I^2 \sim 4\overline{K}/\pi$ corresponding to the maximum distance between two separatrix branches that are generated in the phase space by Equation (7) [27]. We may assess the amplitudes $\delta\varepsilon$ of the energy oscillations of the resonance states that are depicted in Figure 6 as $\delta\varepsilon \approx |\omega(I)| \cdot \max|\delta I|$. The distance $\Delta I_{1,m}$ between two adjacent resonance values $I_{1,m}$ is $\Delta I_{1,m} = 1$, which gives, for the corresponding energy separation, $\Delta\varepsilon = |\omega(I)|\Delta I = |\omega(I)|$ (see Figure 6). Chirikov criterion (11), hence, is reduced to the form

$$K \approx \frac{|\omega(I)| \cdot \max|\delta I|}{|\omega(I)|} = 2\sqrt{\overline{K}/\pi} > 1. \tag{19}$$

The characteristic important feature of the Hamiltonian (12) in the standard map model is the simultaneous overlapping of all resonance energies widths, as shown in Figure 6b. Figure 7 demonstrates the evolution of the rotor trajectories $(I(t), \theta(t))$ in the phase space for some parameter $\overline{K}$ values. It is clearly seen that, initially ($\overline{K} = 0.772$), small islands of instabilities (black areas) increase their size as the parameter $\overline{K}$ increases, forming a "stochastic sea" for large amplitudes ($\overline{K} = 3.972$) of the external force.

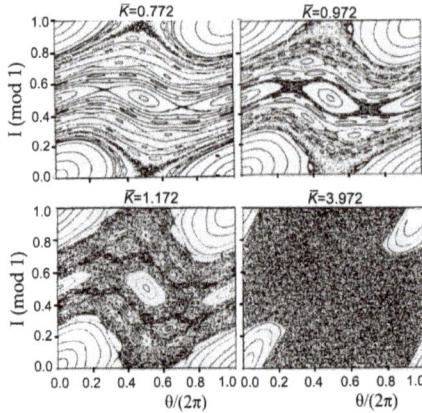

Figure 7. Phase space trajectories of a kicked rotor with Hamiltonian (12) (adopted from [26]).

4.3. Conception of Diffusional Ionization

A significant contribution to the development of the theory of stochastization of quantized systems was made by the authors of Refs. [20,51], who considered the evolution of a bound Rydberg electron with its ionization in an external microwave field. The adaptation of the methods of work [51] in describing the development of dynamic chaos in the act of a single collision (8) is described in [49]. It is shown that, under the influence of a quasimonochromatic internal electric field $E(t)$, nonlinear dynamic resonances can arise due to the coincidence of the overtone $k_0 \omega_\varepsilon$ of the angular frequency ω_ε of motion of RE e_{nl}^- on the Keplerian orbit with charge-exchange frequency $\Delta(R)$ of the internal electron e^- (see Figure 3). As a result, the motion of the RE becomes unstable, and the RE evolution in the energy space takes the character of random walks along the quasi-intersecting "grid" of potential curves (see Figure 4), which opens the possibility of a kinetic description of the RE dynamics

In a series of subsequent works [22,52–54], data on the formation of dynamic nonlinear resonances with a transition to the stage of global chaos for isolated atoms in an external linearly polarized electric field $E(t) = E_0 \cos(\omega_L t + \theta)/2$ were refined. We note that, the widths of the dynamic resonances (10), as well as the effects of stochastic dynamics, are directly related to the matrix elements of the perturbation operators. This conclusion is confirmed by the results of [52], in which the coefficients of the light-induced diffusion equation for a weakly bound electron in an external microwave field are explicitly expressed in terms of the dipole matrix elements for optical transitions. Since the implementation of the diffusion ionization requires finite time, to take into account the regime of dynamic chaos becomes important for slow collisions, which is for a range of thermal and subthermal energies. A transition to the ionization continuum due to a stochastic walk (see Figure 6b) can be described, with good accuracy, in terms of kinetic methods using the distribution function $\overline{f}(\varepsilon, t)$ [22,27] in the bound state region of a Rydberg electron ($\varepsilon < 0$). For calculations, it is more convenient to work with the distribution $f(n, t)$ over the principal quantum number n

$$f(n, t)\, dn = \overline{f}(\varepsilon, t)\, d\varepsilon, \tag{20}$$

using the relation $\varepsilon = -1/(2n^2)$. In this case, the time dependence of $f(n, t)$ is found from the solution of a diffusion equation of the Fokker–Planck type [27]

$$\frac{\partial}{\partial t} f(n, t) = \frac{\partial}{\partial n} D_n \frac{\partial}{\partial n} f(n, t) \tag{21a}$$

with the initial condition that at the moment $t = 0$ the excited electron is at the level n_0, i.e., $f(n, t = 0) = \delta(n - n_0)$ ($\delta(x)$ is the Dirac delta-function). To determine the evolution of $f(n, t)$, Equation (21a) must be supplemented by the following boundary conditions with respect to the quantum variable n at the boundaries of the region of development of stochasticity $nc < n < \infty$:

$$f(n \to \infty, t) = 0, \qquad \frac{\partial}{\partial n} f(n = n_c, t) = 0, \tag{21b}$$

where the quantity n_c is determined from the Chirikov criterion (11), as [22]

$$n_c^4 (n_c \, \omega_L)^{1/3} = 1/(49 E_0) \tag{22}$$

The boundary of the continuum ($n = \infty$) is an absorbing wall for the diffusion flux where $f(n)$ becomes zero. The amplitude E_0 of the microwave field determines the critical value n_c of the principal quantum number below which the electron has regular motion, and the underlying states ($n < n_c$) refer to the deterministic region, since here the energy separation $\Delta \varepsilon = 1/n^3$ of the neighboring levels exceed the widths of the dynamic resonances (see Figure 6a). Thus, the values $n = n_c$ determine the position of the "reflecting wall", on which the diffusion flux vanishes.

According to [22,51], the value of the diffusion coefficient D_n is

$$D_n \cong 0.65 \, E_0^2 n^3 \, \omega_L^{-4/3} \tag{23}$$

From here, the average time, τ_{eff}, required to reach the ionization limit by RE, which begins its diffusion motion with an initial binding energy $\varepsilon = -1/(2n_0^2)$, is determined as [24,49]

$$\tau_{\text{eff}}(n_0) = \frac{\omega_L^{4/3}}{0.65 E_0^2} n_0^2 \left(1 - \frac{n_c}{2 n_0}\right) \tag{24}$$

The distribution $f(n, t)$ makes it possible to find parameters of the diffusion process, such as the average number of jumps from one level to another, experienced by the particle during its diffusion drift to the continuum of energies, the average lifetime of the excited atom, and the degree of its "survival" at a given time interval, i.e., knowledge of $f(n, t)$ relatively simply makes it possible to obtain theoretical data for comparison with the values that were observed in the experiment.

4.4. Diffusional Ionization of Hydrogen Atom in External Field

More detailed information on the processes of diffusional ionization requires numerical calculations. Within the semiclassical approximation, numerical data are extracted from the study of the dynamics of motion (trajectories) of the RE. Analysis of time processes with the strong stochastization of trajectories requires the use of a stable numerical calculation scheme. The corresponding algorithm, which is based on the Floquet technique [55] and geometric integration methods [56,57], was proposed in [54] to find the parameters of atomic systems that were subjected to external periodic fields.

As an example, Figure 8 shows the trajectories of motion and shows the time dependence of the orbital angular momentum **L** in conditions of dynamic chaos development. We take the initial 10P ($n_0 = 10, l_0 = 1$) state of the hydrogen atom in the microwave field with frequency $\omega_L = 3/10^3$ and amplitude E_0, exceeding its threshold value $E_c = 2/(49 n_0^4)$ [22]. Two characteristic initial configurations of the vector $\mathbf{L}_0$ and the Runge-Lenz vector $\mathbf{A}_0$ (directed along the semiaxis of the unperturbed Keplerian orbit [58]) were chosen, corresponding to the maximal changes in the modulus $|\mathbf{L}|$ for cases of two-dimensional ($E_0 = 8, 2E_c$) and three-dimensional ($E_0 = 6, 5E_c$) trajectories. Note that, according to the literature data, the range of values of $n \approx 10$ corresponds to the range of strong interaction of the dipole field of the cluster A_2^+ with the Rydberg electron e_{nl}^- (see Figure 3). Also note that the results of numerical calculations in Figure 8 show a significant change of the orbital momentum **L** in the microwave field under the conditions of development of global chaos, which is in contrast to the main approximation of the authors [22,51], assuming the adiabatic invariance of **L**.

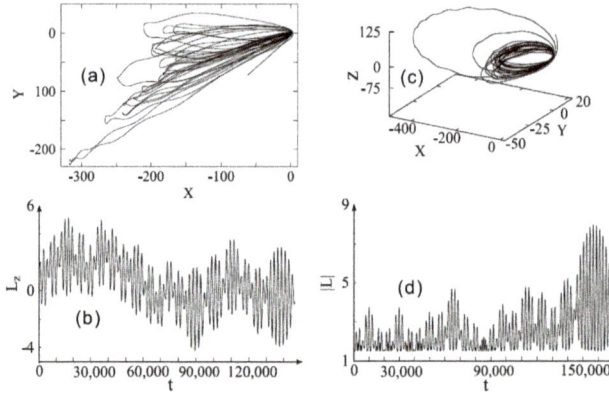

Figure 8. Trajectories of Rydberg electron (frames **a,c**) and the evolution of its angular momentum **L** (frames **b,d**) for the 10P-state ($n_0 = 10, l_0 = 1$) of the hydrogen atom. For two-dimensional (2D) motion of an electron in the $\{X, Y\}$-plane (frame **a**), frame **b** shows the projection L_Z of the momentum **L** on the z-axis, orthogonal to the motion plane $\{X, Y\}$. For three-dimensional (3D) motion (frame **c**), frame **d** shows $|\mathbf{L}|$.

4.5. Diffusional Ionization of the Rydberg Colisional Complex

The diffusion Equation (13) have a number of specific features with respect to the calculations of chaotic motion of RE e_{nl}^- in collisional molecular complexes due to the variation of both the amplitude and the frequency of the internal electric field $E(t)$ as the (adiabatic) internuclear distance R changes (see Figures 3 and 4). At the same time, the RE ionization boundary shifts (see Figure 4), since it is determined by the potential curve $^2\Sigma_g^+$. Figure 9 shows the boundaries of the stochasticity region for RE (see the discussion of Equation (21b)) in the case of collisions of hydrogen A and A** atoms: the lower dashed curve (circles) corresponds to the lower "reflective wall" $n_{min}(R(t))$ ($n_{min} = n_c$ is found from Equation (22)), while the upper dotted curve (crosses) defines the position of the "absorbing wall" $n_{max}(R(t))$. For the Rydberg states, lying above $n_{max}(R)$, the exchange interaction $\Delta(R)$ exceeds their binding energies, so that the internal microwave field results in the fast photoionization of those states. The parameters of the problem are chosen in such a way that no ionization occurs when the RE ($n_0 = 10$) motion is regular (i.e., without any stochatization).

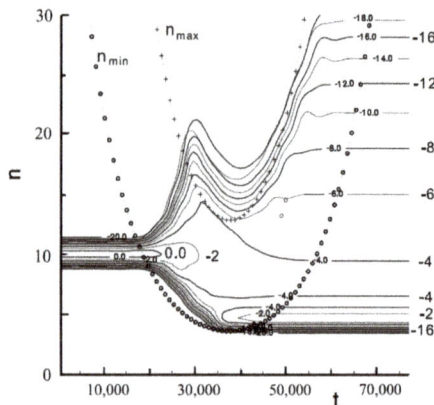

Figure 9. The time evolution of the distribution function $f(n,t)$ of a Rydberg electron e_{nl}^- in a quasimolecular collisional complex A_2^{**} with the impact parameter $\rho = 15$ and collision energy $1.9 \cdot 10^{-3}$ a. u. = 600 K. The initial value $n_0 = 10$.

Figure 9 also shows the results of calculating [47]; the parameters of RE diffusion evolution for a time-varying region of the phase space. The initial distribution, corresponding to the localization of the RE at the energy level with $n = n_0 = 10$, was chosen as the narrow Gaussian distribution $f(n, t = -\infty) = \exp(-(n - n_0)^2)/\sqrt{\pi}$. The principal feature of the present is the realization of the RE diffusion along the energy levels in the act of a single collision. The solid lines in the figure give a map of levels $\ln f(n, t)$—the values of the logarithm are plotted alongside the corresponding curves. The position of the minima of the boundary curves $n_{max/min}(R(t))$ is determined by the turning point with the minimum approach of the nuclei. It is evident that, while the initial value $n_0 = 10$ is located lower than $n_{min}(R(t))$, i.e., lies in the regular motion region, there is no diffusion along the n axis. Diffusion begins to develop after the moment of entry of the initial value $n_0 = 10$ in the stochastic region Ω_{st} (the region lying above the curve $n_{min}(R(t))$). As a result of stochastic diffusion, the Rydberg electron has a finite probability of reaching the ionization boundary and passing to a continuum of energies through the photoionization channel. In this case, ionization is found as the probability of the electron leaving the region of bound states. The validity of such a model was verified on the quantitative level by the example of the chemi-ionization process involving Rydberg atoms of sodium [49].

4.6. Assotiative Ionization Rate Constants

The experimental methods existing in atomic and molecular physics make it possible to compare the ionization parameters that were calculated in the frame of stochastic dynamics with the results of direct measurements performed in atomic beams of different types [59]. Atomic/molecular beams have been widely implemented in practice as convenient sources of particles that are used for investigations in the physics of collisions [60], spectroscopy [61], and the analysis of the interaction of light and matter [62]. In recent decades, a new type of beams, known as "cold beams" [63], has been added to the two classical types of beams, i.e., diffusion [14,64] and supersonic [59] ones. Cold beams are extracted from magneto-optical traps and they have unique prospects for use in nanotechnology [65] due to their extremely narrow divergence angle.

An important parameter that characterizes the efficiency of ionization processes (8) is the rate constants $K(nl, T)$. In experiments, rate constants are found from a measured number of registered charged particles. A theoretical treatment of $K(nl, T)$

$$K(nl, T) = \int \sigma(v_c) \cdot v_c f(v_c, T) dv_c \tag{25}$$

operates with the cross sections $\sigma(v)$ and the distribution function $f(v, T)$ over the relative (impact) velocity $v = v_c$ of colliding atoms of the same mass M:

The velocity distribution $f(v_c, T)$ dependence on the beams source temperature T may be expressed via the characteristic thermal velocity $v_T = (kT/M)^{1/2}$ as $f(v_c, T) = F(v_c/v_T)/v_T$. The function $F(x)$ has quite different profiles in cells, single beams, and crossed beams cases, as demonstrated in Figure 10 [13,64].

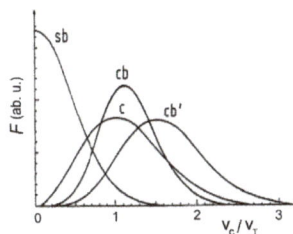

Figure 10. Distribution function *F* in a single beam (sb), crossed beams (cb), counter beams (cb′), and gas cell (c).

Figures 11 and 12 show, following [53,60,66–68], the values of the associative ionization (AI) rate constants $K_{ai}(nl, T)$ obtained under the conditions of a single and two orthogonal beams. The formation of molecular ion occurs upon the collisions of Rydberg sodium atoms $A^{**}(nl)$ excited to nS, nP, or nD states with atoms A in the ground state. Experimental data of Figure 11, frame (a) (dots, $l = 1$), corresponds to crossed beams conditions, T = 600 K [66]. Figure 11 data related to frame (b) (open circles, $l = 1$), frame (c) (open triangles, $l = 2$), and frame (d) (open squares, $l = 2$) were obtained in a single beam of Na atoms, T = 1000 K [67]. Full curves exhibit results of theory [60], accounting for stochastic diffusion effects, while dotted curves correspond to calculations in the frame of regular DSMJ model [43–45].

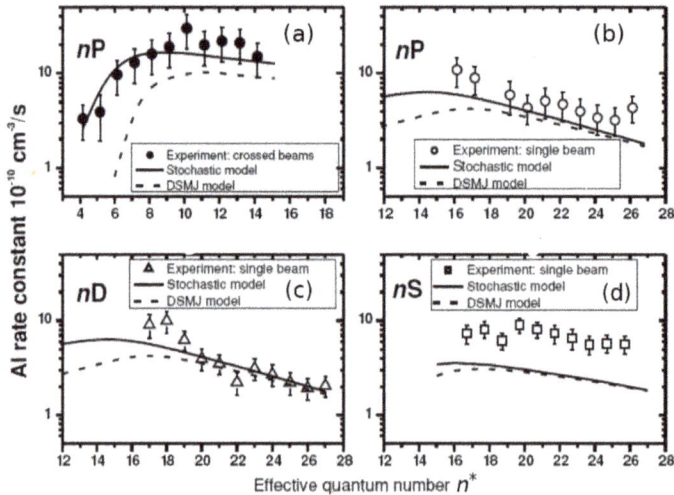

Figure 11. Values of rate constants of associative ionization for collisions in a single beam (frames (**b–d**)) and in two orthogonal crossed beams (frames (**a**)) of sodium atoms excited in nS, nP, and nD Rydberg states [66,67].

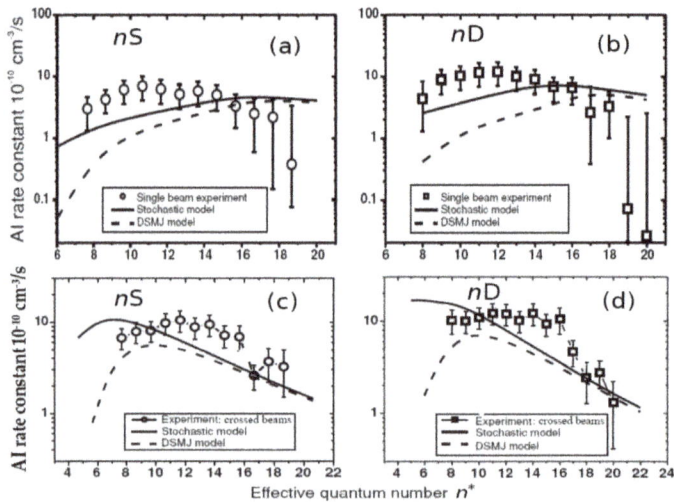

Figure 12. Associative ionization rate constants for Na*(nl) + Na(3s) collisions in a single beam [53] (frames (**a,b**)) and in two orthogonal crossed beams [66] (frames (**c,d**)), T = 600 K.

Figure 12 shows the results of more recent experiments [53,66] on AI rate constants measurements, the authors of which took into account the effects of free electrons escaping due to atoms photoionization by black body radiation.

It can be seen that, in the $4 < n^* < 28$ range of effective quantum numbers, the data of the experiment and the stochastic theory (solid curves) agree with each other (although measurements have large error bars), whereas calculations using the traditional DSMJ model [43–45] (dashed curves) significantly underestimate the corresponding results, particularly for the lower values of n.

4.7. Features of Diffusional Ionization under Conditions of Förster Resonance

In Section 2.2, a significant decrease in the probabilities of Rydberg states radiative decay in Förster resonance (FR) conditions was demonstrated, which is due to the suppression of dipole matrix elements. Since the widths of the nonlinear resonances $\delta\varepsilon$, as shown in [43,52], are directly related to the probabilities of microwave transitions in the excited atom, the blocking of the latter means the blocking of the development of dynamic chaos. The possibility of "controlling" the development of global chaos in the Rydberg diatomic cluster, using the double Stark resonance (or FR) mechanism, was considered in [40,43].

Figure 13 shows the time dependence of the RE binding energy under conditions of the development of diffusional ionization of Rydberg states of the model Sommerfeld atom, which is under the influence of an external microwave field, for different values of the Sommerfeld parameter α (see Equation (5)). It can be seen that the ionization time is significantly prolonged when the FR is realized ($\alpha = 2.81$), which indicates the partial blockage of global stochatization.

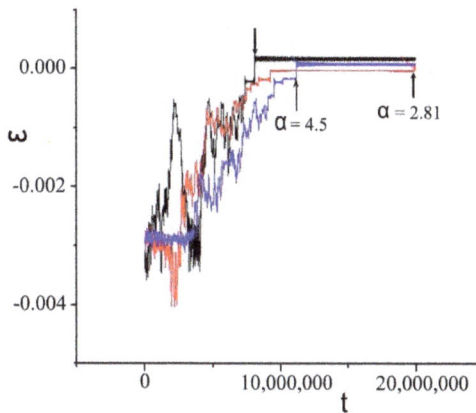

Figure 13. Temporal evolution of binding energy ε of the 13P-state ($l = 1$) of the Rydberg electron in the Sommerfeld atom in an external microwave field of frequency $\omega = 1/13^3$ and amplitude $E_0 = 10E_c$, which exceeds ten times the critical value E_c. The calculations were performed for three values of Sommerfeld parameter: $\alpha = 0$, 2.81, and 4.5. The arrows indicate the moments of ionization. The occurrence of Förster resonance corresponds to $\alpha_{p,s} = 2.81$.

5. Conclusions

Our work presents the results of studies (experiment and theory) of the radiative and collisional kinetics of Rydberg atoms, with their specific features arising from the closeness of highly excited bound states to the energy continuum. The model problems that are considered here reveal significant changes (by orders of magnitude) in the dipole matrix elements values of optical transitions in the vicinity of the Förster resonance, which is an important tool for creating entangled states in the system of cold Rydberg atoms. In particular, the anomalously long lifetimes of the p-series Rydberg states of the sodium atom are explained by the proximity of the Na p- and s-series energy levels structure to the Förster resonance configuration. We have considered the ionization processes of Rydberg atomic complexes in microwave electric fields. Under the multiplicity of the quasi-crossing of energy levels near the ionization continuum, the semiclassical method for taking into account the ionization instability of Rydberg states is discussed. This method is based on the formalism of nonlinear dynamic resonances and the evolution of dynamic chaos in Hamiltonian systems. The results of the numerical modeling of diffusion ionization of atomic hydrogen Rydberg states in an external microwave field are given with a demonstration of the nontrivial evolution of the orbital moment. The possibility of reducing the theoretical analysis of collisional ionization of Rydberg alkali metal atoms to the problem of the stochastic ionization of a Rydberg electron by an "internal" microwave field is shown. This field is induced by the charge exchange processes in the system "ion core of a Rydberg atom plus neutral atom-collision partner". The comparison of the experimental and calculated data on the associative ionization rate constants in "Rydberg sodium atom-normal sodium atom" collision demonstrates the validity of describing the dynamic instability of Rydberg complexes within the framework of the dynamic chaos evolutionary theory for Hamiltonian systems.

Author Contributions: A.N.K. contributed analysis tools; A.A.Z. and V.A.S. analyzed the data with writing-original draft preparation; N.N.B., M.S.D. wrote the paper.

Funding: This research was partially funded by Ministry of Education, Science and Technological Development of the Republic of Serbia grants number III 44002 and 176002, and by the Russian Science Foundation under the grant NO 18-12-00313 in the part regarding the control of Rydberg complexes rate constants via Förster resonance.

Acknowledgments: Authors acknowledge the support of Russian Science Foundation under the grant NO 18-12-00313 in the part regarding the control of Rydberg complexes rate constants via Förster resonance.

Conflicts of Interest: The authors declare no conflict of interest.

References

1. Stebbings, R.F.; Dunning, F.B. (Eds.) *Rydberg States of Atoms and Molecules*; Cambridge University Press: Cambridge, UK, 2011.

2. Gallagher, T.F. *Rydberg Atoms*; Cambridge Monographs on Atomic, Molecular and Chemical Physics; Cambridge University Press: Cambridge, UK, 2005.

3. Jones, M.P.A.; Marcassa, L.G.; Shaffer, J.P. Special issue on Rydberg atom physics. *J. Phys. B At. Mol. Phys.* **2017**, *50*, 060202. [CrossRef]

4. Lim, J.; Lee, H.-G.; Ahn, J. Review of cold Rydberg atoms and their applications. *J. Korean Phys. Soc.* **2013**, *63*, 867–876. [CrossRef]

5. Hofmann, C.S.; Günter, G.; Schempp, H.; Müller, N.L.; Faber, A.; Busche, H.; Robert-de-Saint-Vincent, M.; Weidemüller, M. An experimental approach for investigating many-body phenomena in Rydberg-interacting quantum systems. *Front. Phys.* **2014**, *9*, 571–586. [CrossRef]

6. Pillet, P.; Gallagher, T.F. Rydberg atom interactions from 300 K to 300 K. *J. Phys. B At. Mol. Opt. Phys.* **2016**, *49*, 174003. [CrossRef]

7. Saffman, M.; Walker, T.G.; Mølmer, K. Quantum information with Rydberg atoms. *Rev. Mod. Phys.* **2010**, *82*, 2313. [CrossRef]

8. Ryabtsev, I.I.; Beterov, I.I.; Tretyakov, D.B.; Entin, V.M.; Yakshina, E.A. Spectroscopy of cold rubidium rydberg atoms for applications in quantum information. *Phys.-Uspekhi* **2016**, *59*, 196–208. [CrossRef]

9. Marcassa, L.G.; Shaffer, J.P. Interactions in Ultracold Rydberg Gases. In *Advances in Atomic, Molecular, and Optical Physics*; Arimondo, E., Berman, P.R., Lin, C.C., Eds.; Academic: New York, NY, USA, 2014; Volume 63, pp. 47–133.

10. Shaffer, J.P.; Rittenhouse, S.T.; Sadeghpour, H.R. Ultracold Rydberg molecules. *Nat. Commun.* **2018**, *9*, 1965. [CrossRef] [PubMed]

11. Schlagmüller, M.; Liebisch, T.C.; Engel, F.; Kleinbach, K.S.; Böttcher, F.; Hermann, U.; Westphal, K.M.; Gaj, A.; Löw, R.; Hofferberth, S.; et al. Ultracold Chemical Reactions of a Single Rydberg Atom in a Dense Gas. *Phys. Rev. X* **2016**, *6*, 031020. [CrossRef]

12. Lyon, M.; Rolston, S.L. Ultracold neutral plasmas. *Rep. Prog. Phys.* **2017**, *80*, 017001. [CrossRef] [PubMed]

13. Klyucharev, A.N.; Vujnović, V. Chemi-ionization in thermal-energy binary collisions of optically excited atoms. *Phys. Rep.* **1990**, *185*, 55–81. [CrossRef]
14. Graham, W.G.; Fritsch, W.; Hahn, Y.; Tanis, J.A. *Recombination of Atomic Ions*; Springer Science & Business Media: Berlin, Germany, 2012.
15. Beterov, I.I.; Tretyakov, D.B.; Ryabtsev, I.I.; Entin, V.M.; Ekers, A.; Bezuglov, N.N. Ionization of rydberg atoms by blackbody radiation. *New J. Phys.* **2009**, *11*, 013052. [CrossRef]
16. Hahn, Y. Density dependence of molecular autoionization in a cold gas. *J. Phys. B At. Mol. Opt. Phys.* **2000**, *33*, L655. [CrossRef]
17. Efimov, D.K.; Miculis, K.; Bezuglov, N.N.; Ekers, A. Strong enhancement of Penning ionization for asymmetric atom pairs in cold Rydberg gases: The Tom and Jerry effect. *J. Phys. B At. Mol. Opt. Phys.* **2016**, *49*, 125302. [CrossRef]
18. Gnedin, Y.N.; Mihajlov, A.A.; Ignjatović, L.J.M.; Sakan, N.M.; Srećković, V.A.; Zakharov, M.Y.; Bezuglov, N.N.; Klycharev, A.N. Rydberg atoms in astrophysics. *New Astron. Rev.* **2009**, *53*, 259–265. [CrossRef]
19. Buenker, R.J.; Golubkov, G.V.; Golubkov, M.G.; Karpov, I.; Manzheliy, M. Relativity laws for the variation of rates of clocks moving in free space and GPS positioning errors caused by space-weather events. In *Global Navigation Satellite System-From Stellar Navigation*; Mohamed, A.H., Ed.; In Tech: Berlin, Germany, 2013.
20. Koch, P.M.; van Leeuwen, K.A.H. The importance of resonances in microwave "ionization" of excited hydrogen atoms. *Phys. Rep.* **1995**, *255*, 289–403. [CrossRef]
21. Mitchell, K.A.; Handlay, J.P.; Tighe, B.; Flower, A.; Delos, J.B. Analysis of chaos-induced pulse trains in the ionization of hydrogen. *Phys. Rev. A* **2004**, *70*, 043407. [CrossRef]
22. Krainov, V.P. Ionization of atoms in strong low-frequency electromagnetic field. *J. Exp. Theor. Phys.* **2010**, *111*, 171–179. [CrossRef]
23. Park, H.; Shuman, E.S.; Gallagher, T.F. Ionization of Rb Rydberg atoms in the attractive *ns-np* dipole-dipole potential. *Phys. Rev. A* **2011**, *84*, 052708. [CrossRef]
24. Dashevskaya, E.I.; Litvin, I.; Nikitin, E.E.; Oref, I.; Troe, J. Classical diffusion model of vibrational predissociation of van der Waals complexes Part III. Comparison with quantum calculations. *Phys. Chem. Chem. Phys.* **2002**, *4*, 3330–3340. [CrossRef]
25. Bezuglov, N.N.; Golubkov, G.V.; Klyucharev, A.N. *Ionization of Excited Atoms in Thermal Collisions*; The Atmosphere and Ionosphere: Elementary Processes, Discharges and Plasmoids; Springer: New York, NY, USA; London, UK, 2013; Chapter 1; pp. 1–60.
26. Reichl, L.E. *The Transition to Chaos: Conservative Classical Systems and Quantum Manifestations*; Springer: New York, NY, USA, 2004.
27. Zaslavskii, G.M. *Physics of Chaos in Hamiltonian Systems, 2nd ed*; Imperial College Press: London, UK, 2007.
28. Bezuglov, N.N.; Golubkov, G.V.; Klyucharev, A.N. *Manifestations of "Dynamic Chaos" in Reactions with Participation of Rydberg States*; St. Petersburg State University: St. Petersburg, Russia, 2017. (In Russian)
29. Paris-Mandoki, A.; Gorniaczyk, H.; Tresp, C.; Mirgorodskiy, I.; Hofferberth, S. Tailoring Rydberg interactions via Förster resonances: State combinations, hopping and angular dependence. *J. Phys. B* **2016**, *49*, 164001. [CrossRef]
30. Gianninas, A.; Dufour, P.; Kilic, M.; Brown, W.R.; Bergeron, P.; Hermes, J.J. Precise atmospheric parameters for the shortest-period binary white dwarfs: Gravitational waves, metals, and pulsations. *Astrophys. J.* **2014**, *794*, 35–52. [CrossRef]
31. Bezuglov, N.N.; Borisov, E.N.; Verolainen, Y.F. Distribution of the radiative lifetimes over the excited states of atoms and ions. *Sov. Phys. Uspekhi* **1991**, *34*, 3–29. [CrossRef]
32. Hezel, T.P.; Burkhardt, C.E.; Ciocca, M.; He, L.W.; Leventhal, J.J. Classical view of the properties of Rydberg atoms: Application of the correspondence principle. *Am. J. Phys.* **1992**, *60*, 329–335. [CrossRef]
33. Landau, L.D.; Lifshitz, E.M. *Quantum Mechanics*; Pergamon: Oxford, UK, 1977.
34. Omidvar, K. Semiclassical formula for the radiative mean lifetime of the excited state of the hydrogenlike atoms. *Phys. Rev. A* **1982**, *26*, 3053–3061. [CrossRef]
35. Mack, M.; Grimmel, J.; Karlewski, F.; Sárkány, L.; Hattermann, H.; Fortágh, J. All-optical measurement of Rydberg-state lifetimes. *Phys. Rev. A* **2015**, *92*, 012517. [CrossRef]

36. Tretyakov, D.B.; Beterov, I.I.; Entin, V.M.; Yakshina, E.A.; Ryabtsev, I.I.; Dyubko, S.F.; Alekseev, E.A.; Pogrebnyak, N.L.; Bezuglov, N.N.; Arimondo, E. Effect of photoions on the line shape of the Förster resonance lines and microwave transitions in cold rubidium Rydberg atoms. *J. Exp. Theor. Phys.* **2012**, *114*, 14–24. [CrossRef]
37. Hund, F. *The History of Quantum Theory*; Barnes & Noble Books: New York, NY, USA, 1974; Chapter 11.
38. Bokulich, P.; Bokulich, A. Niels Bohr's generalization of classical mechanics. *Found. Phys.* **2005**, *35*, 347–371. [CrossRef]
39. Delone, N.B.; Goreslavsky, S.P.; Krainov, V.P. Dipole matrix elements in the quasi-classical approximation. *J. Phys. B* **1994**, *27*, 4403. [CrossRef]
40. Arefieff, K.N.; Miculis, N.; Bezuglov, N.N.; Dimitrijević, M.S.; Klyucharev, A.N.; Mihajlov, A.A.; Srećković, V.A. Dynamics Resonances in Atomic States of Astrophysical Relevance. *J. Astrophys. Astron.* **2015**, *36*, 613–622. [CrossRef]
41. Sommerfeld, A. *Atomic Structure and Spectral Lines*; Methuen: London, UK, 1934.
42. Grouzdev, P.F. *Atomic and Ionic Spectra in X-ray and Ultraviolet Region's*; Energoatomizdat: Moscow, Russia, 1982. (In Russian)
43. Zakharov, M.Y.; Bezuglov, N.N.; Klyucharev, A.N.; Matveev, A.A.; Beterov, I.I.; Dulieu, O. Specifics of the stochastic ionization of a Rydberg collision complex with Förster resonance. *Russ. J. Phys. Chem. B* **2011**, *5*, 537–545. [CrossRef]
44. Fermi, E. Sopra lo spostamento per pressione delle righe elevate delle serie spettrali. *Nuovo Cimento* **1934**, *11*, 157–166. [CrossRef]
45. Janev, R.K.; Mihajlov, A.A. Resonant ionization in slow-atom-Rydberg-atom collisions. *Phys. Rev. A* **1980**, *21*, 819–826. [CrossRef]
46. Mihajlov, A.A.; Janev, R.K. Ionisation in atom-Rydberg atom collisions: Ejected electron energy spectra and reaction rate coefficients. *J. Phys. B* **1981**, *14*, 1639. [CrossRef]
47. Duman, E.L.; Shmatov, I.P. Ionization of highly excited atoms in their own gas. *Sov. Phys. JETP* **1980**, *51*, 1061–1065.
48. Srećković, V.A.; Dimitrijević, M.S.; Ignjatović, Lj.M.; Bezuglov, N.N.; Klyucharev, A. The Collisional Atomic Processes of Rydberg Hydrogen and Helium Atoms: Astrophysical Relevance. *Galaxies* **2018**, *6*, 72. [CrossRef]
49. Bezuglov, N.N.; Borodin, V.M.; Kazanskiy, A.K.; Klyucharev, A.N.; Matveev, A.A.; Orlovskii, K.V. Analysis of Fokker-Planck type stochastic equations with variable boundary conditions in an elementary process of collisional ionization. *Opt. Spectrosc.* **2001**, *91*, 19–26. [CrossRef]
50. Zaslavskij, G.M. *Hamiltonian Chaos and Fractional Dynamics*; Oxford Univ. Press: Oxford, UK, 2005.
51. Delone, N.B.; Krainov, V.P.; Shepelyanskii, D.L. Highly-excited atoms in the electromagnetic field. *Sov. Phys. Uspekhi* **1983**, *26*, 551. [CrossRef]
52. Bezuglov, N.N.; Borodin, V.M.; Ekers, A.; Klyucharev, A.N. A quasi-classical description of the stochastic dynamics of a Rydberg electron in a diatomic quasi-molecular complex. *Opt. Spectrosc.* **2002**, *93*, 661–669. [CrossRef]
53. Ryabtsev, I.I.; Tretyakov, D.B.; Beterov, I.I.; Bezuglov, N.N.; Miculis, K.; Ekers, A. Collisional and thermal ionization of sodium Rydberg atoms: I. Experiment for nS and nD atoms with n = 8–20. *J. Phys. B* **2005**, *38*, S17–S35. [CrossRef]
54. Efimov, D.K.; Bezuglov, N.N.; Klyucharev, A.N.; Gnedin, Y.N.; Miculis, K.; Ekers, A. Analysis of light-induced diffusion ionization of a three-dimensional hydrogen atom based on the Floquet technique and split-operator method. *Opt. Spectrosc.* **2014**, *117*, 8–17. [CrossRef]
55. Chu, S.-I.; Telnov, D.A. Beyond the Floquet theorem: Generalized Floquet formalisms and quasienergy methods for atomic and molecular multiphoton processes in intense laser fields. *Phys. Rep.* **2004**, *390*, 1–131. [CrossRef]
56. Hairer, E. *Numeral Geometric Integration*; Universite de Geneve: Geneve, Switzerland, 1999.
57. Kazansky, A.K.; Bezuglov, N.N.; Molisch, A.F.; Fuso, F.; Allegrini, M. Direct numerical method to solve radiation trapping problems with a Doppler-broadening mechanism for partial frequency redistribution. *Phys. Rev. A* **2001**, *64*, 022719. [CrossRef]
58. Landau, L.D.; Lifshitz, E.M. *Mechanics Course of Theoretical Physics Mechanics*; (Nauka, Moscow, 1973); English Tran.; Permagon Press: Oxford, UK; New York, NY, USA; Toronto, ON, Canada, 1976; Volume 1.
59. Ramsey, N.F. *Molecular Beams*, 2nd ed.; Clarendon: Oxford, UK, 1989.

60. Michulis, K.; Beterov, I.I.; Bezuglov, N.N.; Ryabtsev, I.I.; Tretyakov, D.B.; Ekers, A.; Klucharev, A.N. Collisional and thermal ionization of sodium Rydberg atoms: II. Theory for nS, nP and nD states with n= 5–25. *J. Phys. B* **2005**, *38*, 1811–1831. [CrossRef]

61. Sydoryk, I.; Bezuglov, N.N.; Beterov, I.I.; Miculis, K.; Saks, E.; Janovs, A.; Spels, P.; Ekers, A. Broadening and intensity redistribution in the Na(3p) hyperfine excitation spectra due to optical pumping in the weak excitation limit. *Phys. Rev. A* **2008**, *77*, 042511. [CrossRef]

62. Kirova, T.; Cinins, A.; Efimov, D.K.; Bruvelis, M.; Miculis, K.; Bezuglov, N.N.; Auzinsh, M.; Ryabtsev, I.I.; Ekers, A. Hyperfine interaction in the Autler-Townes effect: The formation of bright, dark, and chameleon states. *Phys. Rev. A* **2017**, *96*, 043421. [CrossRef]

63. Porfido, N.; Bezuglov, N.N.; Bruvelis, M.; Shayeganrad, G.; Birindelli, S.; Tantussi, F.; Guerri, I.; Viteau, M.; Fioretti, A.; Ciampini, D.; et al. Nonlinear effects in optical pumping of a cold and slow atomic beam. *Phys. Rev. A* **2015**, *92*, 043408. [CrossRef]

64. Klyucharev, A.N.; Bezuglov, N.N.; Matveev, A.A.; Mihajlov, A.A.; Ignjatović, L.M.; Dimitrijević, M.S. Rate coefficients for the chemi-ionization processes in sodium- and other alkali-metal geocosmical plasmas. *New Astron. Rev.* **2007**, *51*, 547–562. [CrossRef]

65. Tantussi, F.; Mangasuli, V.; Porfido, N.; Prescimone, F.; Fuso, F.; Arimondo, E.; Allegrini, M. Towards laser-manipulated deposition for atom-scale technologies. *Appl. Surf. Sci.* **2009**, *255*, 9665–9670. [CrossRef]

66. Boulmer, J.; Bonanno, R.; Weiner, J. Crossed-beam measurements of absolute rates coefficients in associative ionization collisions between Na*(np) and Na(3s) for $5 \leq n \leq 15$. *J. Phys. B* **1983**, *16*, 3015–3024. [CrossRef]

67. Weiner, J.; Boulmer, J. Associative ionization rate constants as a function of quantum numbers n and l in Na*(np) + Na(3s) collisions for $17 \leq n \leq 27$ and $l = 0$, $l = 1$ and $l \geq 2$. *J. Phys. B* **1986**, *19*, 599–609. [CrossRef]

68. Beterov, I.I.; Tretyakov, D.B.; Ryabtsev, I.I.; Bezuglov, N.N.; Miculis, K.; Ekers, A.; Klucharev, A.N. Collisional and thermal ionization of sodium Rydberg atoms III. Experiment and theory for nS and nD states with n = 8–20 in crossed atomic beams. *J. Phys. B* **2005**, *38*, 4349–4361. [CrossRef]

atoms

MDPI

Article

Spectral Modeling of Hydrogen Radiation Emission in Magnetic Fusion Plasmas

Mohammed Koubiti * and Roshin Raj Sheeba

PIIM laboratory, Aix-Marseille Université and Centre national de la recherche scientifique (CNRS),
F-13397 Marseille, France; roshin-raj.SHEEBA@univ-amu.fr
* Correspondence: mohammed.koubiti@univ-amu.fr

Received: 11 December 2018; Accepted: 1 February 2019; Published: 12 February 2019

check for updates

Abstract: Modeling of the spectral line and continuum radiation emitted by hydrogen isotopes in peripheral regions of magnetic fusion is presented through profiles of the Zeeman-Doppler broadened $H\alpha/D\alpha$ line and those of the Stark broadened high-n Balmer lines extending beyond the series limit for recombining plasmas. The $H\alpha/D\alpha$ line profiles should be modelled while accounting for several populations of neutrals to mimic real situations and analyze experimental data for isotopic ratio determination. On the other side, high-n Balmer lines of hydrogen are used for plasma electron density and temperature diagnostics. Moreover, modelling whole spectra including the continuum radiation contributes to the development of synthetic diagnostics for future magnetic fusion devices for which they can give predictive results through coupling to numerical simulation tools.

Keywords: plasma spectroscopy; Stark broadening; plasma diagnostics; line shape modeling; Zeeman-Doppler broadening; Balmer line series; radiative recombination

1. Introduction

Decades after the first experiments on controlled thermonuclear fusion in tokamaks, and despite the large amounts of scientific and technological progress and the high quality of our scientific knowledge, the path towards a power plant based on this principle has been shortened but there still remains some major problems to overcome. Indeed, several issues need to be solved like the mitigation of disruptions and the retention of tritium by the plasma-facing components or materials (PFCs) and the control of the power and particle loads on these materials as well. For safety reasons, it is crucial to determine the isotopic ratio for deuterium-tritium (D-T) plasmas. However, except for very few D-T discharges in the past and some planned in the near future at JET (Joint European Tokamak), fusion plasma experiments use deuterium gas or sometimes a mixture of hydrogen and deuterium. Therefore, it is crucial to determine as accurate as possible the isotopic ratio H/(H+D) in D-D and H-D plasma experiments in order to fulfill the tight constraints required by the D-T experiments in terms of tritium concentration. Concerning the other major issue of power exhaust, it is mandatory to reduce the huge heat and particle flux loads to acceptable levels because the most advanced target materials do not support power loads in excess of 10–20 MW/m^2. A solution widely used to mitigate power and particle loads consists in the creation of a radiative dense, cold mantle in the divertor which leads to plasma detachment through the volume recombination processes. Detachment has been achieved in many tokamaks like Alcator C-mod [1], DIII-D [2], JET [3,4] and ASDEX Upgrade [5,6]. It is largely admitted that this scenario is the most efficient solution and it has even been foreseen for the operation of ITER (International Thermonuclear Experimental Reactor). Under the conditions of divertor plasma detachment, high-n lines of the Balmer series of hydrogen isotopes extending to (or beyond) the series limit can be observed. These spectra are used to infer both the plasma electron density and the temperature when compared to calculated spectra. Modeling the complete spectra of the radiation emitted in the visible domain by hydrogen or deuterium neutrals for plasma conditions relevant to recombining plasmas is therefore of great interest. This paper is organized as follows. Section 2 is devoted to the modeling of the Hα/Dα line emitted by several neutral populations of hydrogen and deuterium neutrals to mimic the recycling mechanisms taking place in tokamak divertors. Section 3 is focused on the modeling of hydrogen high-n Balmer lines and continuum emission for conditions relevant to recombining plasmas where the electron density is in the range 10^{14}–10^{15} cm^{-3} and the electron temperature is around 1–2 eV or less.

2. The Zeeman–Doppler Hα/Dα Line Profile

In peripheral regions of tokamaks, especially in the divertor and the scrape-off layer (SOL), the Hα/Dα line is one of the most intense lines emitted by plasma in the visible domain. For typical electron densities of about 10^{14} cm^{-3} and temperatures of few eV, the profile of this line is dominated by Doppler broadening in addition to the Zeeman effect which splits and removes the degeneracy of the involved energy levels at n = 2 and n = 3, and introduces an anisotropy due to polarization. To a good approximation, the broadening due to the Stark effect can be neglected and the Hα/Dα line profiles can be modelled by retaining only Zeeman and Doppler effects. The treatment of the Zeeman effect depends, obviously, on the strength of the magnetic field. In addition, energy perturbation due to the Zeeman effect has to be compared to that due to the relativistic corrections which are part of the fine structure. Without any loss of generality, we consider situations with magnetic fields greater than 2 T and assume that the use of the strong field approximation is valid to account for Zeeman effect. Under such conditions, the fine structure effect can be neglected and if observed in a direction perpendicular to the magnetic field direction, one sees the well-known Lorentz triplet of the Balmer-α line composed of an un-shifted π component and two lateral components, known as the σ components, which are equally separated from the unshifted one. The separation between the p and s component depends linearly on the magnetic field strength. It should be noted that in the case of an observation parallel to the magnetic field, only the s components are observed. If the instrumental function of measurement apparatus is weak, each component will show a Gaussian profile due the thermal motion of the neutral emitters. The full width at half maximum (FWHM) of each component is proportional to the square root of the neutral temperature ($\Delta\lambda_{1/2} \propto \sqrt{T}$) according to Doppler broadening theory. However, in divertor tokamaks it is known that several processes contribute to the release of hydrogen or deuterium neutrals. This is known as particle recycling. Indeed, neutrals can be released following dissociative excitation of desorbed molecules, or through reflection as neutrals of impinging ions, or through charge exchange processes. These mechanisms are responsible for the coexistence of several populations of neutrals each with its own velocity distribution function. In addition, in the case of H-D mixtures, the same populations exist for both hydrogen and deuterium. For illustration purposes, as shown in Figure 1, the profiles of the Balmer-α line emitted by the population have a Maxwellian velocity distribution function (VDF) with a temperature of ~3 eV where hydrogen represents 5% and deuterium represents 95%. The chosen temperature roughly represents neutrals released through dissociation of H_2 and D_2 molecules in a tokamak. Detailed information about recycling and the different populations of neutrals in magnetic fusion devices can be found for instance in References [7–9].

A more complex situation is presented in Figure 2 where a second neutral population with a Maxwellian VDF corresponding to a neutral temperature of ~20 eV is added to the one shown in Figure 1. This temperature is an approximation of that of the population of neutrals released through reflection followed by a complete thermalization process due to elastic collisions with plasma ions. Note that this is different from the population of neutrals resulting from a charge exchange process whose temperature is in principle much higher depending on the ion temperature at the location where the charge exchange process takes place. For simplicity, equal contributions to the Balmer-α line from the two neutral populations for both hydrogen and deuterium has been assumed. However, for real situations, where experimental spectra of this line measured in tokamaks are analyzed, generally the three neutrals populations are necessary. The fitting procedures allow us to obtain their temperatures and their relative proportions or contributions, and this is a rather delicate task. For this refer to References [7–9] or to a more recent reference [10], where the fitting is explicitly used to obtain the isotopic ratio H/(H + D).

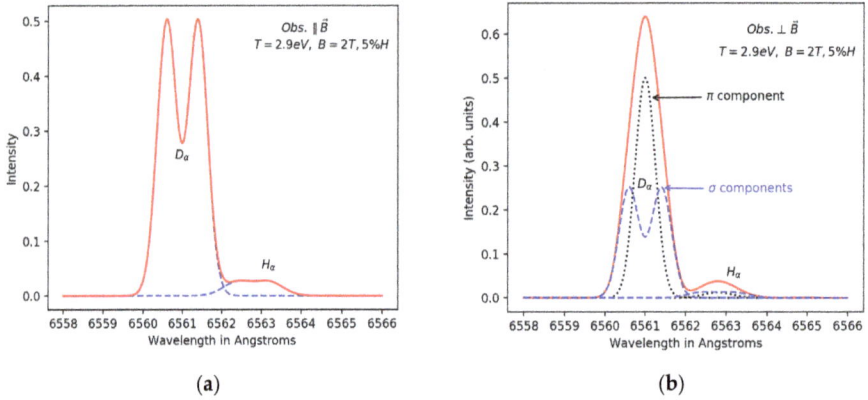

Figure 1. Theoretical profiles of the Balmer-α line emitted by neutrals distributed as 5% of hydrogen and 95% of deuterium atoms, having the same Maxwellian velocity distribution with a temperature of ~3 eV in a plasma where the magnetic field B = 2 T. The Dα/Hα line centers are separated by ~1.8 Å and the Doppler broadening of the Hα is higher than that of the Dα line because of its lower mass. (a) Parallel observation: Only the lateral σ components are seen; (b) perpendicular observation: Both the lateral σ and central π components are shown.

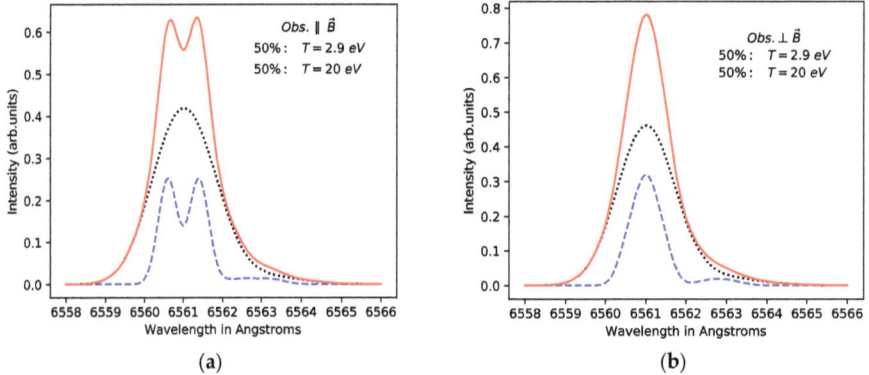

Figure 2. Total profiles (solid red line) of the Hα/Dα line resulting from two neutral populations with equal concentrations but different temperatures: 2.9 eV (blue dashed line) and 20 eV (black dotted line) for B = 2T. A fraction of 5% hydrogen is considered here. (a) Parallel observation; (b) perpendicular observation.

3. The High-n Balmer Lines and Continuum

As mentioned previously, under detachment conditions, i.e., for recombining plasma conditions, e.g., for $n_e = 10^{14}$ cm^{-3} and $T_e = 1$ eV, high-n lines of the Balmer series as well as the continuum radiation can be emitted by hydrogen or deuterium neutrals in divertors of tokamaks. This radiation can therefore be observed and recorded to be used further for divertor diagnostics. The corresponding spectra can allow the extraction of very useful information. Indeed, high-n lines of the Balmer series are very sensitive to Stark broadening and allow us to infer the plasma electron density of the emissive zone. On the other side, the use of the relative intensities of several lines of this series or the continuum slope allow for electron temperature determination. Of course, the determination of these parameters requires the use of theoretical models for line broadening and line shapes retaining all the broadening mechanisms, or at least the major ones. It is therefore necessary to model spectra of both lines and continuum radiation. However, as the Stark broadening increases with the upper quantum number n, beyond a given value n_{IT} called the Inglis-Teller limit [11], adjacent lines overlap and eventually merge into the continuum part of the spectrum leading to an apparent advance of the continuum. One should note that even if this transition region between lines and continuum is not usually used for diagnostics, it should be modelled to obtain a whole or complete spectrum. Therefore, it is clear that from the diagnostic point of view, this discrete-to-continuum transition region is not important. However, from the point of view of synthetic diagnostics which are very important for future devices because of their predictive abilities, the spectrum as a whole plays a very important role and each part is necessary. This is one of the reasons to model a quasi-complete spectrum of the Balmer series say from the H_γ (n = 5) or H_δ (n = 6) to the theoretical Balmer series limit ($\lambda_B^l = 364.6$ nm) and beyond. We present here briefly the background related to the spectra.

3.1. Modeling of the Bound–Bound Transitions

The high-n Balmer lines emitted by hydrogen in a typical recombining plasma are mainly broadened by the Stark effect. The Zeeman effect may be neglected except for a few lines at relatively low densities. The theory of Stark broadening is well established and there exists many codes (either models or numerical simulations) dealing with this. For the calculations shown here, we have used the PPP line shape code [12,13]. For the Stark broadening calculations, a collisional approach known as the impact approximation is used for the plasma electrons contribution while a quasi-static one is used to treat the ion contribution. More precisely, the electronic contribution is calculated using the Griem-Blaha-Kepple GBK model [14] taking into account the frequency dependence of the electronic operator. High order broadening contributions to line broadening such as interference terms in the collision operator and non-binary effects are neglected for the high-n Balmer lines emitted under detachment plasma conditions. For a given transition between levels i and j ($j \rightarrow i$) corresponding to a frequency ν, the intensity is calculated following the expression:

$$I_{j \rightarrow i} = \frac{h\nu}{4\pi} A_{ji} N_j \phi_\nu,$$

where A_{ji} is the Einstein coefficient of the corresponding transition, N_j the population of the upper level j of the transition and ϕ_ν the normalized line shape. The profiles of some Balmer lines are shown in Figure 3a.

3.2. Modeling of the Continuum Radiation

The continuum radiation results from two types of transitions: Bound-free and free-free contributions. For the typical plasma conditions considered here, the radiation due to free-free transitions, known as bremsstrahlung, is negligible in comparison with that due to bound-free transitions. Therefore, for illustration purposes, only the bound-free contribution is retained in the calculations shown here. However, for the complete calculations all contributions are included. The free-bound radiative recombination power density can be written as:

$$\rho_n^R(\nu)d\nu = h\nu n_i n_e f(\varepsilon) v_e \sigma_{rec}(p, \varepsilon)d\varepsilon,$$

where n_i is the ion density, h is the Planck constant, v_e is the electron velocity, $f(\varepsilon)$ is the energy distribution function of free electrons and $\sigma_{rec}(p, \varepsilon)$ is the cross-section for the recombination process [15]. Here p designates the bound state (p = 2, 3, ...). The cross-section, $\sigma_{rec}(p, \varepsilon)$ is obtained using Milne's formula [15], from its relation with the photoionization cross section for the level p, $\sigma_{ion}(p, \nu)$. The expression can be rewritten for the transition to an energy level p, for a Maxwell-Boltzmann electron energy distribution, as:

$$\rho_p^R(\nu) = 2\sqrt{\frac{2}{\pi}\frac{h^4}{m_e^{\frac{3}{2}}c^2}}2p^2\left(\frac{1}{kT_e}\right)^{\frac{3}{2}}\exp\left(-\frac{h\nu - I_H(p)}{kT_e}\right)\sigma_{ion}(p,\nu)\nu^3 n_i n_e$$

where k is the Boltzmann constant, m_e the electron mass, c the speed of light and $I_H(p)$ the ionization potential for level p. The total free-bound contribution is obtained by additions all the contributions from free states to bound states p = 2, 3, ... , as follows:

$$\rho^R(\nu) = \sum_{p=1}^{\infty}\rho_p^R(\nu)$$

Since the continuum is dominated by Balmer radiative recombination (p = 2) and higher (p > 2) radiative recombination contributions are decreasing with increasing p, higher radiative recombination can be neglected. Here we have calculated the contributions up to an energy level of p = 10. The resulting calculation is shown in Figure 3b. Note that the bound-free continuum power density shown here was calculated without taking into account any plasma density effect but for a hypothetical isolated atom, hence the sharp transition seen at the series limit which comes from the recombination threshold.

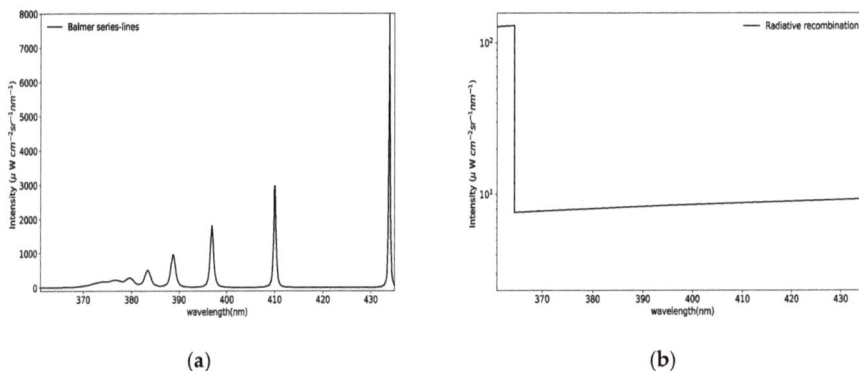

(a)

(b)

Figure 3. (a) Calculated profiles of some Balmer lines are calculated from n = 5 up to level n = 15 and are summed up together for the plasma condition ($n_e = 8 \times 10^{14}$ cm^{-3} and $T_e = 1.1$ eV). (b) Radiative recombination calculation for the same plasma conditions for an isolated atom.

3.3. The Total Spectrum and the Dissolution Approach

To obtain the total spectrum, one should also deal with the discrete-to-continuum transition to combine bound–bound and free–bound contributions. Here we use a dissolution factor approach as discussed in References [16,17] to take into account the merging of lines into the continuum. This method, which is also known as the occupation probability formalism (OPF) [16], is based on two simple suggestions: (i) The potential barrier of the atom is altered by the electric field which adds to the Coulombian field of interaction of the external electron with the core of the atom. For electric microfields exceeding a certain critical value F_c that corresponds to a level above the potential barrier, the energy of that level vanishes. This means that some bound-bound transitions transform partially or totally to free-bound transitions by conserving the oscillator strength density. The probability of energy level j, E_j realization is called the 'dissolution factor' and it is given by:

$$W_j = \int_0^{F_c} P(F)dF,$$

where P(F) is the micro field distribution calculated using [18]. Based on the second suggestion, the radiative recombination continuum should be multiplied with $(1 - W_j)$ to conserve the oscillator strength density. By using an approach like this one, one can take into account the density effect on the atom energy levels. Indeed, with increasing electron densities, the model of the isolated atom becomes less and less valid. The charged particles surrounding the neutral emitter (hydrogen atom) strongly disturb the structure (energy levels and wave functions) of the latter. The dissolution approach states simply that for a given electron density, atomic energy levels which are discrete for an isolated atom cannot be considered as discrete or bound but a combination of free and bound states. A level observed in a pure discrete or bound state has a realization probability of one. In contrast, a level in a pure free state has a realization probability of zero. Note that in the previous expression, j stands for both discrete and continuum states. Numerically speaking, in the emission spectrum calculation, the recombination contribution is extended into lines and dissolution factors are applied at corresponding positions and interpolated. Each line is multiplied by its probability of realization at its upper energy level of the transition and the sum over the line is shown in Figure 4a. Summing up all the contributions together will result in the complete Balmer spectrum calculation as shown in Figure 4b. One can see on this figure that the apparent advance of the continuum which starts at about $\lambda_B^{ap} \approx 370$ nm instead from the theoretical value of 346.6 nm.

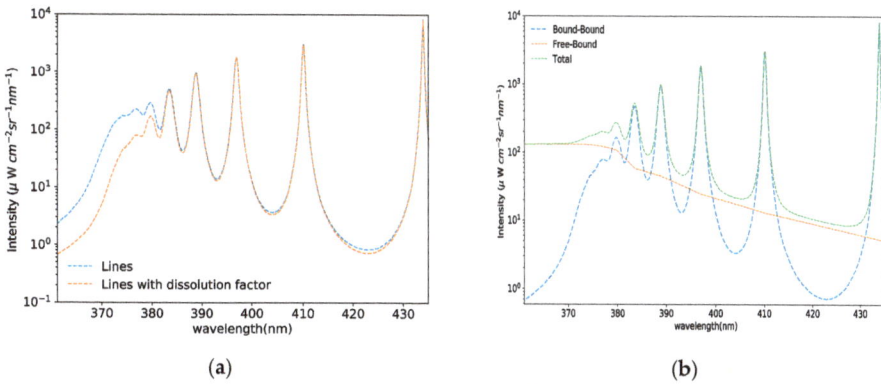

Figure 4. (a) Density effect (dissolution factor approach) on the Balmer lines as computed for $n_e = 8 \times 10^{14}$ cm^{-3} and $T_e = 1.1$ eV. (b) Total spectrum for the same plasma conditions.

4. Discussion and Conclusions

We have shown how the Balmer-α line emission spectra emitted by hydrogen and its isotopes in the peripheral regions of magnetic fusion devices allow the determination of an important quantity: The isotopic ratio. This quantity is crucial for the D-T experiments as the tritium concentration is tightly controlled and should not exceed some value for safety reasons. Of course, the quality of the results depends strongly on both the measurements and on the accuracy of the modeling. In particular, modeling should account for all or at least the major broadening mechanisms as well as the inclusion of all the populations of hydrogen/deuterium neutrals. On the other side, we have seen the spectra of high-n Balmer lines extending to the series limit or beyond it. This can be seen for conditions relevant to recombining plasmas characterized by typical electron densities in the range 10^{14}–10^{15} cm^{-3} and electron temperatures of the order of 1–2 eV. We have mentioned that such spectra can serve for plasma diagnostics. Indeed, comparing such spectra to experimental ones recorded along a given line of sight (LOS) allows the determination of the average electron density along the LOS from the Stark broadening of the discrete lines (bound-bound transitions). The average electron density can also be obtained from the comparison calculated and measured line intensities. However, modeling of such spectra can become more useful by extending its utility beyond diagnostics and using it also for predictive purposes: This is the role of synthetic diagnostics. The modeling presented here is now ready to be coupled to other simulation tools providing plasma parameter spatial distributions for calculating total spectra using these spatial distributions. The theoretical spectra calculated in this way take into account any inhomogeneity of the plasma emission zone along the considered line of sight. The first results of such synthetic spectra considered as synthetic diagnostics will be published elsewhere [19] for conditions relevant to WEST tokamak by using distributions provided by a numerical transport code SolEdge2D-Eirene [20].

Author Contributions: Equal contributions. Spectral calculations of the Balmer series of hydrogen were carried out by R.R.S. Calculations concerning the Ha/Da spectra were done by M.K. Writing of the draft and preparation done by M.K. R.R.S. wrote parts related to synthetic Balmer spectra.

Funding: This research received funding from EUROfusion WPEDU, MST1 and JET1.

Acknowledgments: This work has been carried out within the framework of the EUROfusion Consortium. The views and opinions expressed herein do not necessarily reflect those of the European Commission.

Conflicts of Interest: The authors declare no conflict of interest.

References

1. Pigarov, A.Y.; Terry, J.L.; Lipschultz, B. Study of the discrete-to-continuum transition in a Balmer spectrum from Alcator C-Mod divertor plasmas. *Plasma Phys. Control. Fusion* **1998**, *40*, 2055–2072. [CrossRef]
2. Krasheninnikov, S.I.; Kukushkin, A.S.; Pshenov, A.A. Divertor plasma detachment. *Phys. Plasmas* **2016**, *23*, 055602. [CrossRef]
3. Loarte, A.; Monk, R.; Martin-Solis, J.; Campbell, D.; Chankin, A.; Clément, S.; Davies, S.; Ehrenberg, J.; Erents, S.; Guo, H.; et al. Plasma detachment in JET Mark I divertor experiments. *Nucl. Fusion* **1998**, *38*, 331–371. [CrossRef]
4. Field, A.R.; Balboa, I.; Drewelow, P.; Flanagan, J.; Guillemaut, C.; Harrison, J.R.; Huber, A.; Huber, V.; Lipschultz, B.; Matthews, G.; et al. Dynamics and stability of divertor detachment in H-mode plasmas on JET. *Plasma Phys. Control. Fusion* **2017**, *59*, 095003. [CrossRef]
5. Potzel, S.; Wischmeier, M.; Bernert, M.; Dux, R.; Müller, H.; Scarabosio, A. A new experimental classification of divertor detachment in ASDEX Upgrade. *Nucl. Fusion* **2013**, *54*, 13001. [CrossRef]
6. Potzel, S.; Dux, R.; Müller, H.W.; Scarabosio, A.; Wischmeier, M. Electron density determination in the divertor volume of ASDEX Upgrade via Stark broadening of the Balmer lines. *Plasma Phys. Control. Fusion* **2014**, *56*, 025010. [CrossRef]
7. Kubo, H.; Takenaga, H.; Sugie, T.; Higashijima, S.; Suzuki, S.; Sakasai, A.; Hosogane, N. The spectral profile of the Ha line emitted from the divertor region of JT-60U. *Plasma Phys. Control. Fusion* **1998**, *40*, 1115. [CrossRef]
8. Koubiti, M.; Marandet, Y.; Escarguel, A.; Capes, H.; Godbert-Mouret, L.; Stamm, R.; De Michelis, C.; Guirlet, R.; Mattioli, M. Analysis of asymmetric Dα spectra emitted in front of a neutralizer plate of the Tore-Supra ergodic divertor. *Plasma Phys. Control. Fusion* **2002**, *44*, 261. [CrossRef]
9. Hey, J.; Chu, C.C.; Mertens, P.H.; Brezinsek, S.; Unternberg, B. Atomic collision processes with ions at the edge of magnetically confined fusion plasmas. *J. Phys. B* **2004**, *37*, 2543. [CrossRef]
10. Neverov, V.S.; Kukushkin, A.B.; Stamp, M.F.; Alkseev, A.G.; Brezinsek, S.; von Hellermann, M.; JET contributors. Determination of divertor stray light in high-resolution main chamber Ha spectroscopy in JET-ILW. *Nucl. Fusion* **2017**, *57*, 016031. [CrossRef]
11. Inglis, D.R.; Teller, E. Ionic depression of series limits in one electron spectra. *Astrophys. J.* **1939**, *90*, 439. [CrossRef]
12. Calisti, A.; Mossé, C.; Ferri, S.; Talin, B.; Rosmej, F.; Bureyeeva, L.A.; Lisitsa, V.S. Dynamic Stark broadening as the Dicke narrowing effect. *Phys. Rev. E* **2010**, *81*, 016406. [CrossRef] [PubMed]
13. Ferri, S.; Calisti, A.; Mossé, C.; Rosato, J.; Talin, B.; Alexiou, S.; Gigosos, M.A.; González, M.A.; González-Herrero, D.; Lara, N.; et al. Ion Dynamics Effect on Stark-Broadened Line Shapes: A Cross-Comparison of Various Models. *Atoms* **2014**, *2*, 299–318. [CrossRef]
14. Griem, H.; Blaha, M.; Kepple, P. Stark profile clacultaions for Lyman series lines of one electron ions in dense plasmas. *Phys. Rev. A* **1979**, *19*, 2421. [CrossRef]

15. Goto, M.; Sakamoto, R.; Morita, S. Experimental verification of complete LTE plasma formation in hydrogen pellet cloud. *PPCF* **2007**, *49*, 1163. [CrossRef]

16. D'yachkov, L.G. Smooth transition from spectral lines to a continuum in dense hydrogen plasma. *High Temp.* **2016**, *54*, 5. [CrossRef]

17. Stehle, C.; Jacquemot, S. Line shape in hydrogen opacities. *Astron. Astrophys* **1993**, *271*, 348.

18. Iglesias, C.A. Electronic microfield distributions in strongly coupled plasmas. *Phys. Rev. A* **1983**, *28*, 1168. [CrossRef]

19. Sheeba, R.R.; Koubiti, M.; Calisti, A.; Ferri, S.; Marandet, Y.; Rosato, J.; Stamm, R. Synthetic diagnostics based on hydrogen Balmer series in recombining plasmas for WEST tokamak. *High Density Energy Phys.* **2019**. to be submitted to High Energy Density Physics.

20. Bufferand, H.; Ciraolo, G.; Marandet, Y.; Bucalossi, J.; Ghendrih, P.; Gunn, J.; Mellet, N.; Tamain, P.; Leybros, R.; Fedorczak, N.; et al. Numerical modeling for divertor design of the WEST device with a focus on plasma-wall interactions. *Nucl. Fusion* **2015**, *55*, 053025. [CrossRef]

atoms

MDPI

Article

Developing the Techniques for Solving the Inverse Problem in Photoacoustics

Mioljub Nesic, Marica Popovic and Slobodanka Galovic

Department of Atomic Physics, Vinca Institute for Nuclear Sciences, 11000 Belgrade, Serbia;
maricap@vin.bg.ac.rs (M.P.); bobagal@vin.bg.ac.rs (S.G.)
* Correspondence: mioljub.nesic@vin.bg.ac.rs; Tel.: +381-64-1417-188

Received: 30 November 2018; Accepted: 6 February 2019; Published: 12 February 2019

Abstract: In this work, theoretically/mathematically simulated models are derived for the photoacoustic (PA) frequency response of both volume and surface optically-absorbing samples in a minimum volume PA cell. In the derivation process, the thermal memory influence of both the sample and the air of the gas column are accounted for, as well as the influence of the measurement chain. Within the analysis of the TMS model, the influence of optical, thermal, and elastic properties of the sample was investigated. This analysis revealed that some of the processes, characterized by certain sample properties, exert their dominance only in limited modulation frequency ranges, which are shown to be dependent upon the choice of the sample material and its thickness. Based on the described analysis, two methods are developed for TMS model parameter determination, i.e., sample properties which dominantly influence the PA response in the measurement range: a self-consistent procedure for solving the exponential problems of mathematical physics, and a well-trained three-layer perceptron with back propagation, based upon theory of neural networks. The results of the application of both inverse problem solving methods are compared and discussed. The first method is shown to have the advantage in the number of properties which are determined, while the second one is advantageous in gaining high accuracy in the determination of thermal diffusivity, explicitly. Finally, the execution of inverse PA problem is implemented on experimental measurements performed on macromolecule samples, the results are discussed, and the most important conclusions are derived and presented.

Keywords: photoacoustic; photothermal; inverse problem; thermal memory; minimum volume cell; neural networks; thermal diffusivity; conductivity; linear coefficient of thermal extension

1. Introduction

One of the most plastic and easily understandable definitions of inverse problem was given by professor Mandelis [1]—a field in which one is called upon to reconstruct the cow from the hamburger meat. Indeed, when all the difficulties are taken into considerations, such as ill conditioning, non-linearity, model dependence upon material, experimental range limitations, etc., one truly feels like they are dealing with the impossible. On the other hand, no matter what method of inverse problem solving is opted for, one conclusion seems inevitable—it is necessary to simultaneously develop both the appropriate TMS model (direct solving methods) and the inverse solving procedures (characterization, imaging) in order to obtain optimum results. By reviewing literature regarding TMS models and techniques of inverse solving in photoacoustics, unexplained approximations that could be the limiting factor in determination of sample properties were noticed.

The research in this domain done by our group has taken two directions. From the experimental point of view, it was found that only a narrow bandwidth of frequency measurements has been exploited until now; from our own experience, this is due to the fact that experiential results rarely agree with theoretical predictions over the entire frequency range. Also, the processing of results, almost by rule, considers either amplitude or phase measurements of the signal; never does it account for both of those, simultaneously.

From the theoretical modeling aspect (the aspect of fundamental research), it was found that the influence of finite heat propagation velocity was neglected, as well as the influence of volumetric optical absorption and the possibility of multiple optical reflections. Also, the knowledge of the measurement chain influence can be, in general, considered insufficient, and it plays an important role in the process of obtaining experimental results.

That is why, in the first part of this work, the generalized model of photoacoustic (PA) response was presented and discussed as the basis for the developed inverse solving procedures for PA characterization. Furthermore, two types of inverse problem solving are suggested and analyzed: a self-consistent inverse procedure, and a neural network. Finally the results of the application on experimental results of the first method are presented. At the end, the most important conclusions are derived.

2. Generalized Model of PA Response—Direct Problem Solving

Indirect transmission photoacoustics presumes the use of an air-filled PA cell as the element in which the acoustic signal is created due to the deployment of a monochromatic, amplitude modulated light source: $I = I_0(1 + \cos \omega t)$ upon a sample. Usually, a cylindrical cell is used in combination with a disk-shaped sample of the radius R and the thickness l_s, positioned and fixed in accordance to the "simply supported plate" principle [2]. This sample is exposed, from one side, to the described EM source, while the response is recorded by microphone on the other side, i.e., this is the principle of t transmission gas-microphone configuration, presented in Figure 1a (upper part), while a detailed description is given in Figure 2. The frequency-dependent measurements of PA response are performed using a lock-in amplifier.

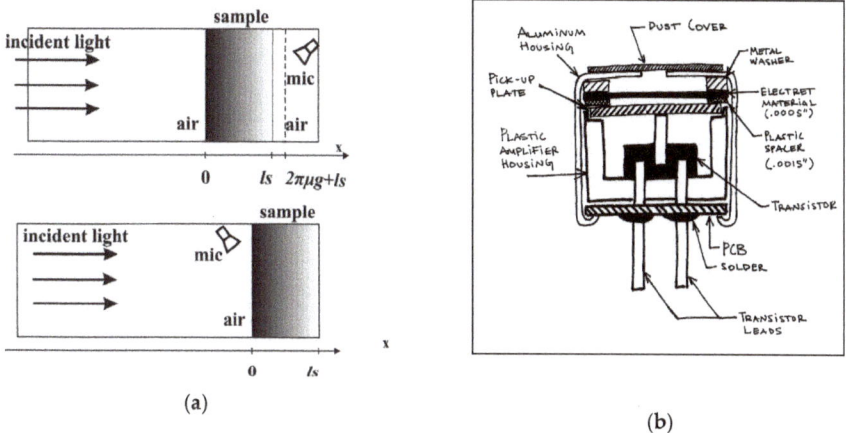

Figure 1. (a) Schematic representation of transmission (up) and reflection (down) configuration in photoacoustics, (b) drawing of a standard electrets microphone (http://www.openmusiclabs.com/wp/wp-content/uploads/2011/03/mic_section_small.jpg, accessed on 29 November 2018).

In transmission PA configuration Figure 1a, the concept of minimum volume cell is used in order to obtain sufficiently high-measured acoustic signal and good signal-to-noise ratio. This means that the microphone chamber itself acts as the interior of the PA cell, as illustrated in Figure 1b [3,4].

Based upon previous research and in accordance with literature defined norms [5], the following designation of thermodynamic properties of the system is introduced:

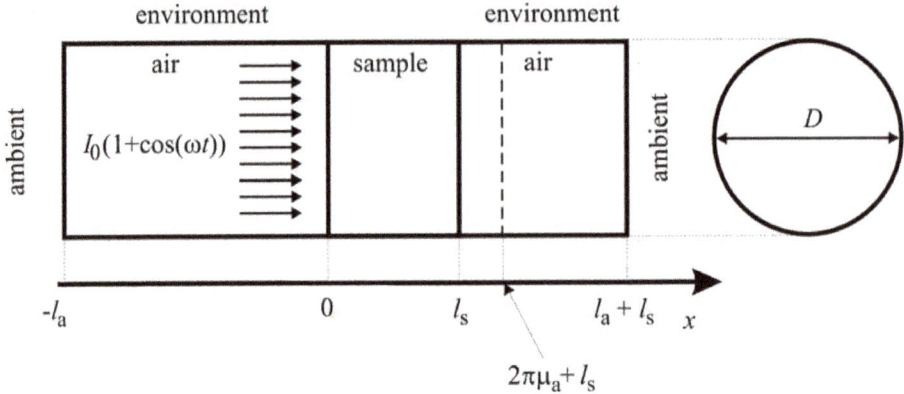

Figure 2. Transmission configuration, taken with approval from [6].

k_i—*thermal conductivity* [W/mK],
D_{Ti}—*thermal diffusivity* [m²/s],
a_T—*thermal coefficient of linear expansion* [K⁻¹],
τ_i—*thermal relaxation time* [s],
and $u_i = \sqrt{D_{Ti}/\tau_i}$—heat propagation velocity [m/s], indexed $i = a,s$ (*air or sample*) and designating the *i*-th medium where it occurs.

Thermal relaxation time and heat propagation velocity are the properties of materials which exist in the generalized theory of heat transfer [6–12]. Explanation attempts regarding the meaning of thermal relaxation time can be found in several papers [13–15]. Avoiding further considerations of the matter (since they go beyond the scope of this work), it is important to note that the investigations in this area are still ongoing and can be approached from different physical viewpoints. As for the stand of our group, we have adopted the most general interpretation of this property, regardless from the material microstructure or thermal energy carriers—thermal relaxation time is the period of time passing between the occurrence of the excitation and the actual change of the heat flux.

Prior to the development of the theoretical/mathematical simulation (TMS) model of the generalized PA response, the following presumptions were introduced:

(a) The cross section of the incident beam is much larger than thesurface area of the sample, thus, planar uniformity of energy distribution justifies the use of one-dimensional (*1D*) approximation [4,8,16–21];

(b) Excitation energy is absorbed within thin surface layer of the sample (the approximation which describes metal samples well, otherwise achieved through the application of thin, opaque absorbent layer) [18,20];

(c) Heat conduction to the surrounding gas (outside of the PA cell) is considered negligible due to its poor thermal conduction properties [18];

(d) Harmonic component alone (of the Fourier transform of the acquired signal) is observed (*lock-in* detection) and frequency characteristics of the PA response are analyzed;

(e) The "thin plate" approximation is applicable since R is much larger than l_s, where the influence of the sample dilatation on the mechanical piston model is negligible and only *thermoelastic* (TE) bending is taken into consideration [4,8,18,19,22,23].

Basing the approach upon literature considerations [4,8,18,19,22,23], the starting expression of the model (for the measured signal directly proportional to the pressure change in the PA cell) can be written in the form:

$$\tilde{p} = \tilde{p}_{th} + \tilde{p}_{ac} \tag{1}$$

where p_{th} denotes the pressure change due to the thermoconducting(*TH*) component of the PA response (the component that originates from the periodic expansion of a thin gas layer closest to the sample in the PA cell), while p_{ac} denotes the pressure change due to the PA component originating from TE vibrations of the sample caused by temperature gradient along symmetry axes of sample (drum effect). These components are then written in the following manner [4], [18], [8]:

$$\tilde{p}_{th} = \frac{\gamma P_0}{l_a T_0} \int_{l_s}^{l_s+2\pi\mu_a} \tilde{\vartheta}_s(l_s) e^{-\tilde{\sigma}_a(x-l_s)} dx, \tag{2a}$$

$$\tilde{p}_{ac} = \frac{3\gamma P_0}{l_a} \alpha_S \frac{R^2}{l_s^3} \int_0^{l_s} \left(x - \frac{l_s}{2}\right) \tilde{\vartheta}_s(x) dx, \tag{2b}$$

where γ annotates the adiabatic coefficient, P_0 is the atmospheric pressure, l_a is the length of the gas column inside the PA chamber, while T_0 stands for the room temperature. Furthermore, α_S is the linear coefficient of thermal expansion of the sample, R is its radius, while l_s annotates its thickness. The symbols $\tilde{\vartheta}_s$ and μ_a represent the complex representative of distribution of dynamic temperature variations across the sample and the thermal diffusion length in the air (gas).

As can be seen from the expressions (1) and (2), the components of the pressure depend on the distribution of the dynamic temperature variation along the sample axis and from the dynamic temperature variation on the unexposed side of the sample (back). We have acquired these values taking into consideration finite heat propagation velocity [9]:

$$\frac{d^2\tilde{\vartheta}_a(x,\omega)}{dx^2} - \tilde{\sigma}_a^2 \tilde{\vartheta}(x,\omega) = 0, \tag{3a}$$

$$\frac{d^2\tilde{\vartheta}_s(x,\omega)}{dx^2} - \tilde{\sigma}_s^2 \tilde{\vartheta}_s(x,\omega) = -\tilde{\sigma}_s \tilde{z}_{cs} S(x) \tag{3b}$$

$$\tilde{q}_i(x,\omega) = -\frac{1}{\tilde{\sigma}_i \tilde{z}_{ci}} \cdot \frac{d\tilde{\vartheta}_i(x,\omega)}{dx}, \quad i = a, s \tag{3c}$$

where q(x) is dynamic heat flux, S(x) represents incident volumetric heat flux which generates perturbations of temperature field, $\tilde{\sigma}_i$ and $\tilde{Z}_{ci}$, are *heat wave vector* and *thermal impedance* of the environment (air or sample), given by:

$$\tilde{\sigma}_i = \frac{1}{\sqrt{D_{Ti}}} \sqrt{j\omega(1+j\omega\tau_i)}, \tag{4a}$$

$$\tilde{Z}_{ci} = \frac{\sqrt{D_{Ti}}}{k_i} \sqrt{\frac{(1+j\omega\tau_i)}{j\omega}}, \tag{4b}$$

Incident volumetric heat flux is calculated as:

$$S(x) = -\frac{dI_{abs}(x)}{dx} \tag{5a}$$

$$I_{abs}(x) = I_0 e^{-\beta x} \tag{5b}$$

$$I_{abs}(x) = I_0(1-R_0)\frac{e^{-\beta x}}{1+Re^{-\beta x}} \tag{5c}$$

The equations for the components of the measured pressure (2a,b), combined with the expressions (3a)–(5c) and solved, become:

$$\tilde{p}_{th} = \frac{\gamma P_0}{l_a T_0} \frac{1}{\tilde{\sigma}_a} e^{2\pi\mu_a} \tilde{\vartheta}_s(l_s) \tag{6a}$$

$$\tilde{\vartheta}_S(l_s) = -\frac{S_0 \beta \tilde{\sigma}_s \tilde{Z}_{cs}}{\beta^2 - \tilde{\sigma}_s^2} \cdot \frac{e^{(\sigma_a - \beta)l_s}\left[(\tilde{r} + \tilde{r}_a)ch(\tilde{\sigma}_s l_s) + (1 + \tilde{r}_a^2)sh(\tilde{\sigma}_s l_s)\right] + (\tilde{r} - \tilde{r}_a)e^{-\sigma_s l_s}}{2\tilde{r}_a ch(\tilde{\sigma}_s l_s) + (1 + \tilde{r}_a^2)sh(\tilde{\sigma}_s l_s)} \cdot \frac{1 - R_0}{1 + R_1 e^{-\beta l_s}},$$
$$\left(\tilde{r} = \frac{\beta}{Z_{cs}}, \tilde{r}_a = \frac{\tilde{Z}_{ca}}{Z_{cs}}\right) \tag{6b}$$

$$\tilde{p}_{ac} = S_0 \frac{6\gamma P_0 R^4}{l_a l_s^2 R_c^2} \alpha_s \frac{\tilde{Z}_{cs}}{\tilde{\sigma}_s^2} \frac{ch(\tilde{\sigma}_s l_s) - \frac{\tilde{\sigma}_s l_s}{2}sh(\tilde{\sigma}_s l_s) - 1}{sh(\tilde{\sigma}_s l_s)} \frac{1 - R_0}{1 + R_1 e^{-\beta l}} \tag{6c}$$

where R_0 and R_1 are, respectively, outer and inner optical reflection coefficient.

If the sample is good optical absorber, heat source becomes surface type and the model given by (6a–c) is reduced to:

$$\tilde{p}_{th} = S_0 \frac{\gamma P_0}{T_0 l_a} \frac{\tilde{Z}_{cs}}{\tilde{\sigma}_a} \frac{1}{sh(\tilde{\sigma}_s l_s)}, \tag{7a}$$

$$\tilde{p}_{ac} = S_0 \frac{6\gamma P_0 R^4}{l_a l_s^3 R_c^2} \alpha_s \frac{\tilde{Z}_{cs}}{(\tilde{\sigma}_s)^2} \frac{ch(\tilde{\sigma}_s l_s) - \frac{\tilde{\sigma}_s l_s}{2}sh(\tilde{\sigma}_s l_s) - 1}{sh(\tilde{\sigma}_s l_s)}. \tag{7b}$$

In the above expressions S_0 stands for the surface heat source, which equals half of the excitation energy intensity, R_c represents the effective radius of the sample [24], and $\omega = 2\pi f$ is radial modulation frequency.

Finally, the PA response is given in the form:

$$\tilde{p}_{ins} = E_0 \frac{1}{1 + jw\tau_e}(\tilde{p}_{th} + \tilde{p}_{ac}) = E_0 \frac{\tilde{p}_{th}}{1 + jw\tau_e}\left(1 + \frac{\tilde{p}_{ac}}{\tilde{p}_{th}}\right). \tag{8}$$

In the expression (8), the influence of the measurement chain is represented through the presence of the element $E_0/(1 + jw\tau_e)$, which can, however, be annulated by diverse normalization procedures [20,23].

When the influence of thermal memory is neglected, the expressions (5a,b) and (7) are reduced to their classic *composite piston* forms [2,4,8,18,19,22,23].

2.1. Multiple Optical Reflections—the Influence of Optical Properties

In thin samples with low optical absorption coefficient an increase of the static component of the PA response temperature variation was calculated as:

$$\Theta_{(0)}^{(s)} = \Theta_{(0)}'^{(s)} \cdot \frac{1}{1 + Re^{-\beta l_s}}. \tag{9}$$

On the other hand, in thin samples ($l_s \sim 10$ μm, optical absorption coefficient $\beta \sim 10^5$ m^{-1}, typical for polymers) with high inner reflection coefficients ($R_1 \sim 0.9$), at low frequencies (100 Hz–10 kHz), temperature variation (the dynamic component of the signal) is significantly increased, more noticeably on the exposed side of the sample ($x = 0$), as presented in Figure 3a.

However, in thick samples (~ 100 μm), as well as in others with low inner reflection coefficients ($R_1 \sim 0.1$), the effect is, surprisingly, the opposite: the temperature variation is decreased compared to the one corresponding to the model which neglects multiple reflections—the effect which can be seen in Figures 3b and 4a,b. Graphic representation of this principle is given in Figure 5.

These considerations present us with the possibility of observing another TMS model parameter—optical coefficient of inner reflection—in the future, but also call for caution; fundamental aspects of heat transfer through various media should be more profoundly studied [25].

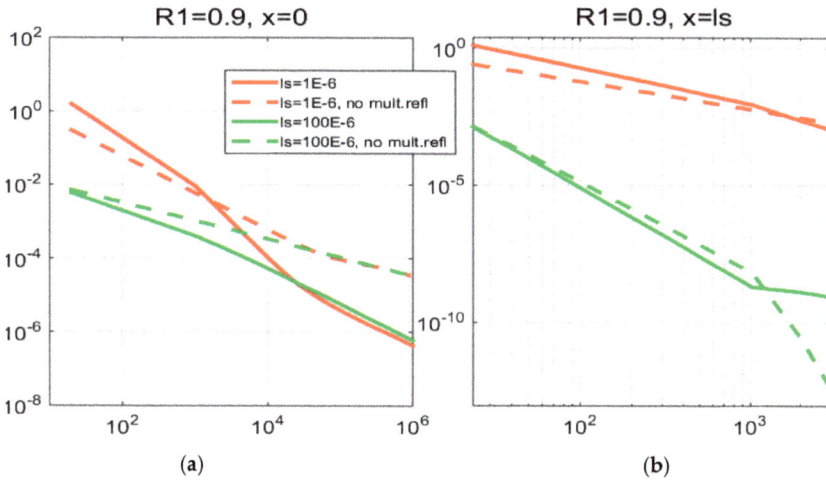

Figure 3. The influence of multiple optical reflections on surface temperature variation of samples with high inner reflection coefficient.

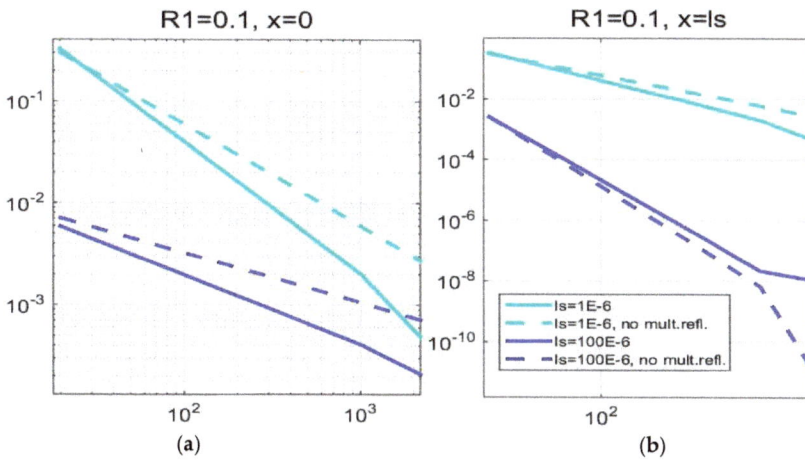

Figure 4. The influence of multiple optical reflections on surface temperature variation of samples with low inner reflection coefficient.

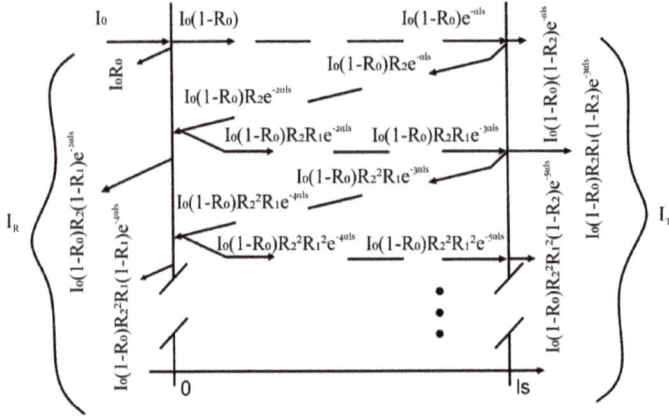

Figure 5. Graphic representation of the principle of multiple optical reflections.

2.2. Thermal Memory Influence

In numerous materials and under certain conditions, the appearance of oscillatory behavior as well as shape changes in both phase and amplitude responses are predicted; however, due to technical limitations of the experiment, these could not be recorded and validated. Instead, theoretical predictions are presented for reference samples (Aluminum, two thickness levels) in Figure 6.

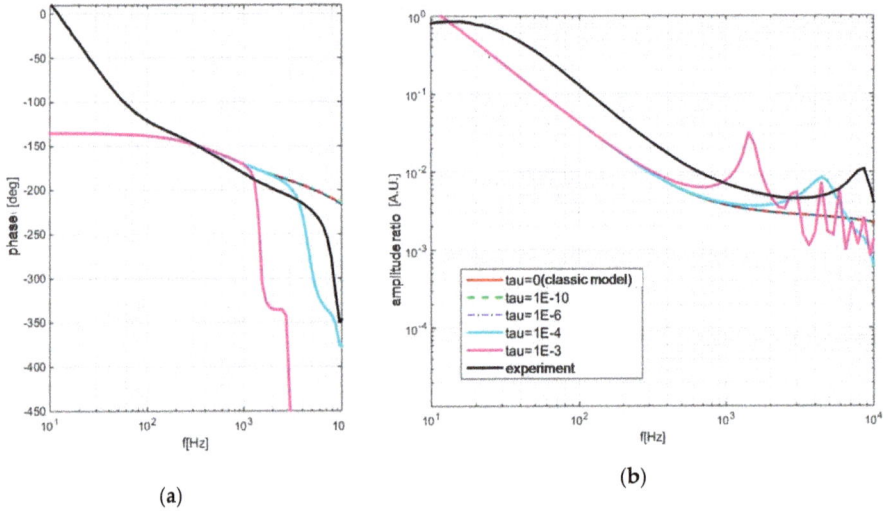

(a)

(b)

Figure 6. *Cont.*

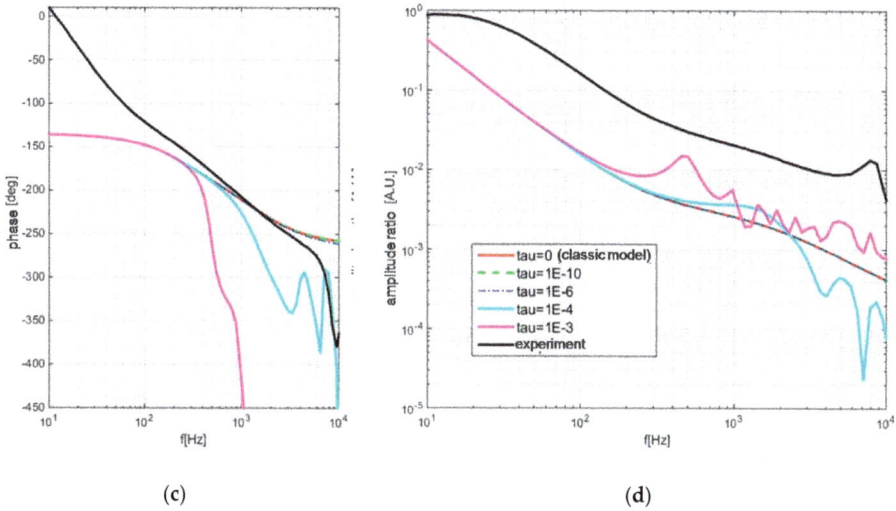

Figure 6. Thermal memory influence on PA response of aluminum samples 100 (**a,b**) and 300 (**c,d**) microns thick: phase is given on the left (**a,c**) and amplitude is on the right (**b,d**).

The most important result of these considerations is the expression which directly links the position of the first peak (the frequency value) with heat propagation velocity through the observed medium [9,10,12,21,26]:

$$f_{1\max} = \frac{1}{2l_s}\sqrt{\frac{D_{Ts}}{\tau_s}} = \frac{1}{2l_s}u_s. \tag{10}$$

2.3. Helmholtz Resonances—the Influence of the Measurement Chain

Resonant peaks observed in this part of frequency domain are, throughout literature, attributed to the influence of measurement chain, although, in measurements, they occur at frequencies lower than expected (frequency characteristic of the microphone, the amplifier, and other electronics) [27–29]. Minimum volume cell has already been observed as an electro-acoustic resonator and it has been modeled with cascade filter array, with transfer function represented as the combination of two Helmholtz resonators:

$$\tilde{p}_u(j\omega) = \tilde{p}(j\omega) \cdot H_V(j\omega) \cdot H_\varepsilon(j\omega). \tag{11}$$

The relation among different elements of the analogous electro-acoustical system and the actual geometrical values of the microphone are given in the following set of expressions [30]:

$$L = \frac{\rho l}{S}, C_i = \frac{V_i}{\rho v^2}, i = V, \varepsilon, \omega_{closed} = v\sqrt{\frac{S}{lV}},$$

$$H_i(j\omega) = \frac{\omega_i^2}{\omega_i^2 - \omega^2 + j\omega\frac{\omega_i}{Q_i}}, \quad \begin{bmatrix} i = V, \varepsilon \\ \omega = 2\pi f \\ \omega_i = 2\pi f_i \end{bmatrix}, \tag{12}$$

while the graphical representation of the analogy is given in Figure 7:

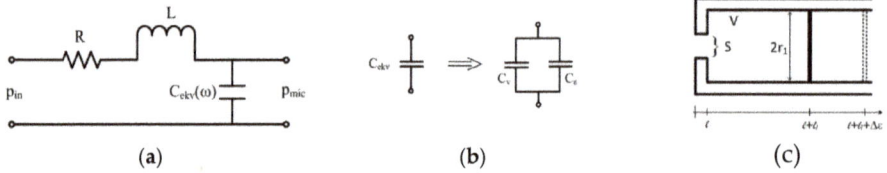

Figure 7. (a) Analogous electrical circuit, (b) capacitance model of the microphone chamber, (c) actual geometry of the microphone chamber.

The results of the application of the model are given in Figure 8:

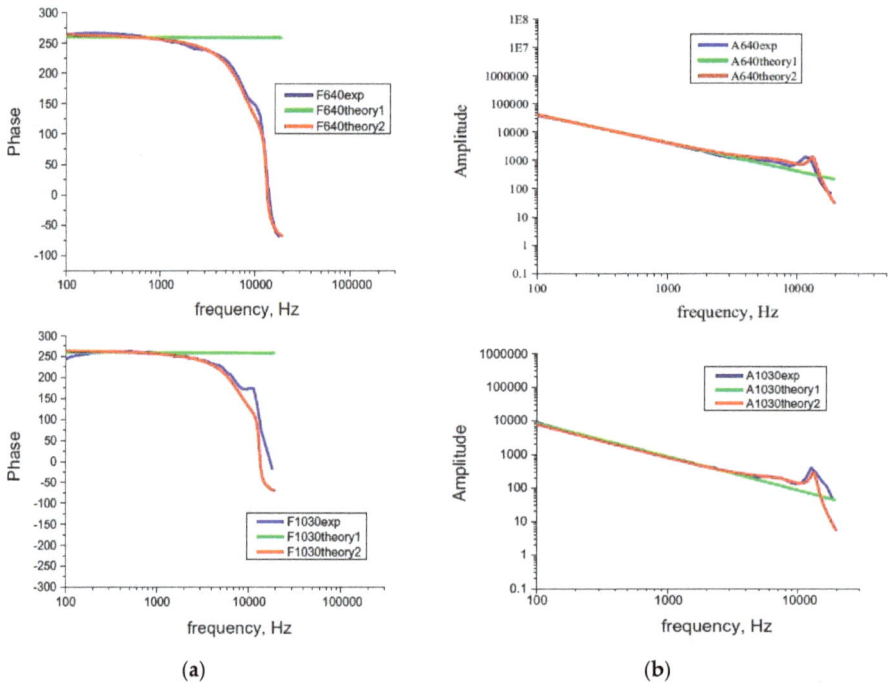

(a)   (b)

Figure 8. The inclusion of Helmholtz resonances in a PA experiment post-processing (**red line**): phase (a) and amplitude response (b), theoretical prediction (**green**), experimental results (**blue**) of HDPE at 640 and 1000 microns.

These considerations not only presented us with the possibility of effectively eliminating microphone influence in our future experiments, but also open the possibility of introducing a novel method of microphone characterization.

3. Techniques for Inverse Solving of PA Response

Based on the described analysis, two methods are developed for TMS model parameter determination, which were applied on numerical experiments:

1. A self-consistent procedure for solving the exponential problems of mathematical physics;
2. A well-trained three-layer perceptron with back propagation, based upon theory of neural networks.

The first method was, consequentially, applied on experimental measurements, with satisfactory results, and published [31] (subsection 2.2.3).

3.1. Self-Consistent Inverse PA Procedure

The idea for the development of the self-consistent inverse procedure for the estimation of thermodynamic parameters originates from theoretical considerations of PA model, where the tendency of phase exhibiting linear dependence upon thermal diffusivity, D_{Ts}, was noticed. The benefit of this approach is that this parameter, when derived from phase data, improves the reliability of multi-parameter fitting done on the rest of the signal (amplitude data). As a matter of fact, analytical methods demonstrated that thermal conductivity, k_s, could not be identified separately, but only as the part of its ratio with linear expansion coefficient, a_s: $\frac{a_T}{k_s}$ [32], which boils the fitting procedure down to only one parameter.

The validity of the idea was demonstrated first by TMS modeling of the problem, i.e., on a numerical experiment, presented in Figures 9 and 10. In Matlab package, the procedure was developed which randomly sets the values of the dataset D_{Ts}, $\frac{a_T}{k_s}$ (in accordance to literature values), and then simulates the PA response at two thickness levels using the given set of parameters with the addition of the certain level of noise. In the next step, the estimation of D_{Ts} is done by the comparison of phase difference data, while the value of $\frac{a_T}{k_s}$ is estimated from the amplitude ratio—both estimates are done by regression analysis: least squares being the method of choice.

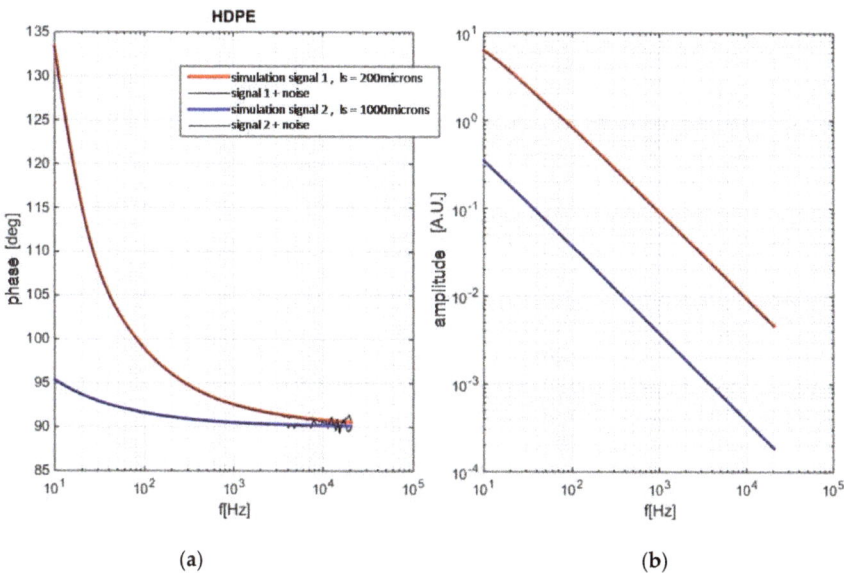

(a)   (b)

Figure 9. Simulated PA response at two thickness levels, with the addition of noise: phase (a) and amplitude (b).

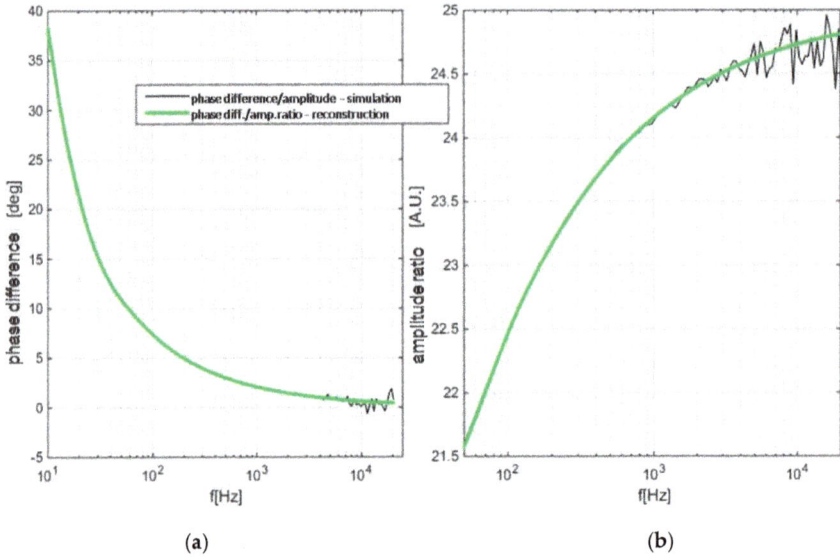

Figure 10. Simulated (**black line**) and reconstructed PA response (**green**), based upon the estimated values of parameters.

The results were quite interesting: after 1000 iterations, the procedure retuned the error for D_{Ts} 0.94% (always under 2%) and 44.15% for $\frac{\alpha_T}{k_s}$ (always above 30 %!). The conclusion was drawn that some parts of the model must be seriously ill-conditioned in the case of soft matter materials.

3.2. The Application of the Neural Network

Finally, a neural network was developed in order to assess the ill-conditioning issue of the inverse problem in photoacoustics of polymers. The type was multilayer perceptron, learning method was back-propagation, and the input parameters: $k_s, D_{Ti}, \alpha_T, l_s$. The material of choice: HDPE. Training was done on 40% of the sample dataset, 10% was used for validation, and testing (reconstruction) was applied on 50% of it. The results, after 10000 simulations were more than satisfying: estimation error for $\frac{\alpha_T}{k_s}$ was as low as 0.71%! As for the accuracy was not uniform: for low and high values of the parameter, the error was noticeable, but still, for the most of the sample set it remained under 2.15%!

However, what was more important than the estimation results, themselves, was the accompanying analytics, which, for the first time, presented the graphical representation of the ill-conditioning of the model, itself! Figure 11 clearly indicates how steep the dependence upon two parameters can be in the case of D_{Ts}.

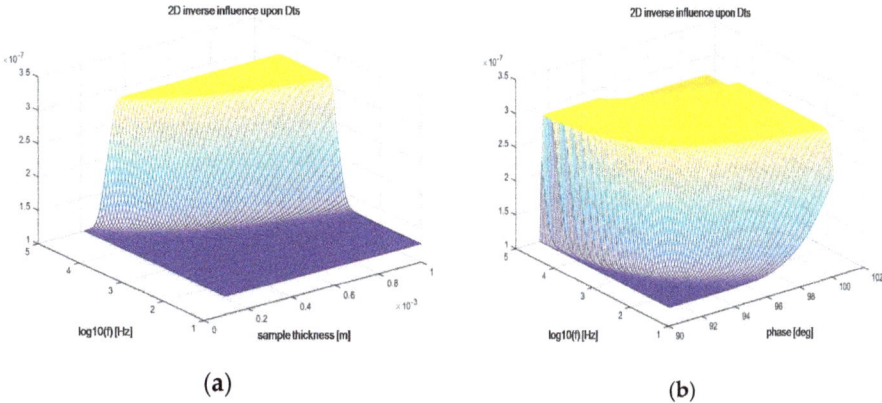

Figure 11. The dependence of D_{Ts} upon two parameters: thickness and modulation frequency (a) and phase and modulation frequency (b).

Finally, a conclusion could be drawn that, in case of soft matter materials such as HDPE, a self-consistent procedure could be more adequate for the estimation of D_{Ti}, while neural network approach clearly stood out when the estimation of $\frac{\alpha_T}{k_s}$ is concerned.

3.3. The Application on Experimental Data

The pioneering paper concerning the application of self-consistent procedure for the estimation of thermodynamic parameters on experimental data was published in 2018 [31]. HDPE samples had been prepared and characterized in advance at "Vinca" Institute for Nuclear Sciences in such a manner that their thickness and chemical or structural composition could not be questioned [31,33–36]. Using methods such as wide angle X-ray diffraction (WAXD) and diffraction scanning calorimetry (DSC), it was proven that regular normalization method (on two levels of thickness) could not be deployed. Also, crystallinity levels were estimated and are presented in Table 1.

Table 1. Crystallinity—functional dependence upon preparation conditions and sample thickness.

χ (%)	200 μm		400 μm	600 μm
	DSC	WAXD	DSC	DSC
Fast Cooled	51.7	50.5	57.4	59.3
Slowly Cooled	73.8	72.5	71.5	70.8

Regression analysis of the difference between theoretical prediction and the experimentally obtained PA response demonstrated that thin samples (200μm) have the potential for differentiating between different levels of crystallinity, as presented in Figure 12.

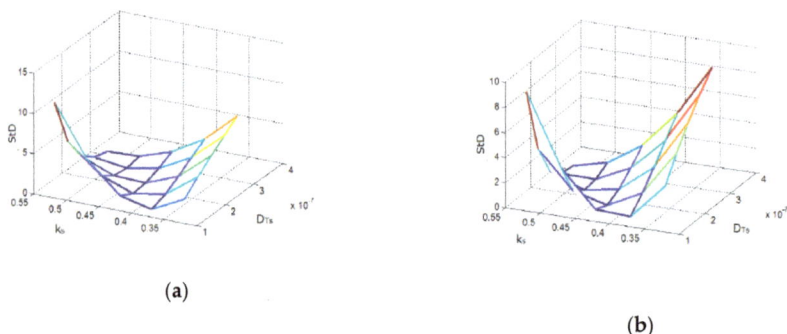

(a)

(b)

Figure 12. Deviation over region of interest as function of (ks, DTs) datasets for 200µm thick samples at two levels of crystallinity: ~70% (**a**) and ~50% (**b**).

Estimated results are presented in Table 2.

Table 2. Estimated values of thermodynamic parameters of HDPE (uncertainty given as half-distance between points).

Thickness [µm]		HDPE—High-Density Polyethylene		
		Fast Cooled	Slowly Cooled	Uncertainty
400, 600	$k_s\left[\frac{W}{m\cdot K}\right]$	0.33	0.33	(±0.02)
	$D_{Ts}\left[\times 10^{-6}\frac{m^2}{s}\right]$	0.313	0.313	(±0.019)
200	$k_s\left[\frac{W}{m\cdot K}\right]$	0.48	0.53	(±0.02)
	$D_{Ts}\left[\times 10^{-6}\frac{m^2}{s}\right]$	0.265	0.313	(±0.019)

Apart from the evident conclusion that rise in crystallinity demonstrated the tendency of D_{Ti}, k_s to increase, one could also say that the decrease in thickness facilitates the process of inverse solving, but also calls for caution when interpreting the dependence upon crystallinity due to the appearance of surface effects.

Another thing worth noticing is the significance of normalization, which was absent in this case (due to the influence of crystallinity) and which is proven to be very important for inverse solving of PA problems.

Finally, the relations among D_{Ti}, k_s and crystallinity amplify the significance of future fundamental heat transfer investigations.

4. Conclusions

The subject of this work is the development of the techniques aiming at solving the inverse problem in photoacoustics. Its mid-term goal is the increase in the number of material properties which can be characterized by PA measurements with a satisfactory level of accuracy, while its long-term goal is the improvement of the methods of PA imaging of different materials, from macromolecule nanostructures and nanoelectronics or nanophotonic devices, to biological tissues.

Within the analysis of the TMS model, the influence of optical, thermal, and elastic properties of the sample were investigated. This analysis revealed that some of the processes, characterized by certain sample properties, exert their dominance only in limited modulation frequency ranges, which are shown to be dependent upon the choice of the sample material and its thickness. In the rest of the range their influence can be neglected, so the TMS model is divided into parts, each corresponding to the appropriate modulation frequency range.

The main conclusions of this progress report are gathered in the form of a bulleted list:

- Generalized model of PA response as the consequence of finite heat propagation velocity was considered and its manifestations—thermal resonances—were described, with potential application in the determination of heat propagation velocity by making use of the location of the first peak;
- The influence of multiple optical reflections on PA response was considered for a specific class of soft matter materials and its potential application, as well as implications regarding fundamental heat transfer were pointed out;
- Minimum volume PA cell was successfully modeled as Helmholtz resonator and innovative applications of PA methods were potentiated;
- Simultaneous use of amplitude and phase measurements was proven to enable the estimation of thermal diffusivity, while difficulties in assessing the ratio of linear expansion coefficient and heat conductivity coefficient pointed out the necessity for the improvement of TMS modeling;
- The application of a neural network on the numerical experiment exposed the necessity for the reconsideration of the thermal piston model in materials with low levels of arrangement (macromolecules, tissue, soft matter);
- The application of self-consistent procedures on the experiment demonstrated the dependence of thermal properties upon thickness and crystallinity.

Author Contributions: Experimental investigation, software modelling and writing—original draft preparation, M.N.; simulations and modelling, M.P.; conceptualization, methodology, formal analysis and writing—review and editing, S.G.

Funding: This research received no external funding.

Acknowledgments: We acknowledge the support of the Ministry of Science of the Republic of Serbia throughout all the explorations done as part of the III45005 project.

Conflicts of Interest: The authors declare no conflict of interest.

Abbreviations

The following abbreviations are used in this manuscript:

TMS	theoretically/mathematically simulated (TMS) models
PA	photoacoustic/photoacoustics
EM	electromagnetic
1D	one-dimensional
HDPE	High-Density Polyethylene
WAXD	wide angle X-ray diffraction
DSC	diffraction scanning calorimetry

References

1. Mandelis, A. Diffusion waves and their uses. *Phys. Today* **2000**, *53*, 29–34. [CrossRef]
2. Todorović, D.M.; Nikolic, P.M. Carrier transport Contribution on Thermoelastic and Electronic Deformation in Semiconductors. In *Progress in Photothermal and Photoacoustic Science and Technology*; Mandelis, A., Hess, P., Eds.; SPIE Press: Washington, DC, USA, 2000; Volume 4, p. 272.
3. Rabasovic, M.D.; Nikolic, M.G.; Dramicanin, M.D.; Franko, M.; Markushev, D.D. Low-cost, portable photoacoustic setup for solid samples. *Meas. Sci. Technol.* **2009**, *20*, 95902. [CrossRef]
4. Perondi, L.F.; Miranda, L.C.M. Minimal-volume photoacoustic cell measurement of thermal diffusivity: Effect of the thermoelastic sample bending. *J. Appl. Phys.* **1987**, *62*, 2955–2959. [CrossRef]
5. Terazima, M.; Hirota, N.; Braslavsky, S.E.; Mandelis, A.; Bialkowski, S.E.; Diebold, G.J.; Miller, R.J.D.; Fournier, D.; Palmer, R.A.; Tam, A. Quantities, Terminology, And Symbols In Photothermal And Related Spectroscopies. *Pure Appl. Chem.* **2004**, *76*, 1083–1111. [CrossRef]
6. Nesic, M.V.; Popovic, M.N.; Stojanovic, Z.; Soskic, Z.N.; Galovic, S.P. Photoacoustic response of thin films: Thermal memory influence. *Hem. Ind.* **2013**, *67*, 139–146. [CrossRef]

7. Markushev, D.D.; Ordonez-Miranda, J.; Rabasović, M.D.; Chirtoc, M.; Todorović, D.M.; Bialkowski, S.E.; Korte, D.; Franko, M. Thermal and elastic characterization of glassy carbon thin films by photoacoustic measurements. *Eur. Phys. J. Plus* **2017**, *132*, 1–9. [CrossRef]

8. Pichardo-Molina, J.L.; Alvarado-Gil, J.J. Heat diffusion and thermolastic vibration influence on the signal of an open photoacoustic cell for two layer systems. *J. Appl. Phys.* **2004**, *95*, 6450–6456. [CrossRef]

9. Galovic, S.P.; Kostoski, D. Photothermal wave propagation in media with thermal memory. *J. Appl. Phys.* **2003**, *93*, 3063–3070. [CrossRef]

10. Nesic, M.V.; Galovic, S.P.; Soskic, Z.N.; Popovic, M.N.; Todorović, D.M. Photothermal thermoelastic bending for media with thermal memory. *Int. J. Thermophys.* **2012**, *33*, 2203–2209. [CrossRef]

11. Galovic, S.P.; Soskic, Z.N.; Popovic, M.N.; Stojanovic, Z.; Cevizovic, D.; Stojanovic, Z. Theory of photoacoustic effect in media with thermal memory. *J. Appl. Phys.* **2014**, *116*, 024901. [CrossRef]

12. Nesic, M.V.; Popovic, M.N.; Galovic, S.P. Influence of thermal memory on the thermoelastic bending component of photoacoustic response. *Hem. Ind.* **2011**, *65*, 219–227. [CrossRef]

13. Gurevich, Y.; Logvinov, G.; Lashkevich, I. Effective thermal conductivity: application to photothermal experiments for the case of bulk light absorption. *Phys. Stat. Solidi (b)* **2004**, *241*, 1286–1298. [CrossRef]

14. Gurevich, Y.; Logvinov, G.; de Rivera, L.N.; Titov, O. Nonstationary temperature distribution caused by bulk absorption of laser pulse. *Rev. Sci. Instrum.* **2003**, *74*, 441–443. [CrossRef]

15. Logvinov, G.N.; Gurevich, Y.G.; Lashkevich, I.M. Surface Heat Capacity and Surface Heat Impedance: An Application to Theory of Thermal Waves. *Jpn. J. Appl. Phys.* **2003**, *42*, 4448–4452. [CrossRef]

16. Rosencwaig, A.; Gerscho, A. Theory of the photoacoustic effect with solids. *J. Appl. Phys.* **1976**, *47*, 64–69. [CrossRef]

17. McDonald, F.A.; Wetsel, G.C. Generalized theory of the photoacoustic effect. *J. Appl. Phys.* **1978**, *49*, 2313–2322. [CrossRef]

18. Rousset, L.; Lepoutre, F.; Bertrand, L. Influence of thermoelastic bending on photoacoustic experiments related to measurements of thermal diffusivity of metals. *J. Appl. Phys.* **1983**, *54*, 2383–2391. [CrossRef]

19. Mansanares, A.M.; Vargas, H.; Galembeck, F.; Buijs, J.; Bicanic, D. Photoacoustic characterization of a two-layer system. *J. Appl. Phys.* **1991**, *70*, 7046–7050. [CrossRef]

20. Balderas-Lopez, J.A.; Mandelis, A. Thermal diffusivity measurements in the photoacoustic open-cell configuration using simple signal normalization techniques. *J. Appl. Phys.* **2001**, *90*, 2273–2279. [CrossRef]

21. Nesic, M.V.; Gusavac, P.; Popovic, M.N.; Soskic, Z.N.; Galovic, S.P. Thermal memory influence on the thermoconducting component of indirect photoacoustic response. *Phys. Scr.* **2012**, *T149*, 14018. [CrossRef]

22. Cao, J. Interferential formulization and interpretation of the photoacoustic effect in multi-layered cells. *J. Phys. D. Appl. Phys.* **2000**, *33*, 200. [CrossRef]

23. Lima, C.A.S.; Lima, M.B.S.; Miranda, L.C.M.; Baeza, J.; Freer, J.; Reyes, N.; Ruiz, J.; Silva, M.D. Photoacoustic characterization of bleached wood pulp and finished papers. *Meas. Sci. Technol.* **2000**, *11*, 504. [CrossRef]

24. Bedoya, A.; Marin, E.; Mansanares, A.M.; Zambrano-Arjona, M.A.; Riech, I.; Calderon, A. On the thermal characterization of solids by photoacoustic calorimetry: Thermal diffusivity and linear thermal expansion coefficient. *Thermochim. Acta* **2015**, *614*, 52–58. [CrossRef]

25. Nesic, M.V.; Popovic, M.N.; Galovic, S.P. The influence of multiple optical reflexions on the photoacoustic frequency response. *Opt. Quantum Electron.* **2016**, *48*, 7. [CrossRef]

26. Novikov, I.A. Harmonic thermal waves in materials with thermal memory. *J. Appl. Phys.* **1997**, *81*, 1067–1072. [CrossRef]

27. Todorović, D.M.; Rabasovic, M.D.; Markushev, D.D. Photoacoustic elastic bending in thin film—Substrate system. *J. Appl. Phys.* **2013**, *114*, 213510.

28. Todorović, D.M.; Rabasovic, M.D.; Markushev, D.D.; Sarajlić, M. Photoacoustic elastic bending in thin film-substrate system: Experimental determination of the thin film parameters. *J. Appl. Phys.* **2014**, *116*, 053506. [CrossRef]

29. Aleksic, S.M.; Markushev, D.K.; Pantic, D.S.; Rabasovic, M.D.; Markushev, D.D.; Todorović, D.M. Electro-Acoustic Influence of the Measuring System on the Photo-Acoustic Signal Amplitude and Phase in Frequency Domain. *Univ. Facta.* **2010**, *14*, 1–13.

30. Popovic, M.N.; Nesic, M.V.; Ciric-Kostic, S.; Zivanov, M.; Markushev, D.D.; Rabasovic, M.D.; Galovic, S.P. Helmholtz Resonances in Photoacoustic Experiment with Laser-Sintered Polyamide Including Thermal Memory of Samples. *Int. J. Thermophys.* **2016**, *37*, 1–9. [CrossRef]

31. Nesic, M.V.; Popovic, M.; Rabasovic, M.; Milicevic, D.; Suljovrujic, E.; Markushev, D.; Stojanovic, Z. Thermal Diffusivity of High-Density Polyethylene Samples of Different Crystallinity Evaluated by Indirect Transmission Photoacoustics. *Int. J. Thermophys.* **2018**, *39*, 24. [CrossRef]

32. Soskic, Z.N.; Ciric-Kostic, S.; Galovic, S.P. An extension to the methodology for characterization of thermal properties of thin solid samples by photoacoustic techniques. *Int. J. Therm. Sci.* **2016**, *109*, 217–230. [CrossRef]

33. Wunderlich, B. *Thermal Analysis of Polymeric Materials*; Springer: Berlin, Germany, 2005.

34. Wunderlich, B.; Cormier, C.M. Heat of fusion of polyethylene. *J. Polym. Sci. Part A-2 Polym. Phys.* **1967**, *5*, 987–988. [CrossRef]

35. Vonk, C.G. Computerization of Ruland's X-ray method for determination of the crystallinity in polymers. *J. Appl. Cryst.* **1973**, *6*, 148–152. [CrossRef]

36. Jokanovic, V. *Instrumental Methods: Key to Understanding Nanotechnologies and Nanomedicine*; Engineering Academy of Serbia: Belgrade, Serbia, 2014.

Article

On the Time Scales of Optical Variability of AGN and the Shape of Their Optical Emission Line Profiles

Edi Bon [1,*], Paola Marziani [2], Predrag Jovanović [1] and Nataša Bon [1,2]

[1]   Astronomical Observatory, Volgina 7, 11060 Belgrade, Serbia; pjovanovic@aob.rs (P.J.); nbon@aob.rs (N.B.)
[2]   National Institute for Astrophysics (INAF), Astronomical Observatory of Padova, IT 35122 Padova, Italy; paola.marziani@inaf.it
*   Correspondence: ebon@aob.rs

Received: 8 December 2018; Accepted: 6 February 2019; Published: 14 February 2019

Abstract: The mechanism of the optical variability of active galactic nuclei (AGN) is still very puzzling. It is now widely accepted that the optical variability of AGN is stochastic, producing red noise-like light curves. In case they were to be periodic or quasi-periodic, one should expect that the time scales of optical AGN variability should relate to orbiting time scales of regions inside the accretion disks with temperatures mainly emitting the light in this wavelength range. Knowing the reverberation scales and masses of AGN, expected orbiting time scales are in the order of decades. Unfortunately, most of monitored AGN light curves are not long enough to investigate such time scales of periodicity. Here we investigate the AGN optical variability time scales and their possible connections with the broad emission line shapes.

Keywords: AGN; black holes; gravitational waves; binary black holes; quasars

1. Introduction

Active galactic nuclei (AGN) are very strong and variable emitters [1]. It is widely believed that the AGN patterns correspond to red noise-like curves [2,3], as a result of unpredicted processes of fluctuations of their accretion disks (AD). In some cases, these variations appear periodic, with periodogram peak significance jumping above the red noise levels (see for example, [4]). Unfortunately, most AGN monitoring campaign time intervals are not sufficiently long for detecting periodicity in optical domain of spectra, since expected orbiting time scales in AGN ADs that would affect the optical part of the spectrum are of the order of years and decades. Luckily, significant periodicities are detected in several AGN which are extensively monitored for sufficiently long time, such as OJ287 [4,5], NGC 4151 [6–8], NGC 5548 [9,10], Ark 120 [11], 3c273 [12,13].

It is expected that in galaxy mergers, their cores should eventually end up close to each other, and get gravitationally bounded in the close-orbiting system of supermassive black holes (SMBHs). Such configurations are called supermassive binary black hole (SMBBH) systems, with period of several to tens of years resulting from orbiting timescales of the system.

Some numerical simulations of SMBBHs show expected periodic behaviors of their light curves [14–17].

1.1. Variability Time Scales and Amplitudes

Main variability time scales of AGN are light crossing (order of days to months), orbiting (years to decades), and sound speed (order of hundreds of years) [18].

The amplitude of variation is different in each object. The variability of radio loud and radio quite types seems to be more or less similar, within the one sigma of difference [19]. The variability is most significant in case of some "changing look"(CL) objects, such as NGC 5548 [9], NGC 4151 [7], and some other examples see, (e.g., [20,21]). Some of them appear to be periodic, such as for example NGC 5548, NGC 4151, and a few more periodic CL candidates; see [7,9,22].

It is not yet clear what triggers such variability that can lead to changing type. Investigating the periodic CL case of NGC 5548 in the context of Eigenvector 1, showing that besides that this object is mainly Pop B1, it also varies in long term observations lead to conclusion that the main driver of the variability could be connected to the accretion rate changes, and the obscuration effects [23]. The evidence of intrinsic obscuration within the broad line region itself, as also found for a recent very short epoch in monitoring of this object [24]. How these two mechanisms combined could produce periodical variability is still not clear.

1.2. Periodic Variability of AGN

Mechanisms proposed to explain the periodic emission variability of AGN are: jet/outflow precession, disk precession, disk warping, orbiting of spiral arms, flares, and other kinds of instabilities orbiting within the accretion disk, tidal disruption event, the existence of a SMBBH system in their cores see, [9,25–32].

AGN are very hard to prove to be periodic [3]. Therefore, standard methods such as Fourier and Lomb-Scargle [33,34] may show peaks of high looking significance but the derived p-value may not be valid [3,9,35]. There are a few historic AGN light curves spanning over 100 years (monitored first as variable stars, before recognized to be distant galaxies) found to show significant periodicity of order of years to decades see (e.g., [4–6,13]).

1.3. Supermassive Black Hole Binaries

A probable explanation of periodic variability is the possibility of SMBBH, since such scenario should be able to produce most significant variations of accretion rates, as well as gas ejections and obscuration effects.

The orbiting variability time scales in AGN are of an order of several decades. Unfortunately, it is hard to find cases of AGN light curves with several repeating patterns [4,5,9,10,29,30,36], needed for the clear detection of periodicity, above the red noise level, since the length of current AGN monitoring campaigns are of the order of orbiting time scales. Therefore, as an indicator of orbiting effects we could be tracing the broad emission line shifts, and if the periodicity in radial velocity curves is the same, it could indicate that the mechanism which drives both curves could be linked to the orbiting within the broad line region [7,9–11,31,37]. Radial velocity curves are harder to obtain due to even shorter records of spectral observations, and therefore only a few candidates are detected, such as NGC 4151 with a 15.9 year periodicity [6,7], NGC 5548 with a ≈15 year periodicity [9,10], and Ark 120 with a ≈20 year periodicity [11]. It is interesting that current results of all these cases show that the radial velocity curves of the red side of broad emission lines show larger amplitude shift with more significant indication of periodicity, as in case of NGC 3516 [38], where the same effect is observed in red wing of Fe Kα line. However, this is expected, since that the gravitational effects are more significant on the red side of the line due to gravitational redshift effect [39], and it could be seen in magneto hydrodynamic simulations of eccentric SMBBH systems [27].

Periodic light curve candidates also include a blazar OJ 287, with 11.5-year period, as the most famous SMBBH candidate [4,5], the quasar PG 1302-102 (6 year period, [29]), the blazar PG 1553+113 (2 year period, [40]), a 13- and 21-year periodicity found in 3C 273 [12,13].

The light variations may not be the only direct effect of the moving SMBHs. The effects of SMBBH system orbital motion should be also affecting the jet bending, which could be used to model the SMBBH system properties with jet observations (see, for example [41–44]) Unfortunately, quasi-periodicities in the jet emission can be induced by intrinsic oscillatory disk instabilities that can mimic periodical behavior.

Most significant effects to light curves are expected from orbiting SMBHs. Simulations of such emission produced in SMBBH systems is very complex, and therefore mainly tested for simplified configurations, such as comparable mass SMBHS, and nearly circular orbits [45,46]. Only a few simulations are currently available for the eccentric high-mass ratio systems [27,28,47,48].

1.4. Broad Emission Line Shape Connection with the Variability Time Scales

The connection of AGN variability with broad emission line shape changes was suggested in several papers see, e.g., [23,38,49–51]. The orbiting gas is expected to be in some form of flattened distribution [52–56] that could be surrounded with isotropic gas component [57–60].

Orbiting gas properties are closely related to the mass of the central supermassive body, gas distribution, and inclination of the system. Therefore, the shape of broad emission lines could give us some constraint of the gas configuration and the central BH mass.

Recently, one such model was proposed [23], where they investigated the connection between the variability time scales of active galactic nuclei (AGN) optical light curves with the shapes of their broad emission line profiles. Knowing that the perturbing region orbital period is related to the SMBH mass and to the radial distance of the perturbing region [38,51], they propose that the variability time scale of the optical continuum light curves could be connected with the perturbing region located at the part of a disk seen as the inner and outer radius for the optical broad Balmer lines. Using the accretion disk model [52] as a tool to measure the disk size and parameters (like the inclination angle), and connecting it with the variability time scales of the AGN light curves, they were able to obtain similar masses of the central BHs, as from reverberation campaigns. From virial mass and the orbiting calculated with this model it is possible to calculate the inclination of the emitting gas orbiting plane. In [23] the obtained inclination angle from their method agreed with the inclination angle obtained by the disk model fitted in the broad emission line profiles. This could indicate that the inner and outer radii of an accretion disk might be indeed connected with the AGN variability time scales.

In case the variability time scales relate to the orbiting time scales of SMBBH systems, then identification of such systems could be very important for gravitational wave (GW) observations [61]. In the epochs of last orbits, the gas could be squeezed producing super-Eddington outflows [61].

2. Method

Knowing that the light curves of a continuum at 5100 Å and broad Hβ emission line are highly correlated [62] may indicate the same origin of their variability. Therefore, the source that drives the variability could leave a trace in the shapes of their broad lines. Analysis of variation time scales may give us valuable information about why they vary the way they do, while the line profiles could provide us with the information about the kinematic parameters of the variability drive (such us the radii where the source of variation is located).

To investigate the variability patterns, we used periodicity analysis of optical continuum light curves, with hope to find periodic or quasi-periodic variability patterns, with time scales corresponding to the orbital time scales within the region where the optical light could be originating inside the AD. If we could detect signatures in the broad emission line profiles, which could be produced by the effects of the same phenomena that drives the variability of the same time scale for that peculiar periodic pattern, then we could be able to determine dynamical properties of such AGN. In case we could identify more than one significant period in their light curves that could be linked to the radius of amplified ring emission in the broad emission line shape, then for each ring-radius pair we should expect to obtain the same mass (or at least very close value) of the central BH mass using a Kepler's laws.

To test these assumptions, as a first step we model synthetic line emission of an orbiting gas in the flat, disk-like gas distribution, assuming that photo-ionization processes produce the emission line from that region, that we could approximate with the accretion disk emission model [39,49,52–54,56,63,64].

By matching the disk model with the line profile, we determine the inclination, and inner and outer radii. In the shapes of BELs sometimes could be found small bump-like features that cannot be modeled with such disk models, assuming emissivity parameter that would correspond to photo-ionization (and assuming a negligible mass of a perturbed region in the ring compared to the central SMBH).

Simulated profiles of AD emission usually have characteristic two peaks in the core of the line, while the wings are broadened due to relativistic effects. The two peaks are usually blended by the isotropic emission component, located away from the AD, which is present in majority of AGN spectra [57–60,65,66]. Only in a very small number (less than 1%) of objects the two peaks are clearly recognized see e.g., [54,67].

Model of AD with Amplified Thin Ring Emission

The AD model is an idealization of emission with assumption of homogeneous AD that may not be the case, and therefore is not sufficient to describe all features in observed profiles, like for e.g., small bumps in the wings of the line profiles that are often present. To describe these additional features (like bumps in the broad Hβ emission line profiles) we assumed additional amplified emission component located in thin rings inside the AD. We model the thin ring emission assuming it is located somewhere in the interval from 100–4000 R$_g$ (which we vary in the model with a step of 100 R$_g$), assuming the ring width of 100 R$_g$. We add the contribution of each ring profile to the AD model profile as an additive combination of rings profiles, with its intensity multiplied by a scaling factor (see Figures 1–3).

Assuming that the time scale of perturbed disk (cooling time, shock wave progression, or anything that produced additional emission from that ring) is significantly longer than the orbital time scale, then we could approximate that the quasi-periodic variations found in optical light curves correspond to the orbiting of some features inside of the AD at radii that we can associate with amplified thin rings that we can locate by matching the emission lines and synthetic modeled profiles. Their radii are measured in units of gravitational radii Rg, since from the AD model we cannot obtain the information about the central mass. By connecting each period to thin ring-radius with the assumption that the shorter period corresponds to closer ring, and the outer ring corresponds to a longer period, we could be able to calculate the mass of the central SMBH using the Kepler's third law for a circular orbit:

$$P = 2\pi\sqrt{\frac{r^3}{GM}} = \frac{2\pi GM\zeta^{3/2}}{c^3},\tag{1}$$

where $r = \zeta(r_g)$ is the ring-radius in gravitational radii and P is the circular orbital period of the orbiting region at such radii see, [23,50,51].

We decided to use the relativistic ray-tracing AD model[1] see [39,64,68], to build a new, more complex model, of an AD assuming the amplified thin ring emission. We assumed that the origin of the variability patterns, detected in quasi-periods, could be traced back in the broad line profiles (seen as amplified thin rings emission inside of AD at different radii). Making a connection of periods and radii, using the Equation (1), could allow us to open a new window into dynamical properties of AGN.

Our model assumes the emission of an AD and ring models as a linear combination of contributions to the line profiles with the same parameters as in the parent AD, which are preserved in ring models. To construct our model, we simulated the grid of models of the Hβ line profiles using the code which includes both special relativistic and general relativistic effects on radiation from the accretion disk around SMBH see e.g., [68]. This AD model is based on the ray-tracing method in the Kerr metric [63,64], for different values of inner and outer radii and inclinations of rings in AD. The emissivity index was fixed as q $= -2$, assuming the emissivity law to be $\sim r^q$, as expected for the case of photo-ionization. The model is then constructed using a previous match of the AD profile to the emission line, as a starting point. Then the scaled contributions of the thin ring profiles are added to the AD profile until bumpy features in observed spectra are described with the synthetic spectrum.

---

[1] We tested several different AD models [39,52,56,63,64,68] for a line fit, and found that obtained inclinations were practically the same regardless of the model used.

Beside the fact that the shape of the line is fitted more realistic then with a simple AD model, we are also obtaining a valuable information about the radii in the disk, where the emission is amplified.

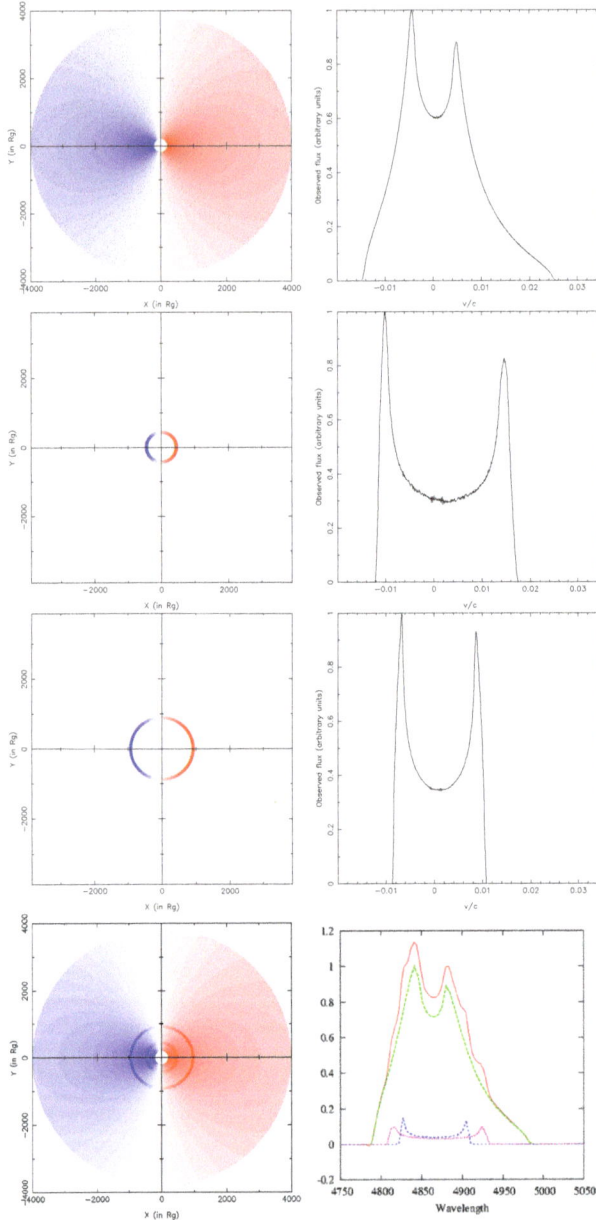

Figure 1. Construction of a synthetic profile as a linear combination of components: AD (R_{inn} = 100 R_g, R_{out} = 8000 R_g, top panel right) and two additional synthetic ring profile models (R_{inn1} = 500 R_g, R_{out1} = 600 R_g, R_{inn} = 1000 R_g, R_{out2} = 1100 R_g mid right panels respectively). The resulting emission line profile is presented as right bottom panel as sum of contribution of AD and two ring profiles with intensity scaled by a factor of 0.1. Representations of disk and ring models are presented on the right panels.

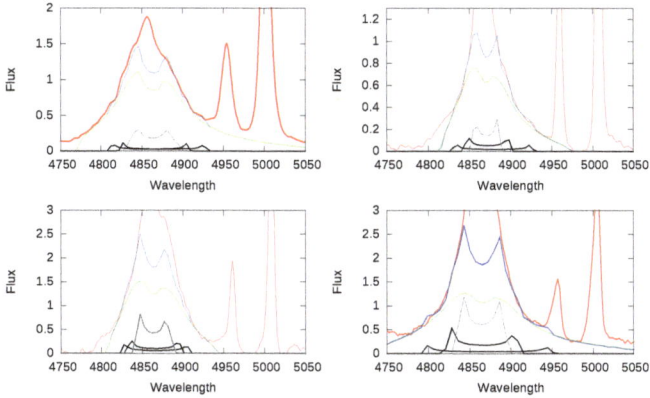

Figure 2. A few examples of AD + ring model matching the broad Hβ emission line profiles.

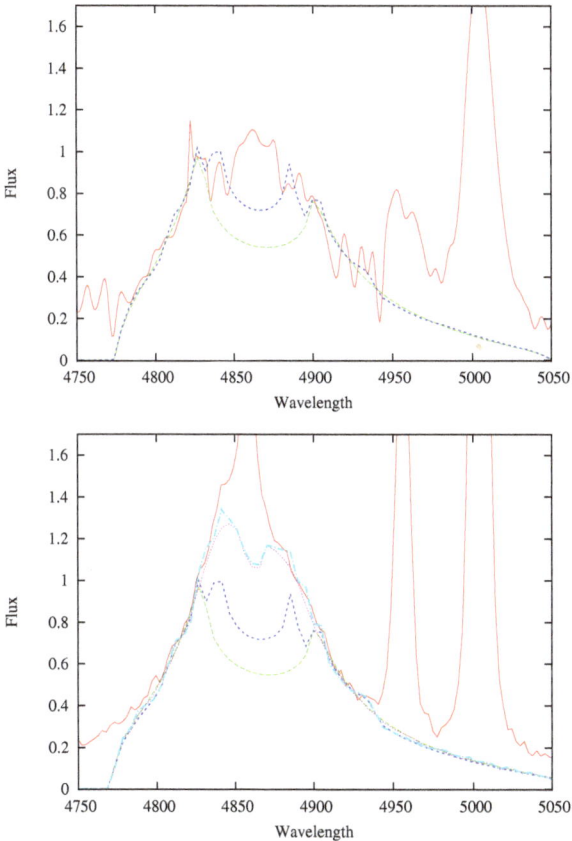

Figure 3. A few examples of AD + ring model matching the monitoring broad Hβ emission line profiles of a single object in low (**top** panel) and high (**bottom** panel) state. Both profiles are matched with models with same parameters, except for the outer radius, which is 1500 R$_g$ for the low state profile (**top**) and 8000 R$_g$ for the higher state line profile (**bottom** panel). The model indicates that the difference between these two states is in ionization of outer BLR, with radii above 1500 R$_g$.

One example of how the profiles are constructed is presented in Figure 1. As could be seen, by adding small contributions of ring profiles to the AD line shape, we make a model for the same inclination. In this peculiar case, the inclination angle is assumed to be 16 degrees. The AD was calculated with R_{inn} = 200 R_g and R_{out} = 4000 R_g. The velocity distribution over the surface of AD in ray-tracing model is presented on the left, while the line shape binned from it is presented on the right panel. For the rings we assumed the same inclination, emissivity law, and the disk shift. We simulated two ring emission profiles, assuming $Rinn_1$ = 500 R_g to $Rout_1$ = 600 R_g and $Rinn_2$ = 1000 R_g to $Rout_2$ = 1100 R_g, and added them to the AD profile assuming the multiplying factor of 0.1. The resulting profile could be seen in the bottom panel of Figure 1.

For the periodicity analysis we used standard Lomb-Scargle method [33,34]. Obtained periods are fitted afterwards with linear combination of sine functions, as an additional test of results.

3. Results

To test this hypothesis, we selected a sample presented in [62], where all spectra and light curves were publicly available. Also, for the periodicity analysis, beside light curves from [62], we used additional light curves from Catalina Real-Time Transient Survey see [29,30,36]. We used only single epoch spectra instead of averaged, to avoid smoothing by averaging. In some cases, we used spectra of the same objects, from [69] paper, due to better S/N.

Using the radii measured from line profile matching to the model, we measure the ring radii that we hoped to be connected to the variability quasi-periods, as a measurement of variability time scales. Assuming that the orbital time scale is the only match to time scale of variability patterns seen in these light curves, we combine measured radii and periods, and derived mass assuming circular Keplerian orbit of this brighter region positioned in the ring of the AD. We test the model by calculating masses of the central BH, with expectation that the obtained results for masses, for each pair of period and radius, should be equal, or at least of similar or close values.

We measured ring radii in units of gravitational radii (R_g) from the line profiles, and detected two or more significant periods, with a threshold at 400 days period to avoid effects of Earth's orbital period of 1 year. Unfortunately, expected orbital period in optical part of the AD for typical AGN of $M \sim 10^8$ $M_\odot$ is about one year. Therefore, any amplification in the AD profile under 300 R_g was also avoided, since it would correspond to such time scales, or even shorter periods.

Here we do not consider any details about what produces the hot ring region in the AD, and we are fully aware that the measured periods are significant above the standard, white noise levels, but may not appear significant compared to the red noise AR curves [2,3]. We were mainly interested in measuring time scales of orbital periods assuming that the variability patterns [19] in the light curves could be induced by the orbital time scales.

Examples of model in the line profile matched to single epoch spectral profiles are presented in Figure 2. PG 0052 in low and high state is presented as an example of an object spectroscopically monitored. The low- and high-state profiles are matched with the same parameters, except for the contribution of the outer part of the disk with upper limit of 1500 R_g. This radius in this case is the one we connected with the longer period, while the shorter period is paired with radius of 1000 R_g (of the amplified ring making bumps at the core of the profile next to the narrow OIII [4959] line, see Figure 3). These bumps are even better recognized in higher S/N spectrum of the same object presented at the top right panel of the Figure 2, where one can see the observed spectrum in red, model of an AD+rings (blue), just AD model (green), and each of the additional ring contributions (black) that are paired with periods. The outer part rings are presented with gray line[2]. The model indicates that the difference

---

[2] This outer contribution would correspond to much longer period that we could not be able to detect with such short monitoring intervals of several years, in case it could be connected to any periodicity. These distant radii are probably excited due to reverberation processes.

between these two states of this monitoring is in ionization of outer BLR, with radii above 1500 R_g. Measuring this radius, and pairing it with characteristic periodicity can allow us to calculate the mass, assuming that we know the inclination from the fit of the model, as well.

Therefore, assuming that ring radii are connected to quasi-periods obtained from light curves, we calculate central BH mass, for each pair of period-radius. Results of calculated masses are presented on the plot against FWHM in the Figure 4. Calculated masses from each period–radius pair result with similar value of mass (see Figure 4), that is, what we should expect if the hypothesis of the model is correct. Therefore, obtaining close values of mass estimates is justifying the model hypothesis.

Also, getting practically identical values for the mass is very important for the error estimates, showing that this method therefore may be more precise than the reverberation mapping technique.

We note that the obtained masses with our method are mainly in a good agreement with previous results obtained from reverberation mapping results [62]. This could be seen in Figure 5, where we presented the ratio between masses obtained with our method and with the reverberation mapping method given in [62]. The ratio with reverberation of masses obtained from mean profiles are given in orange, while the ratio with masse from RMS profiles are given in blue. These differences of masses obtained with our method and reverberation mapping method (see Figure 5) could be due to the fact that in reverberation mapping technique, the inclinations of each objects are not taken into account when calculating masses. The assumption that FWHM is not corrected for the inclination, or not weighted properly for the contributions of virial and isotropic component, may influence the "f" factor in the virial formula that was assumed to be the same for all objects. Also, the luminosity measurements were assumed to be connected with the BLR radius, which is another approximation in reverberation mapping techniques, and which may add additional systematic effects as well. We note that the largest discrepancy is seen on parts of the plot that is most affected by the inclination effect, see Figure 5 for FWHM under 2000 km/s, and above 8000 km/s, since the "f" factor in virial formula was assumed to correspond to the averaged inclination angle in all sources.

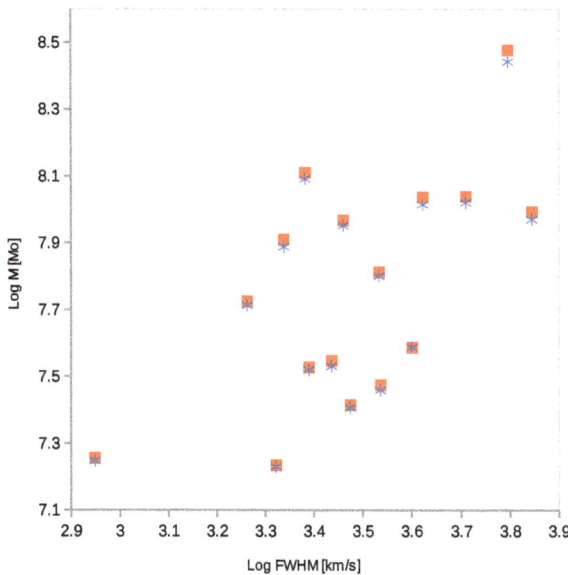

Figure 4. The Full Width at Half Maximum of Hβ line versus mass of central SMBH, calculated for each radius-period pair (orange squares and blue asterisks).

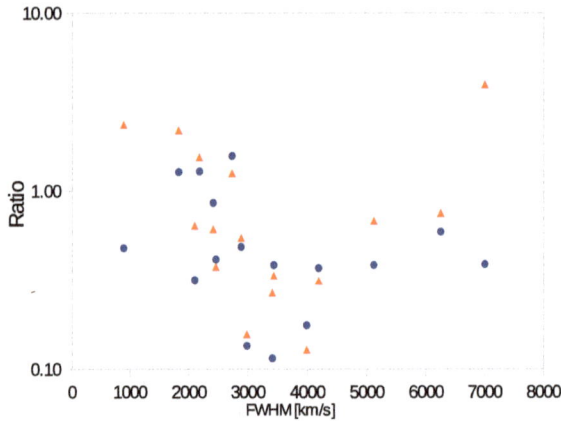

Figure 5. The comparison of SMBH masses obtained in this paper with previous reverberation mapping results from Kaspi et al. [62]. Blue full circles represent the ratio of masses from this work and the masses from the reverberation results obtained using FWHM of averaged line profiles, while the orange triangles correspond to ratio with reverberation masses from the RMS profiles.

4. Possible Interpretations

Possible interpretations of periodicities are discussed in many works [7,9–11,30,31,35,36,50,51,54,70]. Assuming circular orbits in the disk as we did here, we suggest that possible source of periodicity should be in the AD, amplifying an emission contribution at that radius. We are aware that at such radii, the standard models of thermal emission of AD [71,72] shows that the temperature of the disk is relatively low, under 1000 K, which is not sufficient enough for the photo-ionization mechanism to produce optical broad emission lines, or to significantly contribute to the optical continuum flux, without additional emission mechanism, like shocks [50,54,70], hot spot [49,51], secondary orbiting object on a circular orbit around the central SMBH with additional accretion mechanism that is sufficient to produce significant contribution to the continuum and the line emission see e.g., ([73–75], where their MHD simulations show a fast forming of intermediate mass BH's in AD on a circular orbit). It is expected that in such cases the voids or gaps would be formed see, for e.g., [73,76] in the AD (similar to the planet formation in the stellar disks), with pilling up of matter at the outer border of the gap ring see, ([73,76], and the references therein) that may be the region that could be associated with the amplified ring emission that we modeled here.

5. Conclusions

We simulated an AD emission profiles with amplified thin ring regions and matched them with the observed profiles of the broad Hβ emission lines, of PG quasars from the sample selected in [62]. With periodicity analysis, we find significant periodicity from the available optical light curves. We pair each ring profile with the periodicity. Using Kepler's third law, we then calculate central SMBH masses.

Our results show:

(1) The model of an AD with amplified thin ring emission could describe the observed Hβ emission line profiles.

(2) Masses calculated from each pair of period and thin ring, result with very similar values (with the discrepancy of less than 20%), justifying the model initial hypothesis and indicating that the features fitted in the line profiles are not some random noisy patterns, but features that could be connected to the real physical properties of the BLR emission.

(3) The obtained masses are with the same order of magnitude as obtained from the reverberation mapping method, with the discrepancy that could be addressed to the lack of use of inclination angle correction in the reverberation mapping analysis.

(4) Even though, we do not go into a detailed interpretation of the origin of periodic variability, this method could be used on single epoch spectra of central mass BH determination, combined with long term photometric monitoring.

We plan to extend the sample (Bon et al. 2019, in preparation), selecting more AGN with long term monitoring data.

Author Contributions: Conceptualization, E.B. and P.M.; Methodology, E.B., P.J., P.M. and N.B.; Software, P.J., N.B., P.M. and E.B.; Validation, E.B., P.M., N.B. and P.J.; Formal Analysis, E.B., N.B. and P.J.; Investigation, E.B., P.M. and N.B.; Data Curation, E.B., P.M., N.B. and P.J.; Writing—Original Draft Preparation, E.B. and P.M.; Writing—Review and Editing, E.B. and P.M.; Visualization, E.B. and P.J.; Supervision, E.B. and P.M.; Project Administration, E.B.

Funding: This research was funded by the Ministry of Education, Science and Technological Development of the Republic of Serbia, through the project 176003 "Gravitation and the large-scale structure of the Universe" and 176001 "Astrophysical spectroscopy of extragalactic objects". We also wish to acknowledge the COST Action CA16104 "GWverse", supported by COST (European Cooperation in Science and Technology).

Acknowledgments: This research is supported by the Ministry of Education, Science and Technological Development of the Republic of Serbia, through the project 176003 "Gravitation and the large-scale structure of the Universe" and 176001 "Astrophysical spectroscopy of extragalactic objects". We also wish to acknowledge the COST Action CA16104 "GWverse", supported by COST (European Cooperation in Science and Technology).

Conflicts of Interest: The authors declare no conflict of interest.

References

1. Gaskell, C.M.; Klimek, E.S. Variability of Active Galactic Nuclei from Optical to X-ray Regions. *Astron. Astrophys. Trans.* **2003**, *22*, 661–679. [CrossRef]
2. Vaughan, S.; Uttley, P. Detecting X-ray QPOs in active galaxies. *Adv. Space Res.* **2006**, *38*, 1405–1408. [CrossRef]
3. Vaughan, S.; Uttley, P.; Markowitz, A.G.; Huppenkothen, D.; Middleton, M.J.; Alston, W.N.; Scargle, J.D.; Farr, W.M. False periodicities in quasar time-domain surveys. *Mon. Not. R. Astron. Soc.* **2016**, *461*, 3145–3152. [CrossRef]
4. Bhatta, G.; Zola, S.; Ostrowski, M.; Winiarski, M.; Ogłoza, W.; Dróżdż, M.; Siwak, M.; Liakos, A.; Kozieł-Wierzbowska, D.; Gazeas, K.; et al. Detection of Possible Quasi-periodic Oscillations in the Long-term Optical Light Curve of the BL Lac Object OJ 287. *Astrophys. J.* **2016**, *832*, 47. [CrossRef]
5. Sillanpaa, A.; Haarala, S.; Valtonen, M. J.; Sundelius, B.; Byrd, G. G. OJ 287-Binary pair of supermassive black holes. Provided by the SAO/NASA Astrophysics Data System. *Astrophys. J.* **1988**, *325*, 628–634. [CrossRef]
6. Oknyanskij, V.; Lyuty, V. Optical Variability of NGC 4151 During 100 Years. *Peremennye Zvezdy Prilozhenie* **2007**, *28*, 7.
7. Bon, E.; Jovanović, P.; Marziani, P.; Shapovalova, A.I.; Bon, N.; Jovanović, V.B.; Borka, D.; Sulentic, J.; Popović, L.Č. The First Spectroscopically Resolved Sub-parsec Orbit of a Supermassive Binary Black Hole. *Astrophys. J.* **2012**, *759*, 118. [CrossRef]
8. Guo, D.-F.; Hu, S.-M.; Tao, J.; Yin, H.X.; Chen, X.; Pan, H.J. Optical monitoring of the Seyfert galaxy NGC 4151 and possible periodicities in its historical light curve. *Res. Astron. Astrophys.* **2014**, *14*, 923. [CrossRef]
9. Bon, E.; Zucker, S.; Netzer, H.; Marziani, P.; Bon, N.; Jovanović, P.; Shapovalova, A.I.; Komossa, S.; Gaskell, C. M.; Popović, L.Č.; et al. Evidence for Periodicity in 43 year-long Monitoring of NGC 5548. *Astrophys. J. Suppl.* **2016**, *225*, 29. [CrossRef]
10. Li, Y.R.; Wang, J.M.; Ho, L.C.; Lu, K.-X.; Qiu, J.; Du, P.; Hu, C.; Huang, Y.-K.; Zhang, Z.-X.; Wang, K.; et al. Spectroscopic Indication of a Centi-parsec Supermassive Black Hole Binary in the Galactic Center of NGC 5548. *Astrophys. J.* **2016**, *822*, 4. [CrossRef]
11. Li, Y.-R.; Wang, J.-M.; Zhang, Z.-X.; Wang, K.; Huang, Y.-K.; Lu, K.-X.; Hu, C.; Du, P.; Ho, L.C.; Bai, J.-M.; et al. A Possible 20-Year Periodicity in Long-term Variations of the Nearby Radio-Quiet Active Galactic Nucleus Ark 120. *arXiv* **2017**, arXiv:1705.07781.

12. Yuan, Y.-H.; Fan, J.H.; Tao, J.; Qian, B.-C.; Constantin, D.; Xiao, H.-B.; Pei, Z.-Y.; Lin, C. Optical monitoring of BL Lac object S5 0716+714 and FSRQ 3C 273 from 2000 to 2014. *Astron. Astrophys.* **2017**, *605*, A43. [CrossRef]

13. Babadzhanyants, M.K.; Belokon, E.T. New evidence of the reality of a 13-year period in the optical variability of the quasar 3C 273 and its correlation with observed parameters of the parsec-scale radio jet. *Astron. Rep.* **1993**, *37*, 127.

14. Artymowicz, P.; Lubow, S.H. Dynamics of binary-disk interaction. 1: Resonances and disk gap sizes. *Astrophys. J.* **1994**, *421*, 651–667. [CrossRef]

15. Roedig, C.; Sesana, A.; Dotti, M.; Cuadra, J.; Amaro-Seoane, P.; Haardt, F. Evolution of binary black holes in self gravitating discs. Dissecting the torques. *Astron. Astrophys.* **2012**, *545*, A127. 6361/201219986. [CrossRef]

16. D'Orazio, D. J.; Haiman, Z.; MacFadyen, A. Accretion into the central cavity of a circumbinary disc. *Mon. Not. R. Astron. Soc.* **2013**, *436*, 2997–3020. [CrossRef]

17. MacFadyen, A.I.; Milosavljević, M. An Eccentric Circumbinary Accretion Disk and the Detection of Binary Massive Black Holes. *Astrophys. J.* **2008**, *672*, 83. [CrossRef]

18. Netzer, H. *The Physics and Evolution of Active Galactic Nuclei*; Cambridge University Press: Cambridge, UK, 2013.

19. Marziani, P.; Bon, E.; Grieco, A.; Bon, N.; Dultzin, D.; Del Olmo, A.; D'Onofrio, M. Optical variability patterns of radio-quiet and radio-loud quasars. In *New Frontiers in Black Hole Astrophysics, Proceedings of the International Astronomical Union, IAU Symposium, Baton Rouge, LA, USA, 16 October 2017*; Cambridge University Press: Ljubljana, Slovenia, 2017; Volume 324, p. 243. [CrossRef]

20. MacLeod, C.L.; Ross, N.P.; Lawrence, A.; Goad, M.; Horne, K.; Burgett, W.; Chambers, K.C.; Flewelling, H.; Hodapp, K.; Kaiser, N.; et al. A systematic search for changing-look quasars in SDSS. *Mon. Not. R. Astron. Soc.* **2016**, *457*, 389–404. [CrossRef]

21. Oknyansky, V.L.; Gaskell, C.M.; Huseynov, N.A.; Lipunov, V.M.; Shatsky, N.I.; Tsygankov, S.S.; Gorbovskoy, E.S.; Mikailov, K.M.; Tatarnikov, A.M.; Buckley, D.A.H.; et al. The curtain remains open: NGC 2617 continues in a high state. *Mon. Not. R. Astron. Soc.* **2017**, *467*, 1496–1504. [CrossRef]

22. Oknyansky, V.L.; Malanchev, K.L.; Gaskell, C.M. Changing-look Narrow-Line Seyfert 1s? In Proceeding of the POS: Revisiting Narrow-Line Seyfert 1 Galaxies and Their Place in the Universe, Padova, Italy, 9–13 April 2018.

23. Bon, N.; Bon, E.; Marziani, P. Exploring possible relations between optical variability time scales and broad emission line shapes in AGN. *Front. Astron. Space Sci.* **2018**, *5*, 3. [CrossRef]

24. Pei, L.; Fausnaugh, M.M.; Barth, A.J.; Peterson, B.M.; Bentz, M.C.; De Rosa, G.; Denney, K.D.; Goad, M.R.; Kochanek, C.S.; Korista, K.T.; et al. Space Telescope and Optical Reverberation Mapping Project. V. Optical Spectroscopic Campaign and Emission-line Analysis for NGC 5548. *Astrophys. J.* **2017**, *837*, 131. [CrossRef]

25. Komossa, S.; Burwitz, V.; Hasinger, G.; Predehl, P.K.J.S.; Kaastra, J.S.; Ikebe, Y. Discovery of a Binary Active Galactic Nucleus in the Ultraluminous Infrared Galaxy NGC 6240 Using Chandra. *Astrophys. J. Lett.* **2003**, *582*, L15. [CrossRef]

26. Komossa, S. Observational evidence for binary black holes and active double nuclei. *Mem. Soc. Astron. Ital.* **2006**, *77*, 733.

27. Bogdanović, T.S.; Britton, D.; Sigurdsson, S.; Eracleous, M. Modeling of Emission Signatures of Massive Black Hole Binaries. I. Methods. *Astrophys. J. Suppl.* **2008**, *174*, 455. [CrossRef]

28. Bogdanović, T.; Eracleous, M.; Sigurdsson, S. Emission lines as a tool in search for supermassive black hole binaries and recoiling black holes. *New Astron. Rev.* **2009**, *53*, 113–120. [CrossRef]

29. Graham, M.J.; Djorgovski, S.G.; Stern, D.; Glikman, E.; Drake, A.J.d; Mahabal, A.A.; Donalek, C.; Larson, S.; Christensen, E. A possible close supermassive black-hole binary in a quasar with optical periodicity. *Nature* **2015**, *518*, 74. [CrossRef] [PubMed]

30. Graham, M.J.; Djorgovski, S.G.; Stern, D.; Drake, A.J.; Mahabal, A.A.; Donalek, C.; Glikman, E.; Larson, S.; Christensen, E. A systematic search for close supermassive black hole binaries in the Catalina Real-time Transient Survey. *Mon. Not. R. Astron. Soc.* **2015**, *453*, 1562–1576. [CrossRef]

31. Popović, L.Č. Super-massive binary black holes and emission lines in active galactic nuclei. *New Astron. Rev.* **2012**, *56*, 74–91. [CrossRef]

32. Komossa, S. Tidal disruption of stars by supermassive black holes: Status of observations. *J. High Energy Astrophys.* **2015**, *7*, 148–157. [CrossRef]

33. Lomb, N.R. Least-squares frequency analysis of unequally spaced data. *Astrophys. Space Sci.* **1976**, *39*, 447–462. [CrossRef]

34. Scargle, J.D. Studies in astronomical time series analysis. II - Statistical aspects of spectral analysis of unevenly spaced data. *Astrophys. J.* **1982**, *263*, 835–853. [CrossRef]

35. Bon, E.; Marziani, P.; Bon, N. Periodic optical variability of AGN. *New Front. Black Hole Astrophys.* **2017**, *324*, 176–179. [CrossRef]

36. Charisi, M.; Bartos, I.; Haiman, Z.; Goad, M.; Horne, K.; Burgett, W.; Chambers, K.C.; Flewelling, H.; Hodapp, K.; Kaiser, N.; et al. A population of short-period variable quasars from PTF as supermassive black hole binary candidates. *Mon. Not. R. Astron. Soc.* **2016**, *463*, 2145–2171. [CrossRef]

37. Begelman, M.C.; Blandford, R.D.; Rees, M.J. Massive black hole binaries in active galactic nuclei. *Nature* **1980**, *287*, 307. [CrossRef]

38. Iwasawa, K.; Miniutti, G.; Fabian, A.C. Flux and energy modulation of redshifted iron emission in NGC 3516: Implications for the black hole mass. *Mon. Not. R. Astron. Soc.* **2004**, *355*, 1073–1079. [CrossRef]

39. Bon, N.; Bon, E.; Marziani, P.; Jovanović, P. Gravitational redshift of emission lines in the AGN spectra. *Astrophys. Space Sci.* **2015**, *360*, 41. [CrossRef]

40. Ackermann, M.; Ajello, M.; Albert, A.; Atwood, W.B.; Baldini, L.; Ballet, J.; Barbiellini, G.; Bastieri, D.; Becerra Gonzalez, J.; Bellazzini, R.; et al. Multiwavelength Evidence for Quasi-periodic Modulation in the Gamma-Ray Blazar PG 1553+113. *Astrophys. J.* **2015**, *813*, L41. [CrossRef]

41. Kun, E.; Karouzos, M.; Gabányi, K.É.; Britzen, S.; Kurtanidze, O.M.; Gergely, L.Á. Flaring radio lanterns along the ridge line: Long-term oscillatory motion in the jet of S5 1803+784. *Mon. Not. R. Astron. Soc.* **2018**, *478*, 359–370. [CrossRef]

42. Kun, E.; Frey, S.; Gabányi, K.É.; Britzen, S.; Cseh, D.; Gergely, L.Á. Constraining the parameters of the putative supermassive binary black hole in PG 1302–102 from its radio structure. *Mon. Not. R. Astron. Soc.* **2015**, *454*, 1290–1296. [CrossRef]

43. Kun, E.; Gabányi, K.É.; Karouzos, M.; Britzen, S.; Gergely, L.Á. A spinning supermassive black hole binary model consistent with VLBI observations of the S5 1928 + 738 jet. *Mon. Not. R. Astron. Soc.* **2014**, *445*, 1370–1382. [CrossRef]

44. Britzen, S.; Roland, J.; Laskar, J.; Kokkotas, K.; Campbell, R.M.; Witzel, A. On the origin of compact radio sources-The binary black hole model applied to the gamma-bright quasar PKS 0420-014. *Astron. Astrophys.* **2001**, *374*, 784–799. [CrossRef]

45. Farris, B.D.; Duffell, P.; MacFadyen, A.I.; Haiman, Z. Characteristic signatures in the thermal emission from accreting binary black holes. *Mon. Not. R. Astron. Soc.* **2015**, *446*, L36–L40. [CrossRef]

46. Cuadra, J.; Armitage, P.J.; Alexander, R.D.; Begelman, M.C. Massive black hole binary mergers within subparsec scale gas discs. *Mon. Not. R. Astron. Soc.* **2009**, *393*, 1423–1432. [CrossRef]

47. Miranda, R.; Muñoz, D.J.; Lai, D. Viscous hydrodynamics simulations of circumbinary accretion discs: Variability, quasi-steady state and angular momentum transfer. *Mon. Not. R. Astron. Soc.* **2017**, *466*, 1170–1191. [CrossRef]

48. Hayasaki, K.; Saito, H.; Mineshige, S. Binary Black Hole Accretion Flows From a Misaligned Circumbinary Disk. *Aesthetic. Surg. J.* **2013**, *65*, 86. [CrossRef]

49. Jovanović, P.; Popović, L.Č.; Stalevski, M.; Shapovalova, A.I. Variability of the Hβ Line Profiles as an Indicator of Orbiting Bright Spots in Accretion Disks of Quasars: A Case Study of 3C 390.3. *Astrophys. J.* **2010**, *718*, 168. [CrossRef]

50. Gezari, S.; Halpern, J.P.; Eracleous, M. Long-Term Profile Variability of Double-peaked Emission Lines in Active Galactic Nuclei. *Astrophys. J. Suppl.* **2007**, *169*, 167. [CrossRef]

51. Newman, J.A.; Eracleous, M.; Filippenko, A.V.; Halpern, J.P. Measurement of an Active Galactic Nucleus Central Mass on Centiparsec Scales: Results of Long-Term Optical Monitoring of Arp 102B. *Astrophys. J.* **1997**, *485*, 570. [CrossRef]

52. Chen, K.; Halpern, J.P. Structure of line-emitting accretion disks in active galactic nuclei—ARP 102B. *Astrophys. J.* **1989**, *344*, 115–124. [CrossRef]

53. Chen, K.; Halpern, J.P.; Filippenko, A.V. Kinematic evidence for a relativistic Keplerian disk—ARP 102B. *Astrophys. J.* **1989**, *339*, 742–751. [CrossRef]

54. Eracleous, M.; Halpern, J.P. Doubled-peaked emission lines in active galactic nuclei. *Astrophys. J. Suppl.* **1994**, *90*, 1–30. [CrossRef]

55. Gavrilović, N.; Bon, E.; Popović, L.Č.; Prugniel, P. Determination of Accretion Disc Parameters in the Case of Five AGN with Double-peaked Lines. *AIP Conf. Proc.* **2007**, *938*, 94–97. [CrossRef]
56. Jovanović, P.; Popović, L.Č. Observational effects of strong gravity in vicinity of supermassive black holes. *Fortschritte der Physik* **2008**, *56*, 456–461. [CrossRef]
57. Bon, E.; Popović, L.Č.; Ilić, D.; Mediavilla, E. Stratification in the broad line region of AGN: The two-component model. *New Astron. Rev.* **2006**, *50*, 716–719. [CrossRef]
58. Bon, E. The Disk Emission in Single Peaked Lines for 12 AGNs. *Serb. Astron. J.* **2008**, *177*, 9–13. [CrossRef]
59. Bon, E.; Popović, L.Č.; Gavrilović, N.; La Mura, G.; Mediavilla, E. Contribution of a disc component to single-peaked broad lines of active galactic nuclei. *Mon. Not. R. Astron. Soc.* **2009**, *400*, 924–936. [CrossRef]
60. Bon, E.; Gavrilović, N.; La Mura, G.; Popović, L.Č. Complex broad emission line profiles of AGN - Geometry of the broad line region. *New Astron. Rev.* **2009**, *53*, 121–127. [CrossRef]
61. Barack, L.; Cardoso, V.; Nissanke, S.; Sotiriou, T.P.; Askar, A.; Belczynski, C.; Bertone, G.; Bon, E.; Blas, D.; Brito, R.; et al. Black holes, gravitational waves and fundamental physics: A roadmap. *arXiv* **2018**, arXiv:1806.05195.
62. Kaspi, S.; Smith, P.S.; Netzer, H.; Maoz, D.; Jannuzi, B.T.; Giveon, U. Reverberation Measurements for 17 Quasars and the Size-Mass-Luminosity Relations in Active Galactic Nuclei. *Astrophys. J.* **2000**, *533*, 631. [CrossRef]
63. Fanton, C.; Calvani, M.; de Felice, F.; Cadez, A. Detecting Accretion Disks in Active Galactic Nuclei. *Publ. Astron. Soc. Jpn.* **1997**, *49*, 159–169. [CrossRef]
64. Čadež, A.; Fanton, C.; Calvani, M. Line emission from accretion discs around black holes: The analytic approach. *New Astron.* **1998**, *3*, 647–654. [CrossRef]
65. Popović, L.Č.; Mediavilla, E.; Bon, E.; Ilić, D. Contribution of the disk emission to the broad emission lines in AGNs: Two-component model. *Astron. Astrophys.* **2004**, *423*, 909–918. [CrossRef]
66. Popović, L.Č.; Mediavilla, E.G.; Bon, E.; Stanić, N.; Kubičela, A. The Line Emission Region in III Zw 2: Kinematics and Variability. *Astrophys. J.* **2003**, *599*, 185. [CrossRef]
67. Strateva, I.V.; Strauss, M.A.; Hao, L.; Schlegel, D.J.; Hall, P.B.; Gunn, J.E.; Li, L.-X.; Ivezić, Ž.; Richards, G.T.; Zakamska, N.L. Double-peaked Low-Ionization Emission Lines in Active Galactic Nuclei. *Astron. J.* **2003**, *126*, 1720. [CrossRef]
68. Jovanović, P. The broad Fe Kα line and supermassive black holes. *New Astron. Rev.* **2012**, *56*, 37–48. [CrossRef]
69. Marziani, P.; Sulentic, J.W.; Zamanov, R.; Calvani, M.; Dultzin-Hacyan, D.; Bachev, R.; Zwitter, T. An Optical Spectroscopic Atlas of Low-Redshift Active Galactic Nuclei. *Astrophys. J. Suppl.* **2003**, *145*, 199. [CrossRef]
70. Chakrabarti, S.K.; Wiita, P.J. Variable emission lines as evidence of spiral shocks in accretion disks around active galactic nuclei. *Astrophys. J.* **1994**, *434*, 518–522. [CrossRef]
71. Shakura, N.I.; Sunyaev, R.A. Black holes in binary systems. Observational appearance. *Astron. Astrophys.* **1973**, *24*, 337–355.
72. Czerny, B.; Hryniewicz, K. The origin of the broad line region in active galactic nuclei. *Astron. Astrophys.* **2011**, *525*, L8. [CrossRef]
73. McKernan, B.; Ford, K.E.S.; Kocsis, B.; Lyra, W.; Winter, L.M. Intermediate-mass black holes in AGN discs - II. Model predictions and observational constraints. *Mon. Not. R. Astron. Soc.* **2014**, *441*, 900–909. [CrossRef]
74. Lin, D.N.C. Star/Disk Interaction in the Nuclei of Active Galaxies. In *Emission Lines in Active Galaxies: New Methods and Techniques, Proceedings of the IAU Colloquium 159: Emission Lines in Active Galaxies: New Methods and Techniques, Shanghai, China, 17–20 June 1996*; Astronomical Society of the Pacific: San Francisco, CA, USA, 1997; Volume 113, p. 64.
75. Lin, D.N.C.; Papaloizou, J.C.B. Theory of Accretion Disks II: Application to Observed Systems. *Annu. Rev. Astron. Astrophys.* **1996**, *34*, 703–747. [CrossRef]
76. McKernan, B.; Ford, K.E.S.; Kocsis, B.; Haiman, Z. Ripple effects and oscillations in the broad Fe Kα line as a probe of massive black hole mergers. *Mon. Not. R. Astron. Soc.* **2013**, *432*, 1468–1482. [CrossRef]

Article

Cavity-Enhanced Photodetachment of H⁻ as a Means to Produce Energetic Neutral Beams for Plasma Heating

Christophe Blondel [1,*,†], David Bresteau [2,†,‡] and Cyril Drag [1,†]

[1] Laboratoire de Physique des Plasmas, École Polytechnique, Centre National de la Recherche Scientifique, Université Paris-Sud, Sorbonne Université, Route de Saclay, F-91128 Palaiseau CEDEX, France
[2] Laboratoire Aimé-Cotton, Centre National de la Recherche Scientifique, Université Paris-Sud, Université Paris-Saclay, Bâtiment 505, F-91405 Orsay CEDEX, France
* Correspondence: christophe.blondel@lpp.polytechnique.fr
† These authors contributed equally to this work.
‡ Current address: LIDyL, Commissariat à l'Énergie atomique et aux Énergies Alternatives, Centre de Saclay, F-91191 Gif-sur-Yvette CEDEX, France.

Received: 8 February 2019; Accepted: 23 February 2019; Published: 1 March 2019

check for updates

Abstract: Neutral beam injection, for plasma heating, will supposedly be achieved, in ITER, by collisional detachment of a pre-accelerated D⁻ beam. Collisional detachment, however, makes use of a D_2-filled neutralisation chamber, which has severe drawbacks, including the necessity to set the D⁻-ion source at −1 MV. Photodetachment, in contradistinction, would have several advantages as a neutralisation method, including the absence of gas injection, and the possibility to set the ion source close to the earth potential. Photodetachment, however, requires a very high laser flux. The presented work has consisted in implementing an optical cavity, with a finesse greater than 3000, to make such a high illumination possible with a state-of-the-art CW (continuous-wave) laser. A 1.2 keV ^{1}H⁻-beam (only 20 times slower than the 1 MeV ^{2}D⁻ ion beams to be prepared for ITER) was photodetached with more-than-50% efficiency, with only 24 W of CW laser input. This experimental demonstration paves the way for developing real-size photoneutralizers, based on the implementation of refolded optical cavities around the ion beams of neutral beam injectors. Depending on whether the specifications of the laser power or the cavity finesse will be more difficult to achieve in real scale, different architectures can be considered, with greater or smaller numbers of optical refoldings or (inclusively) optical cavities in succession, on the beam to be neutralised.

Keywords: photodetachment; magnetically confined fusion; neutral beam injection; plasma heating; optical cavity amplification

---

1. Introduction

The history of fast D^0 neutral beam generation for plasma heating has followed three technical ways in succession. Electron capture by accelerated D^+ ion was historically the first one. Collisional detachment of accelerated D⁻ ions was then developed to overcome the decrease of electron-capture probability at higher acceleration voltages. It is the procedure now implemented on ITER, but with a limited efficiency that will probably not be sustainable for industrial developments.

Photodetachment of accelerated D⁻ ions has been considered since the 1980s as a very promising technique [1]. Its energy cost, namely the circa 1 eV energy of every absorbed photon, is very low when compared to the 1 MeV kinetic energy of every produced neutral atom. Photons are insensitive to electric fields, and the photodetachment zone does not need any gas input, which makes it possible to set it at a high (positive) voltage and keep the D$^-$ ion source close to the earth potential, with innumerable advantages as concerns electrical engineering.

The photodetachment process can never (but asymptotically) be complete. One can, however, get as close to 100% efficiency as one wants: undetached ions are still present in the beam and having them photodetached is only a matter of increasing the illumination. This stands in contrast with the collisional technique, where most of the D$^-$ ions that have not been successfully channelled to making a D^0 beam have been either lost due to spurious collisions or further and irremediably stripped to D$^+$ ions.

Implementation of the photodetachment solution has been hindered, however, by the huge light flux required, e.g., to detach a 1 MeV ion beam, say with a 1 cm width: this would mean several MW of light power [2].

Having that power out of a laser source was something no laser promoter could even dream of in the 20th century. Meanwhile, it was imagined that such a power could be produced inside a laser resonator, and that the ion beam could be sent to cross the laser cavity directly [3]. Along these principles, a H$^-$ beam could be produced inside the extended cavity of a YAG laser, with a neutralization efficiency of about 5 % [4].

Amplification in an external optical cavity makes optimization of the cavity finesse easier, at the expense of having the cavity length permanently tuned to a multiple of the laser wavelength [5]. Atomic or molecular physics experiments have used that way to enhance the photoelectron signal [6,7], but, again, with a detachment efficiency that was never greater than a few %. An essential difficulty, for better efficiencies, has been the availability of continuous-wave lasers both powerful and spectrally narrow enough to be used for cavity injection.

The situation changed recently, with the advent of doped fibre-based laser amplifiers, which makes spectrally narrow CW lasers with powers greater than 10 W commercially available. The present communication reports on a demonstration experiment that has shown efficient photodetachment of a H$^-$ ion beam, at a reduced scale as concerns both the ion beam kinetic energy (1.2 keV instead of 1 MeV) and its diameter (a millimetre instead of several centimetres), but with a cavity finesse and intracavity light fluxes that are already of the same order of magnitude as what will be required, eventually, for full-scale implementation.

2. Experimental Set-Up

2.1. Ion Beam

The experimental set-up has already been described elsewhere [8]. The hydrogen ion beam is produced by a cesium sputtering ion source [9]. Neutral-atom time-of-flight measurements, following pulsed photodetachment by a frequency-tripled Nd:YAG laser, make it possible to check that the admixture of O$^-$ ions, which may be present due to oxidization of the cathode but do not detach at a 1 μm wavelength, will not lead to an underestimation of the photodetachment efficiency. H$^-$ ions travel along the ion-beam machine with a 1.2 keV kinetic energy, i.e., a velocity of about 4.8×10^5 m s^{-1}. The residual pressure in the chamber where the ions are brought to cross the laser beam, is a few 10^{-5} Pa.

H$^-$ anions are illuminated by the laser beam, inside an intra-vacuum optical cavity (Figure 1), and some of them are photodetached. The photon momentum, about 6.2×10^{-28} kg m/s at the wavelength λ = 1064 nm, is several orders of magnitude smaller than the ion momentum, about 8.0×10^{-22} kg m/s, which makes the deviation due to photon absorption negligible. The residual ion beam is de-merged from the produced H^0 neutral beam by a transverse electric field, a few cm downstream of the interaction region. A Beam Imaging Solutions® BOS-25 "Beam observation system" (Logmont, CO, USA) can be used to bring both the ion and neutral beams to observation. A Faraday cup can also be used to monitor the residual ion current and measure the photodetachment efficiency. The undetached ion beam current was typically 1 nA.

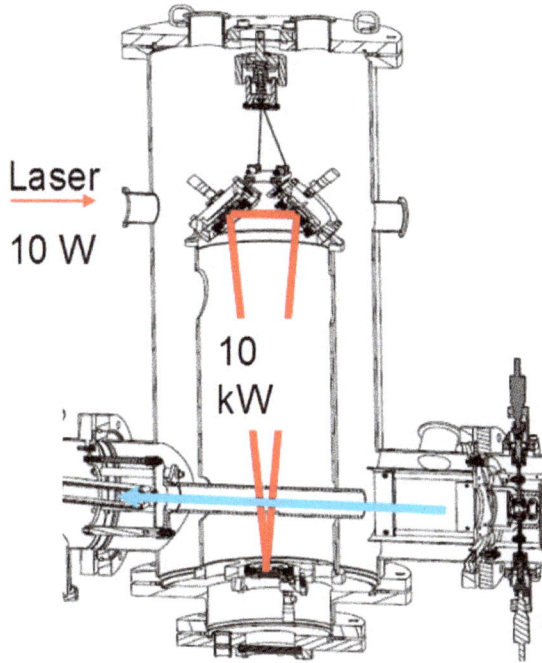

Figure 1. View of the experimental set-up. The optical cavity is suspended inside the vacuum chamber so as to reduce its coupling to external vibrations. The triangular optical circuit is about 1 m long. The laser beam is injected into the cavity horizontally, and (unshown) photodiodes and/or powermeters can be set to measure transmitted and reflected intensities, respectively, to get a measurement of the cavity finesse and amplification factor. The intensity is multiplied by a factor of 1000 inside the cavity, with respect to the input laser power, which makes typically a 10 kW laser beam circulate between the mirrors, with only 10 W of input power. Before reaching the laser beam, the ions pass (right to left) through an adjustable diaphragm made of two vertical plates mounted on micrometer screws, so as to make a quantitative measurement possible of the detachment rate, for ions that actually cross the intracavity optical mode. The neutral beam produced continues straightforward to be either detected by an electron multiplier or visualized by a beam imager made of microchannel plates and a phosphor screen. The intensity of the residual anion beam can be monitored by subsequent diversion of the remaining ions by deflection plates towards a Faraday cup.

2.2. Laser System

The *Azur Light Systems* (ALS) laser [10] consists of a single-mode ytterbium doped *NKT Photonics* Koheras Y10 fibre laser (Talence, France), followed by an ALS-IR-10-USF amplifier (Talence, France). Its maximum output is 25 W and its spectral bandwidth about 10 kHz, with a ±2 GHz tunability achieved via an internal piezoelectric actuator of the *NKT* oscillator. An electro-optic phase-modulator is set between the oscillator and the amplifier, and used for error-signal generation, when the laser light is sent into the cavity and partially reflected, using the Pound–Drever–Hall (PDH) method [11].

2.3. Optical Cavity and Light Storage

2.3.1. Geometry

The optical cavity is a triangular ring cavity, as shown by Figure 1, with an optical length $L \simeq 1$ m. It is equipped with high-reflectivity commercial mirrors: two (upper) plane mirrors set at a 43° angle of incidence, and a concave (lower) mirror with a radius of curvature $R \simeq 5$ m, set on a piezoelectric mount.

The waist w_0 of the TEM$_{00}$ mode is about 713 µm at the wavelength $\lambda = 1064$ nm. It is located at the centre of symmetry of the cavity, between the two 43° plane mirrors. Diffraction, however, does not make the beam much broader all along the cavity circuit. The calculated value of w is 743 µm, at the positions where the ion beam crosses the laser beam. The cavity frame is suspended inside the vacuum chamber by soft springs, in order to make it as insensitive as possible to external vibrations. The cavity can be rotated, as a whole, around a vertical axis to make it possible for the ion beam to pass twice through the laser beam, as shown in Figure 1.

2.3.2. Cavity Finesse and Amplification Factor

Defining θ as the multiplication factor of the amplitude after one round trip in the cavity, $\Theta = |\theta|^2$ is the corresponding multiplication factor of the intensity. Light accumulation inside the cavity leads to an integrated amplitude gain $(1 - \theta)^{-1}$. Cavity resonance occurs every time the cavity length L is an integer multiple of the wavelength λ, which makes the argument of θ an integer multiple of 2π, with θ close to 1. The finesse F of the cavity is defined as the ratio of the spectral interval between two adjacent resonances to the full width at half-maximum of every intensity resonance. Within an excellent approximation, F is given by

$$F \simeq \pi \frac{\sqrt{|\theta|}}{1 - |\theta|}. \tag{1}$$

The intensity amplification factor Γ, defined as the ratio of the internal intensity to the input intensity, at resonance, is given by

$$\Gamma = \frac{T_1}{(1 - |\theta|)^2} \tag{2}$$

with T_1 the transmission factor of the input mirror.

In the well-known case of a symmetrical lossless two-mirror cavity, the amplification factor Γ is just equal to F/π. With three mirrors and, in an even more general case, with losses other than the mirror transmissions (which determine both input and output couplings of the cavity), the ratio can take different values. Optimizing the finesse or the amplification factor are not subject to all identical constraints. Essential differences can be conveniently described by a simultaneous plot of Γ and F/π, as functions of T_1 and S, with S the energy fraction lost, for each intracavity round trip, for all other reasons than partial transmission back through the input mirror. The plot is done in Figure 2, assuming that the input mirror is a perfect one, i.e., with a reflection factor $R_1 = 1 - T_1$. The finesse F, which only depends on the total intracavity losses, then becomes a function of $T_1 + S$ only, which makes the contour lines of constant F parallel straight lines drawn on the developable F/π surface. The Γ and F/π surfaces intersect on the line $S = T_1$. Factor Γ can go up to $2 \times F/\pi$, if S can be made negligible with respect to T_1. In actual situations, however, a constraint is more often that S losses are imposed, and one has to find the best-matching input mirror. The maximum amplification factor, with such a constraint, is met with an input mirror such that $T_1 = S$, which brings us back to the median line, with a ratio $\Gamma/F = 1/\pi$ only. As a matter of fact, the working point of the present experiment, shown in Figure 2, even corresponds to a slightly lower ratio, with $F \simeq 3600$ and $\Gamma \simeq 900$.

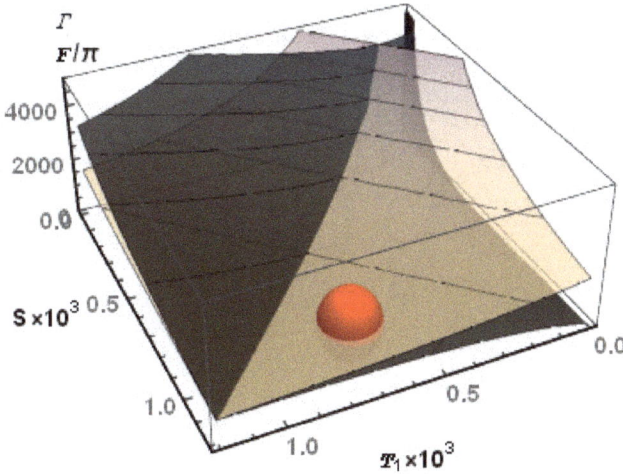

Figure 2. Two-dimensional chart of the finesse F (light) and amplification factor Γ (dark), as functions of the input mirror transmission T_1 and other intracavity losses S, assuming that the intracavity reflection factor of the input mirror is just equal to $1 - T_1$ (no other losses). The finesse has been divided by π to make it conspicuous that $T_1 = S$ is the place where $\Gamma = F/\pi$. When S happens to be much smaller than T_1, the Γ/F ratio can go up to $2/\pi$. However, in every section of constant S, the maximum value of Γ is found on the $T_1 = S$ diagonal, with a Γ/F ratio of only $1/\pi$. The ellipse shows the working point for the present study.

The input mirror was an *Opto4u* one-inch-in-diameter 45° mirror, the reflectivity of which is given by the manufacturer to be > 0.992 (resp. > 0.997) in p (resp. s) polarisation, i.e., with the electric field parallel (resp. perpendicular) to the incidence plane. We have measured its transmission and found it to be $6.4(3) \times 10^{-4}$ ($2.0(2) \times 10^{-5}$, respectively). More details on the optical components of the cavity have been published elsewhere [8]. As a matter of fact, mirror reflection factors happen to be significantly greater than just told by their minimum specifications, but the experiment also shows that intracavity attenuation does not only result from outward mirror transmission. Absorption, scattering and diffraction make factor S a little greater than 10^{-3}, whatever the polarization. Having a transmission factor $T_1 = 6.4(3) \times 10^{-4}$ at the input was thus slightly too little, meaning that the cavity was under-coupled. The even lower coupling coefficient with the other polarization was definitely too low to take advantage of the better finesse expected in the s case. Unfortunately, the time devoted to the experiment was too limited to order new, better-suited optical components. All experiments have thus been carried out in p polarization. Obvious solutions nevertheless exist to improve the cavity finesse and get higher amplification factors.

The intracavity power P can be determined either as the ratio of the output at one or the other secondary mirrors, provided that their transmission factor has been measured, or by monitoring the reflected intensity at the input mirror in a dynamical regime, scanning the cavity length. When this length L is scanned fast enough to go through a resonance in a time shorter than the cavity storage time, a "ringing" phenomenon occurs, as shown by Figure 3, which consists of fast oscillations of the reflected power, following every resonance (the intensities transmitted by the secondary mirrors exhibit similar phenomena). If scanning is not made so fast that the time lapse between resonances gets shorter than the cavity damping time, the observed intensity oscillations can be fitted by an analytical formula [8], which gives a means to evaluate the cavity finesse without depending on light flux calibration.

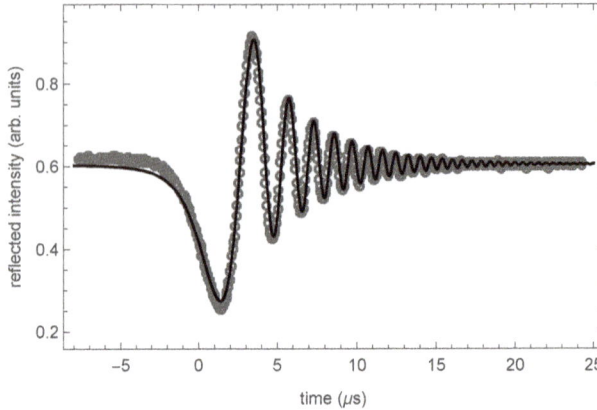

Figure 3. Intensity reflected by the input cavity mirror as a function of time, for a scanning velocity of one of the cavity mirrors $175(2)\,\mu\mathrm{m\,s}^{-1}$. The time derivative of the varying cavity length L can be directly measured by extending the scan to times large enough to make resonances appear in succession, which they do every time L has changed exactly by λ. The fitting curve here corresponds to a finesse $F = 3681$. Dynamic reflection at the input mirror does not only give information about the cavity finesse, but on the transmission coefficient T_1 of the input mirror too. The result here is $T_1 = 6.0(1) \times 10^{-4}$.

For input powers greater than 12 W, it was observed that the increase of the transmitted power ceased to be a linear function of the input. This had not been observed when the cavity had been tested at atmospheric pressure, and may be due to thermal effects. As a consequence, despite the cavity amplification factor Γ having been confirmed, both by finesse and power measurements to be about 900 at low intensities, the maximum intracavity power did not exceed 14 kW, for the input power 24 W.

3. Results

A series of images recorded with the "beam observation system" is displayed on Figure 4, which shows that, at the maximum illumination, the neutral beam has reached an intensity similar to the intensity of the ion beam (notwithstanding the possibility of a slight sensitivity difference between the two species).

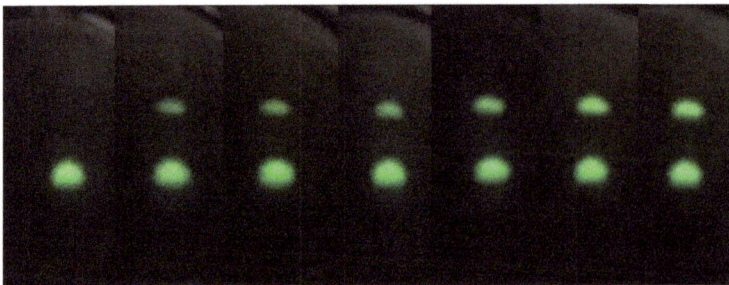

Figure 4. Visual monitoring of neutral beam production. The undetached ions are deflected by an electric field weak enough not to send them out of the detector, leading to simultaneous observation of both the ion (below) and neutral (above) beams. The beams are offset by 6.2 mm. The intracavity light power was 0, 1.3, 1.45, 2, 3, 6.5 and 11.5 kW (it being understood that the ions cross twice a beam of that power), from left to right.

More quantitative measurements are carried out with the Faraday cup. A plot of the beam photodetachment rate η as a function of the intracavity power is shown in Figure 5, together with what numerical modelling predicts for three possible offsets of the ion beam with respect to the cavity mode. The experimental results appear quite compatible with the value $3.6 \times 10^{-21} \text{ m}^2$ of the cross-section most widely reported by theorists [12–14] or the most recent experimental value [15] if one admits a medium 250 µm offset. The larger value $4.5 \times 10^{-21} \text{ m}^2$, which was found to be the most probable in the 2014 measurement [12] (although with a large ±14% uncertainty) would also be compatible with the observation, assuming a $\delta = 400$ µm offset.

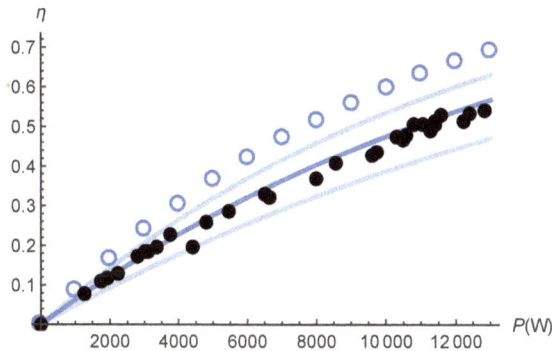

Figure 5. Photodetachment ratio η, measured as the attenuation of the negative ion beam when the laser is switched on, as a function of the intracavity power P. The black dots are the experimental points. The continuous lines correspond to what can be predicted by theory, for a $3.6 \times 10^{-21} \text{ m}^2$ cross-section and three different hypotheses for the offset δ of the two laser beams, the one with respect to the other, in the direction perpendicular to the ion beam: zero, 250 and 400 µm, from top to bottom, respectively. The open circles ○ show the photodetachment ratios that would be obtained for twice heavier, hence $\sqrt{2}$ times slower D$^-$ ions, for the most probable $\delta = 250$ µm value.

4. Discussion

We have shown that optical-cavity enhancement can make photodetachment an actual means of neutralizing the greater part of a H$^-$ beam, and produce a neutral beam efficiently. Up-scaling remains to be done as concerns the beam diameter and ion velocity, but the demonstration already dealt with cavity finesses of the order of magnitude to be met in real-size neutral beam injectors. The power fluxes reached, due to intra-cavity amplification, were already of the order of magnitude of the fluxes to be dealt with, when more power is stored in larger cavities, with larger beam diameters. The corresponding thermal effects may explain the reduction of the finesse observed at higher intensities. Mechanical compensation of these effects was not implemented, as it will be easier on larger mirrors. Testing such a compensation at a reduced scale would thus be an unnecessary challenge.

The use of a ring cavity, instead of a stationary-wave one, appears highly recommendable, since it does not add to the difficulty and prevents the light reflected at the input mirror to go back to the laser. Beam-refolding makes it possible for the intracavity light beam to pass several times on the ion beam [16] and future designs should not be limited to the tested three-mirror triangular scheme. The maximum practical number of mirrors or refoldings will be determined by the reflectivity of the mirrors (the better the reflectivity, the larger the possible refolding index) and the smallest allowable spectral width of a cavity mode (the longer the cavity, the smaller the spectral width).

Investigations are carried out, on the other hand, on the possibility of having a H$^-$ or a D$^-$ beam efficiently photodetached by incoherent light recycling, in a so-called "photon cell" [17]. Both ways appear worthy of being investigated further. The pros and cons of coherent intracavity amplification vs. incoherent intensity summation have thus to be balanced:

Table 1 shows that the current state of the art makes both solutions potentially equivalent, as for the numbers of advantages and drawbacks. A typical illustration of that balance is that the amplification obtained by amplitude summation, of the order of the squared finesse F^2, gets counterbalanced by the weakness of the transmission of the input mirror, necessarily of the order of $1/F$. Nevertheless, some of the listed difficulties could prove easier to overcome, in the one or the other solution.

Table 1. The lesser ($\oplus$), greater ($\ominus$) or similar difficulties ($\odot$) of coherent amplitude summation in an optical cavity vs. incoherent intensity summation in a multipass cell. The orders of magnitude are estimated for a real-size neutral beam injector, with a kinetic energy of 1 MeV per D^0 atom, assuming that the width of the ion beam could be reduced to 1 cm.

Parameter	Resonant Cavity	Multipass
Summation carried out	$\oplus$ Amplitude summation	$\ominus$ Intensity summation
Input coupling	$\ominus 1/F$	$\oplus 1$
Laser required	$\ominus$ Locked single mode	$\oplus$ Multimode admitted
Laser power required	$\odot$ 1 kW	$\odot$ 500 kW
Laser power available in 2018	$\odot$ 100 W	$\odot$ 100 kW
Price in 2018	$\oplus$ ca. 100 k€	$\ominus$ ca. 2 M€
Dissipated power	$\oplus$ a few kW	$\ominus$ > 1 MW
Intended amplification	$\ominus \times5000$	$\oplus \times100$
Required stability	$\ominus \delta L \ll \lambda$	$\oplus$ a few degrees' angular accuracy
Spatial filtering	$\ominus$ TEM$_{00}$ filtering	$\oplus < 3°$ divergence

The major drawbacks of the optical cavity scheme are that resonance makes it necessary for the input laser to be spectrally narrower than a resonant mode of the cavity and for the laser wavelength to be continuously locked on a sub-multiple of the cavity length. This is a fundamental difficulty that cannot disappear, but making tunable lasers spectrally narrow has been the bread-and-butter business of laser technology for nearly 50 years.

Incoherent summation by the multiplication of non-overlapping laser beams in a stable cavity may appear more tentative. Distributed reflections, in that case, however spread over a larger area of the mirrors, which makes the geometrical stability of the ray pattern more sensitive to spherical aberrations. Coherently packing all reflections together in the TEM$_{00}$ mode, on the axis, should make it easier to keep these aberrations low and reach higher reflectivities, with smaller-diameter mirrors. Accordingly, one can expect the necessary power to be put into a resonant cavity to remain smaller, and the whole set-up, based on lower diameter mirrors, to remain cheaper. As a result, a coherent cavity would also be advantageous as concerns the power consumption and lost energy to be disposed of.

5. Conclusions

Photoneutralization of a negative-ion beam has been achieved with more-than-50% efficiency, thanks to the use of an optical cavity, at a reduced scale as concerns the beam diameter and ion velocity. The orders of magnitude of both the cavity finesse and light flux on the cavity mirrors, however, are already the same as those required in future real scale neutral beam injectors. As for the laser power needed to develop those injectors, laser technology gets closer and closer everyday to providing the few, spectrally narrow, kW of continuous laser power, suitable for cavity injection. All components of future photodetachment-based D^0 injectors are thus to become available in the near future, in a realistic scenario [18]. The development of a real-scale experiment, where a D^- beam of several centimetres would be detached in a cavity after acceleration to 1 MV, would thus appear highly recommendable, without prejudice against the parallel development of multipass cells, which can appear equally promising, for the implementation of photodetachment in future neutral beam injectors.

Author Contributions: Conceptualization, C.B. and C.D.; methodology, C.B. and C.D.; software, C.B. and D.B.; validation, D.B. and C.D.; formal analysis, C.B., D.B. and C.D.; investigation, D.B. and C.D.; resources, D.B. and C.D.; data curation, D.B. and C.D.; writing original draft preparation, D.B.; writing review and editing, C.B.; visualization, C.B., D.B. and C.D.; supervision, C.B.; project administration, C.B.; funding acquisition, C.B.

Funding: This research was funded by the Agence nationale de la recherche (ANR) grant number ANR-13-BS04-0016-01 and by the Euratom research and training programme 2014–2018 grant number 633053. The views and opinions expressed herein do not necessarily reflect those of the European Commission.

Acknowledgments: This work was supported by grant ANR-13-BS04-0016-01 of the French Agence Nationale de la Recherche. It was also carried out within the framework of the Eurofusion consortium and has received funding from the Euratom research and training programme 2014–2018 under Grant No. 633053. The views and opinions expressed herein do not necessarily reflect those of the European Commission.

Conflicts of Interest: The authors declare no conflict of interest.

References

1. Fink, J.H.; Alessi, J.G. Neutralizer options for high energy H^- beams. *AIP Conf. Proc.* **1987**, *158*, 618–630.
2. Fink, J.H. Photodetachment technology. *AIP Conf. Proc.* **1984**, *111*, 547–560. [CrossRef]
3. Vanek, V.; Hursman, T.; Copeland, D.; Goebel, D.M.; Prelec, K. Technology of a laser resonator for the photodetachment neutralizer. *AIP Conf. Proc.* **1984**, *111*, 568–586.
4. Van Zyl, B.; Utterback, N.G.; Amme, R.C. Generation of a fast atomic hydrogen beam. *Rev. Sci. Instrum.* **1976**, *47*, 814–819. [CrossRef]
5. Chaibi, W.; Blondel, C.; Cabaret, L.; Delsart, C.; Drag, C.; Simonin, A. Photoneutralization of Negative Ion Beam for Future Fusion Reactor. *AIP Conf. Proc.* **2009**, *1097*, 385–394. [CrossRef]
6. Ervin, K.M.; Ho, J.; Lineberger, W.C. A study of the singlet and triplet states of vinylidene by photoelectron spectroscopy of $H_2C = C^-$, $D_2C = C^-$, and $HDC = C^-$. Vinylidene–acetylene isomerization. *J. Chem. Phys.* **1989**, *91*, 5974–5992. [CrossRef]
7. Kim, J.B.; Wenthold, P.G.; Lineberger, W.C. Ultraviolet Photoelectron Spectroscopy of *o*-, *m*-, and *p*-Halobenzyl Anions. *J. Phys. Chem. A* **1999**, *103*, 10833–10841. [CrossRef]
8. Bresteau, D.; Blondel, C.; Drag, C. Saturation of the photoneutralization of a H^- beam in continuous operation. *Rev. Sci. Instrum.* **2017**, *88*, 113103. [CrossRef] [PubMed]
9. Available online: http://www.pelletron.com/ (accessed on 27 February 2019).
10. Guiraud, G.; Traynor, N.; Santarelli, G. High-power and low-intensity noise laser at 1064 nm. *Opt. Lett.* **2016**, *41*, 4040–4043. [CrossRef] [PubMed]
11. Black, E.D. An introduction to Pound–Drever–Hall laser frequency stabilization. *Am. J. Phys.* **2001**, *69*, 79–87. [CrossRef]
12. Vandevraye, M.; Babilotte, P.; Drag, C.; Blondel, C. Laser measurement of the photodetachment cross section of H^- at the wavelength 1064 nm. *Phys. Rev. A* **2014**, *90*, 013411. [CrossRef]
13. Scott, M.P.; Kinnen, A.J.; McIntyre, M.W. Photon collisions with atoms and ions within an intermediate-energy R-matrix framework. *Phys. Rev. A* **2012**, *86*, 032707. [CrossRef]
14. McLaughlin, B.M.; Stancil, P.C.; Sadeghpour, H.R.; Forrey, R.C. H^- photodetachment and radiative attachment for astrophysical applications. *J. Phys. B At. Mol. Opt. Phys.* **2017**, *50*, 114001. [CrossRef]
15. Génévriez, M.; Urbain, X. Animated-beam measurement of the photodetachment cross section of H^-. *Phys. Rev. A* **2015**, *91*, 033403. [CrossRef]
16. Kovari, M.; Crowley, B. Laser photodetachment neutraliser for negative ion beams. *Fusion Eng. Des.* **2010**, *85*, 745–751. [CrossRef]
17. Popov, S.; Atlukhanov, M.G.; Burdakov, A.V.; Ivanov, A.; Kasatov, A.; Kolmogorov, A.V.; Vakhrushev, R.V.; Ushkova, M.Y.; Smirnov, A.; Dunaevsky, A. Neutralization of negative hydrogen and deuterium ion beams using non-resonance adiabatic photon trap. *Nucl. Fusion* **2018**, *58*, 096016. [CrossRef]
18. Simonin, A.; Achard, J.; Achkasov, K.; Bechu, S.; Baudouin, C.; Baulaigue, O.; Blondel, C.; Boeuf, J.P.; Bresteau, D.; Cartry, G.; et al. R&D around a photoneutralizer-based NBI system (Siphore) in view of a DEMO Tokamak steady state fusion reactor. *Nucl. Fusion* **2015**, *55*, 123020.

Article

Electron-Induced Chemistry in the Condensed Phase

Jan Hendrik Bredehöft

Institute for Applied and Physical Chemistry, University of Bremen, Leobener Str.5, 28359 Bremen, Germany;
jhbredehoeft@uni-bremen.de; Tel.: +49-421-218-63201

Received: 30 November 2018; Accepted: 1 March 2019; Published: 4 March 2019

Abstract: Electron–molecule interactions have been studied for a long time. Most of these studies have in the past been limited to the gas phase. In the condensed-phase processes that have recently attracted attention from academia as well as industry, a theoretical understanding is mostly based on electron–molecule interaction data from these gas phase experiments. When transferring this knowledge to condensed-phase problems, where number densities are much higher and multi-body interactions are common, care must be taken to critically interpret data, in the light of this chemical environment. The paper presented here highlights three typical challenges, namely the shift of ionization energies, the difference in absolute cross-sections and branching ratios, and the occurrence of multi-body processes that can stabilize otherwise unstable intermediates. Examples from recent research in astrochemistry, where radiation driven chemistry is imminently important are used to illustrate these challenges.

Keywords: low-energy electrons; electron–molecule interactions; astrochemistry

1. Introduction

In many physical processes that involve ionizing radiation, chemical reactions take place, sometimes unbeknownst to the experimenter. Ionizing radiation is any form of electromagnetic or particulate radiation that has sufficient energy to overcome a substance's ionization potential, thus knocking out an electron from an atomic or molecular orbital and forming a cation. Often, some care is taken to make sure that the formed cations do not interfere with the intended outcome of a process, but the electron is just as often neglected. In dilute gas phase or vacuum, where mean free paths are long, this treatment is certainly warranted, but when number densities are higher, like in cold plasmas or in solid-state processes, the reactions caused by these secondary electrons can no longer be neglected. Especially so, since estimates put the number of secondary electrons generated per MeV of energy lost at around 50,000 [1]. The energy of all these electrons is typically some few eV, which makes them extremely potent at triggering chemical reactions. Depending on the exact energy of the electron, its interaction with a molecule can trigger three main principal processes, which produce reactive species that can go on to form bigger and more complex chemical structures:

Above the ionization threshold, the impinging electron will knock out one other electron from the molecule in a process called electron impaction ionization (EI). This forms a molecular cation. If the original molecule was in a singlet or triplet state, the formed cation will be a duplet (a radical). The energy dependence of this process sees a steady rise from onset at the ionization threshold to a plateau between roughly 50 to 100 eV and a slow decrease towards higher energies. This decrease is caused by the ever shorter interaction times between the molecule and increasingly faster electrons. At energies well above the ionization threshold, the formed cation often breaks up into smaller fragments. This happens in a characteristic way depending on energy and molecular structure. This effect is utilized in mass spectrometry, where the characteristic fragmentation patterns of molecules at a standardized ionization energy of 70 eV are widely available in databases. Data on the ionization potential can likewise be found for a huge range of compounds. Data on the energy dependence of both fragmentation (appearance energies) and total ionization cross-sections are also available, but to a much smaller extent.

At energies below the ionization threshold, the formation of cationic species is usually not possible. There are however two other mechanisms that play a role here. The first one is neutral dissociation (ND), where inelastic scattering of the impinging electron leads to excitation of the targeted molecule and its subsequent breakup into smaller uncharged fragments. This process is believed to follow a similar energy dependence as EI, albeit with lower total cross-sections and an onset energy that is lower than the ionization threshold. There is however very little data on the fragmentation pathways or energy dependence of ND processes, since the formed species carry no charge and are thus very hard to collect, detect and identify experimentally.

The third primary interaction process of electrons and molecules that can be used to drive chemical reactions is electron attachment (EA). EA can happen at specific energies, where the molecule can capture the impinging electron into either an unoccupied molecular orbital or into a dipole-bound state. These energies are called resonances and are specific to the targeted molecule. The formed radical anion state is much less stable than the cations formed by EI, since the simple detachment of the electron leads back to the original molecule. If the electron affinity of the molecule is positive, it is possible that the hyperpotential surface of the newly formed state is dissociative, which leads to bond cleavage. This process is called dissociative electron attachment (DEA). Usually one of the fragments retains the extra electron, while the other part remains with the radical site. The bond cleavage is specific to the resonance, which provides a tool for selective bond cleavage in larger molecules [2]. Some data on the energy dependence of EA/DEA cross-sections are available, at least for small and simple molecules.

In all three cases, highly reactive (intermediate) species are formed, which can then react with a neighboring molecule, to form larger and more complex species from simple starting materials. These electron-induced reactions have increasingly attracted academic attention in recent years, be it in the context of technical applications like Focused Electron Beam Induced Deposition (FEBID) [3] and curing of polymers [4], as well as the driving force for reactions in astrochemical settings [5,6]. The transfer of the available gas phase data to processes in the condensed phase, however, is not always quite as straightforward as one might hope. In the following paragraphs, I will present a few recent examples of astrochemical research from my lab that highlight the typical challenges faced by experimenters, when trying to apply gas-phase data to condensed-phase problems.

2. Results and Discussion

2.1. The Ionization Potential

The most important parameter for electron-induced chemistry is very often the ionization potential, *IE*, of a substance. Values for most chemical compounds can be found in the literature or in databases such as the NIST Chemistry Webbook [7]. There exist a number of theoretical approximations that describe the energy dependence of the ionization cross-section, starting from *IE*. The most commonly used one is the Binary Encounter-Bethe (BEB) model [8,9], which takes a number of molecular parameters as input, all of which are quite easily accessible by quantum chemical calculations. Together, the *IE* as onset and σ_{BEB} as the energy dependence, give an accurate prediction of the ionization probability by electron impact. For energies that are not greater than *IE* by more than a few eV, no significant fragmentation of the molecular compound is observed. This is referred to as 'soft ionization'. If the formed cation undergoes a chemical reaction, *IE* and σ_{BEB} should give an accurate prediction of the energy dependence of product yields.

Figure 1 depicts the calculated formation cross-section of ethylamine from the net reaction

$$C_2H_4 + NH_3 \rightarrow C_2H_5NH_2,$$

which is triggered by electron impact ionization of either ethylene (C_2H_4) or ammonia (NH_3), as well as experimentally obtained data of the amount of ethylamine actually produced. Upon inspection of the data, two things are very clear: (1) There is considerable product formation at energies below the predicted threshold; and (2) product formation reaches saturation between 10 and 12 eV, even though there is a predicted increase at higher energies.

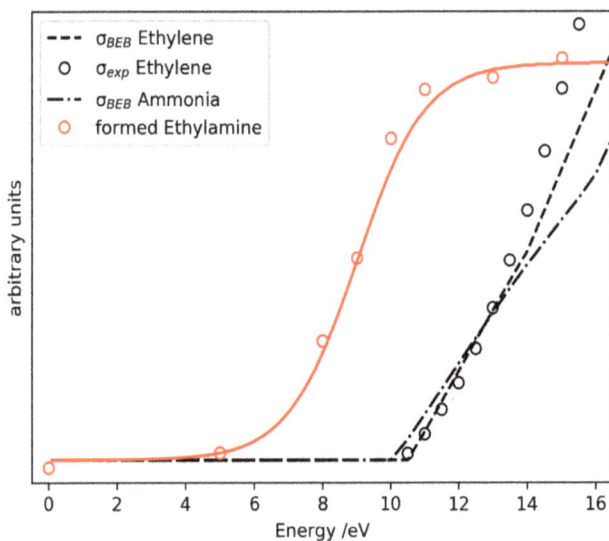

Figure 1. Predicted (black lines) electron impact cross-sections for ethylene and ammonia. Experimental cross-sections for ethylene (black open circles) are taken from Rapp and Englander-Golden [10]. The amount of ethylamine actually formed by electron irradiation of a 1:1 mixture of ammonia and ethylene is shown in red circles. The red line only serves as guide to the eye. Experimental data taken from Böhler et al. [11].

The second observation is explained more easily. The very simple kinetic model used here is based on the assumption of initial rates, i.e., that use of starting materials is negligible. It also neglects any product degradation by the electron beam. The first observation, the shift in energy, however, is not explained by such omissions in the modeling. The data used for the prediction was obtained in the gas phase, where produced cations do not feel any influence of other molecules. In the condensed phase, the energy of the cation is lowered by the polarization of the surrounding medium, which leads to a lowered ionization potential. This is an effect universally observed in experiments with ions in the condensed phase. The shift is usually around 2 eV [12–14].

The gas phase data for ionization potentials, although widely available, is not directly applicable to condensed-phase chemistry, because of the energy shift that ensues when ions are stabilized by polarization of a matrix.

2.2. DEA Cross-Sections

When dealing with electrons with an energy below the ionization threshold of a substance (gas phase *IE* minus approximately 2 eV), the most discussed process that leads to chemical reactions is DEA. The formed radical anions are highly reactive and the specificity with which bonds can be cleaved allows for very precise reaction control. This makes DEA a powerful tool in a chemist's toolbox. Fortunately, there is some data available on EA and subsequent fragmentation channels, at least for small molecules like water [15], ammonia [16], carbon monoxide [17], and carbon dioxide [18] Here, the same caveats as in *IE* data apply: The process forms an ionic species, which is stabilized by polarization and thus the energy at which these processes are observed is lowered with respect to the gas phase. The predictions of fragmentation patterns and especially about where the charge and the radical site end up, nevertheless are very valuable when trying to untangle a reaction mechanism.

Rawat et al. [19] measured absolute DEA cross-sections in ammonia and found two energies at which DEA occurs. At the lower of the resonances, centered around 5.5 eV, cross-sections for the formation of NH_2^- (and by extension H*) were determined to around 1.6×10^{-18} cm^2, while the formation of H$^-$ and NH_2^* was observed with a cross-section of 2.3×10^{-18} cm^2. At the higher resonance, centered around 10.5 eV, the cross-sections were 1×10^{-19} cm^2 for NH_2^- and 5×10^{-19} cm^2 for H$^-$. This data helped a lot in understanding the formation of formamide (H_2NCHO) from the electron irradiation of mixed CO:NH_3 ices in a reaction like,

$$CO + NH_3 \rightarrow H_2NCHO.$$

The energy dependence of formamide production shows a resonance with a maximum between 8 and 9 eV (Figure 2). This resonant shape and product formation observed at energies as low as 6 eV pointed to DEA as the initial electron–molecule interaction process.

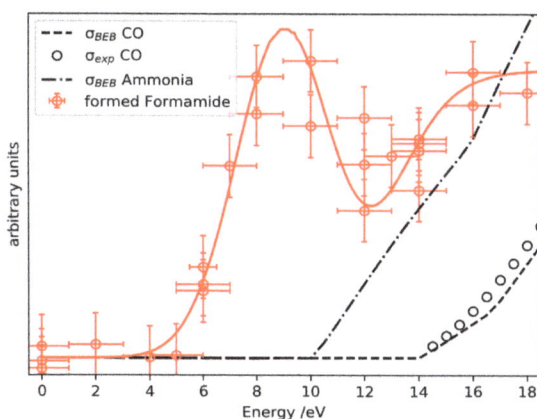

Figure 2. Predicted (black lines) electron impact cross-sections for carbon monoxide and ammonia. Experimental cross-sections for carbon monoxide (black open circles) are taken from Rapp and Englander-Golden [10]. The amount of formamide actually formed by electron irradiation of a 1:1 mixture of ammonia and carbon monoxide is shown in red circles. The red line only serves as a guide to the eye. Experimental data taken from Bredehöft et al. [20].

The reaction could proceed via either the NH_2^* radical or the H* radical attaching to carbon monoxide. The formed intermediate radicals $^*H_2NCO$ or $^*HCO$ could then go on to form formamide after reaction with another molecule of ammonia. However, neither of the two radicals could be observed experimentally, since both are very short-lived species thanks to their extremely high reactivity. In cases like this, the reaction mechanism can often be inferred by looking at the side-products of the reaction. If the reaction proceeds via the $^*H_2NCO$ radical, one side-product would be formed by addition of another amino-group (–NH_2) rather than an –H. This would form the molecule urea (($H_2N)_2CO$). If the reaction proceeds via the $^*HCO$ radical, the corresponding side-product would form by addition of –H rather than –NH_2, leading to formaldehyde (H_2CO) (see Figure 3). The gas phase data by Rawat et al. [19] predicts a ratio of 5:1 urea:formaldehyde. The experiment, however, shows absolutely no trace of urea at all, while formaldehyde is formed in about the same quantity as formamide. This is a very strong indication that the reaction proceeds via the channel that is less favorable in the gas phase. This could be due to a perturbation of the electron structure with respect to the gas phase, which is caused by close proximity to other molecules. In the case of water ice, it has been shown that DEA energies can shift with film thickness and temperature and thus film structure, and can even be higher than in the gas phase [21]. These changes to electronic structure can very well have an influence on electron affinity of the fragments and thus on branching ratios for anion formation.

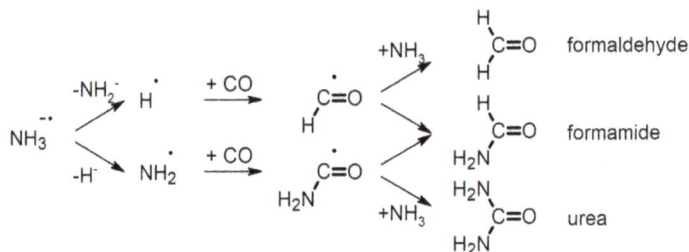

Figure 3. Possible reaction pathways from ammonia to formamide as well as the two expected side-products formaldehyde and urea.

Data on DEA in the gas phase can thus predict which reaction channels are possibly open in condensed-phase chemistry. They need to be corrected in terms of energy due to stabilization of ions and absolute cross-sections seem to be no reliable indication of which channels are actually most active in condensed-phase chemistry, but they do give a prediction of what possible reaction products could look like.

2.3. Prediction of Possible Reaction Routes

In the last example of this paper, the reaction of ethylene and water to form ethanol,

$$C_2H_4 + H_2O \rightarrow C_2H_5OH,$$

shall be described. It is the analogous reaction to the formation of ethylamine from Section 2.1, and above the ionization threshold, the reaction indeed proceeds in much the same way, with only the NH_3 replaced by H_2O. Below the ionization threshold, however, there is also some formation of ethanol, especially at energies below 4 eV. Figure 4 depicts the experimental data as well as some σ_{BEB}-based predictions for the formation via EI above the ionization threshold.

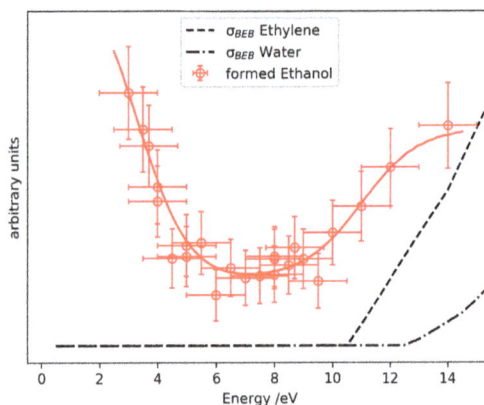

Figure 4. Predicted (black lines) electron impact cross-sections for ethylene and water. The amount of ethanol actually formed by electron irradiation of a 1:1 mixture of ethylene and water is shown in red circles. The red line only serves as a guide to the eye. Experimental data taken from Warneke et al. [22].

There is obviously some significant process at work at very low energies. Since this was not observed in the case of ethylene + ammonia, but is observed in ethylene + water, ostensibly DEA to water must be responsible. DEA cross-sections of water were reported by Curtis et al. [23]. Their data, unfortunately, shows the lowest resonance centered around 6.5 eV, with absolutely nothing happening below 5 eV. While this needs to corrected for stabilization of the ion, a shift of around 4 eV cannot be explained in this way, ruling out DEA to water as the initiating step in the reaction. DEA cross-sections to C_2H_4 have also been reported, by Szymańska et al. [24], but in much the same way as in water, no anion formation was observed below 6 eV. The only process that is known to occur is a non-dissociative electron attachment to water at around 1.5 eV. It produces the short-lived $C_2H_4^{-*}$ radical anion, which quickly decays by auto-detachment (loss of the electron). It is not readily apparent why this process should lead to product formation in the case of ethanol, but not in the case of ethylamine. This conundrum was resolved by looking at the chemistry of the $C_2H_4^{-*}$ radical anion. It is a strong base and will quickly abstract a proton from a nearby water molecule when embedded in a water ice matrix. The formed ethyl radical $^*C_2H_5$ cannot decay back and is thus available for driving the reaction to ethanol. This is not possible with the much less acidic ammonia. This type of process can of course not be captured by gas phase experiments, where great care is taken to eliminate contaminants such as water, and where molecular beams are tuned so as to exclude interactions between molecules.

3. Conclusions

When endeavoring to investigate condensed-phase chemistry triggered by electron–molecules interactions, often the only help in interpreting experimental data is found in data gathered on electron–molecule interactions in the gas phase. Some data, like ionization potentials are easily available, while others, like data on neutral dissociation cross-sections are pretty much non-existent. While being of great help, these gas phase data:

1. must be corrected for the energy shift due to stabilization of ions in a matrix,
2. should be taken with a great deal of caution with regards to absolute cross-sections and thus branching ratios, and
3. are utterly unable to predict the formation of reactive species that relies on interaction between molecules as well as with electrons.

Nevertheless, data that is of limited use is still much better than no data at all.

Funding: This research received no external funding.

Conflicts of Interest: The author declares no conflict of interest.

References

1. *International Commission on Radiation Units and Measurement*; ICRU Report 31; ICRU: Washington, DC, USA, 1979.
2. Ptasinska, S.; Denifl, S.; Scheier, P.; Illenberger, E. Bond- and Site-Selective Loss of H Atoms from Nucleobases by Very-Low-Energy Electrons (<3 eV). *Angew. Chem. Int. Ed.* **2005**, *44*, 6941–6943. [CrossRef]
3. Utke, I.; Gölzhäuser, A. Small, Minimally Invasive, Direct: Electrons Induce Local Reactions of Adsorbed Functional Molecules on the Nanoscale. *Angew. Chem. Int. Ed.* **2010**, *49*. [CrossRef] [PubMed]
4. Swiderek, P.; Jolondz, E.; Bredehöft, J.H.; Borrmann, T.; Dölle, C.; Ott, M.; Schmüser, C.; Hartwig, A.; Danilov, V.; Wagner, H.-E.; et al. Modification of Polydimethylsiloxane Coatings by H_2 RF Plasma, $Xe2^*$ Excimer VUV Radiation, and Low-Energy Electron Beams. *Macromol. Mater. Eng.* **2012**, *297*, 1091–1101. [CrossRef]
5. Öberg, K.I. Photochemistry and Astrochemistry: Photo-chemical Pathways to Interstellar Complex Organic Molecules. *Chem. Rev.* **2016**, *116*, 9631–9663. [CrossRef] [PubMed]
6. Mason, N.J.; Nair, B.; Jheeta, S.; Szymańska, E. Electron induced chemistry: A new frontier in astrochemistry. *Faraday Discuss.* **2014**, *168*, 235–247. [CrossRef] [PubMed]

7. National Institute of Standards and Technology. Mass Spectra. In *NIST Chemistry WebBook, NIST Standard Reference Database Number 69*; Linstrom, P.J., Mallard, W.G., Eds.; National Institute of Standards and Technology: Gaithersburg, MD, USA, 2016; p. 20899. Available online: http://webbook.nist.gov (accessed on 30 November 2018).

8. Kim, Y.-K.; Rudd, M.E. Binary-encounter-dipole model for electron-impact ionization. *Phys. Rev. A* **1994**, *50*, 3954–3967. [CrossRef] [PubMed]

9. Hwang, W.; Kim, Y.-K.; Rudd, M.E. New model for elctron-impact ionization cross sections of molecules. *J. Chem. Phys.* **1996**, *104*, 2956–2966. [CrossRef]

10. Rapp, D.; Englander-Golden, P. Total Cross Sections for Ionization and Attachment in Gases by Electron Impact. I. Positive Ionization. *J. Chem. Phys.* **1965**, *43*, 1464–1479. [CrossRef]

11. Böhler, E.; Bredehöft, J.H.; Swiderek, P. Low-energy electron-induced hydroamination reactions between different amines and olefins. *J. Phys. Chem. C* **2014**, *118*, 6922–6933. [CrossRef]

12. Arumainayagam, C.R.; Lee, H.L.; Nelson, R.B.; Haines, D.R.; Gunawardane, R.P. Low-energy electron-induced reactions in condensed matter. *Surf. Sci. Rep.* **2010**, *65*, 1–44. [CrossRef]

13. Balog, R.; Langer, J.; Gohlke, S.; Stano, M.; Abdoul-Carime, H.; Illenberger, E. Low energy electron driven reactions in free and bound molecules: From unimolecular processes in the gas phase to complex reactions in a condensed environment. *Int. J. Mass Spectrom.* **2004**, *233*, 267–291. [CrossRef]

14. Bald, I.; Langer, J.; Tegeder, P.; Ingólfsson, O. From isolated molecules through clusters and condensates to the building blocks of life. *Int. J. Mass Spectrom.* **2008**, *277*, 4–25. [CrossRef]

15. McConkey, J.W.; Malone, C.P.; Johnson, P.V.; Winstead, C.; McKoy, V.; Kanik, I. Electron impact dissociation of oxygen-containing molecules—A critical review. *Phys. Rep.* **2008**, *466*, 1–103. [CrossRef]

16. Bhargava Ram, N.; Krishnakumar, E. Dissociative electron attachment resonances in ammonia: A velocity slice imaging based study. *J. Chem. Phys.* **2012**, *136*, 164308. [CrossRef] [PubMed]

17. Nag, P.; Nandi, D. Fragmentation dynamics in dissociative electron attachment to CO probed by velocity slice imaging. *Phys. Chem. Chem. Phys.* **2015**, *17*, 7130–7137. [CrossRef] [PubMed]

18. Itikawa, Y. Cross Sections for Electron Collisions with Carbon Dioxide. *J. Phys. Chem. Ref. Data* **2002**, *31*, 749–767. [CrossRef]

19. Rawat, P.; Prabhudesai, V.S.; Rahman, M.A.; Bhargava Ram, N.; Krishnakumar, E. Absolute cross sections for dissociative electron attachment to NH_3 and CH_4. *Int. J. Mass Spectrom.* **2008**, *277*, 96–102. [CrossRef]

20. Bredehöft, J.H.; Böhler, E.; Schmidt, F.; Borrmann, T.; Swiderek, P. Electron-Induced Synthesis of Formamide in Condensed Mixtures of Carbon Monoxide and Ammonia. *ACS Earth Space Chem.* **2017**, *1*, 50–59. [CrossRef]

21. Simpson, W.C.; Orlando, T.M.; Parenteau, L.; Nagesha, K.; Sanche, L. Dissociative electron attachment in nanoscale ice films: Thickness and charge trapping effects. *J. Chem. Phys.* **1998**, *108*, 5027–5034. [CrossRef]

22. Warneke, J.; Wang, Z.; Swiderek, P.; Bredehöft, J.H. Electron-induced hydration of an alkene: Alternative reaction pathways. *Angew. Chem. Int. Ed.* **2015**, *54*, 4397–4400. [CrossRef] [PubMed]

23. Curtis, M.G.; Walker, I.C. Dissociative Electron Attachment in Water and Methanol (5–14 eV). *J. Chem. Soc. Faraday Trans.* **1992**, *88*, 2805–2810. [CrossRef]

24. Szymańska, E.; Mason, N.J.; Krishnakumar, E.; Matias, C.; Mauracher, A.; Scheier, P.; Denifl, S. Dissociative electron attachment and dipolar dissociation in ethylene. *Int. J. Mass. Spectrom.* **2014**, *365–366*, 356–364. [CrossRef]

atoms

MDPI

Review

Interrelationship between Lab, Space, Astrophysical, Magnetic Fusion, and Inertial Fusion Plasma Experiments

Mark E. Koepke

Department of Physics and Astronomy, West Virginia University, Morgantown, WV 26506-6315, USA; mark.koepke@mail.wvu.edu; Tel.: +1-304-293-4912

Received: 1 December 2018; Accepted: 3 February 2019; Published: 11 March 2019

check for updates

Abstract: The objectives of this review are to articulate geospace, heliospheric, and astrophysical plasma physics issues that are addressable by laboratory experiments, to convey the wide range of laboratory experiments involved in this interdisciplinary alliance, and to illustrate how lab experiments on the centimeter or meter scale can develop, through the intermediary of a computer simulation, physically credible scaling of physical processes taking place in a distant part of the universe over enormous length scales. The space physics motivation of laboratory investigations and the scaling of laboratory plasma parameters to space plasma conditions, having expanded to magnetic fusion and inertial fusion experiments, are discussed. Examples demonstrating how laboratory experiments develop physical insight, validate or invalidate theoretical models, discover unexpected behavior, and establish observational signatures for the space community are presented. The various device configurations found in space-related laboratory investigations are outlined.

Keywords: laboratory plasma; astrophysical plasma; fusion plasma; lasers; stars; extragalactic objects; spectra; spectroscopy; scaling laws

---

1. Introduction

Many advances in understanding geospace, heliospheric, and astrophysical plasma phenomena are linked to insight derived from theoretical modeling and/or laboratory plasma experiments [1–3]. Geospace plasma physics includes space weather during periods of magnetic storms, substorms, and geomagnetic quiet; nonlinear plasma behavior such as structure evolution in turbulence and particle transport, fluid and kinetic instabilities, wave–particle interactions, ionospheric-magnetospheric-auroral coupling; solar-wind interaction with magnetospheres; and solar-corona heating. Heliospheric physics is concerned with investigating the interaction of the Sun's heliosphere with the local interstellar medium, as well as with the origin and evolution of the Sun and solar wind, the magnetospheres of the Earth and outer planets, and low-energy cosmic rays. High-energy (HE) and high-energy-density (HED) astrophysics is the study of electromagnetic radiation, from ultra-energetic cosmic phenomena ranging from black holes to the Big Bang, and of ionized matter at ultra-high pressure (~1 Mbar to 1000 Gbar, i.e., 1 million to 1 trillion Earth-surface atmospheres of pressure), density, and temperature (i.e., stored energy in matter $>10^{10}$ J/m^3, e.g., solid-density material at 10,000 K (~1 eV)) for which observations are made in the extreme-ultraviolet (EUV), X-ray, and gamma-ray bands. Examples of places where these HED conditions occur are in Earth's, Jupiter's, and Sun's core and inside igniting Inertial Confinement Fusion (ICF) implosions (~250 Gbar). Figure 1 illustrates the wide ranges of density and temperature naturally occurring in geospace, in the heliosphere, and in astrophysical environments and artificially occurring in larger magnetic and inertial fusion devices, as well as in smaller university-scale laboratory devices.

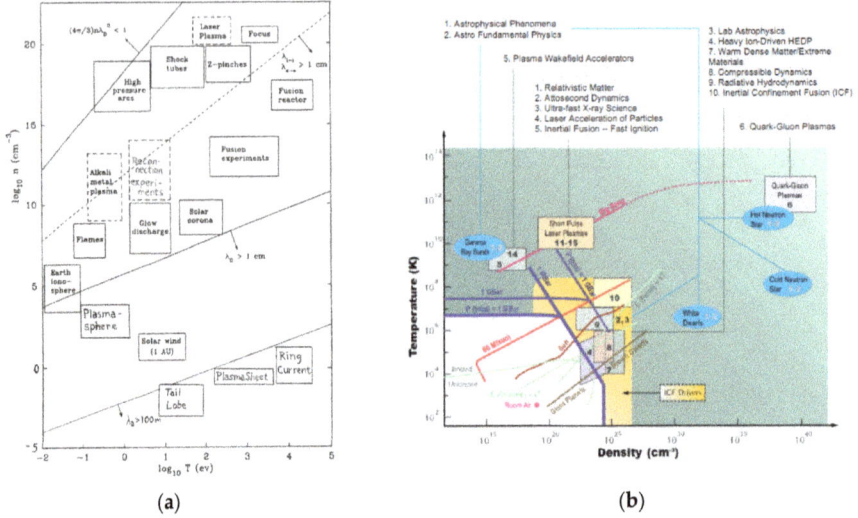

Figure 1. Typical parameters of lab and nature's plasmas: Cartesian representation of density and temperature range associated with (**a**) heliospheric (10^2–10^9 K) [2] and (**b**) astrophysical (10^2–10^{14} K) plasma [4]. Upper-right region corresponds to pushing parameters toward nuclear fusion ignition.

Modern facilities and new experimental techniques have provided access to novel plasma regimes, such as those associated with the ultra-relativistic, beam-in-plasma interaction at the SLAC National Accelerator Laboratory, the ultra-cold, highly correlated plasmas being studied at NIST, University of Maryland, European Organization for Nuclear Research (CERN), and University of Michigan, and the low-temperature micro-discharges that simultaneously share aspects of the solid, liquid, and plasma states while being created to explore novel plasma chemistry in numerous academic and government research laboratories. Plasma source, boundary conditions, and configuration geometry affect space and lab plasma and processes. In geospace cases, the major plasma domains of the magnetospheric boundary layer, bow shock, plasmasphere, plasma sheet boundary layer, polar caps, and lobe region constitute examples of important sources, interfacial layers, and configuration geometry. In astrophysical cases, the major plasma domains associated with plasma jet direction, shock front interface, and coherent radiation- and particle-beam boundaries are examples of the source, gradient, and directional factors. The unexpected new phenomena that now frame the cutting edge of discovery plasma science elevate these new regimes to a common priority in fundamental and applied research that is better illustrated by a circular plasma-parameter space (Figure 2) rather than the cartesian plasma-parameter space familiar in the quest for fusion.

Circumnavigating the range of temperatures, densities, and magnetic fields shown in Figures 1 and 2 are a number of pervasive themes in plasma behavior that can be characterized in terms of universal processes that are, at least partially, independent of the specific process being investigated. Some of these processes are well understood and predictable, whereas many others are neither. Advances for which (a) lab experiments helped explain phenomena and processes, (b) lab data influenced the interpretation of space data, and (c) laboratory validation of atomic and molecular spectroscopic processes contributed to telescopic probing of distant events are itemized in Sections 2–4. As identified in the Plasma 2010 report of the National Academies [5], six critical plasma processes define the research frontier.

- Explosive instability in plasmas
- Multiphase plasma dynamics
- Particle acceleration and energetic particles in plasmas

- Turbulence and transport in plasmas
- Magnetic self-organization in plasmas
- Correlations in plasmas

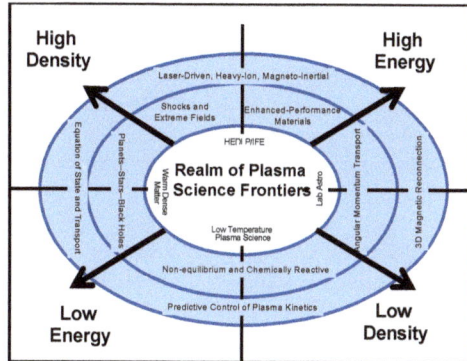

Figure 2. Phenomenological regimes depicted on a circular energy–density range. Angular dispersion of regimes emphasizes their unbiased priority assignment in discovery science.

2. Geospace–Lab Interrelationship

Fälthammar [1] describes the previous generation of lab-magnetospheric plasma interrelations as exploratory measurements in simulated configurations. Partial scale-model experiments, such as Birkeland's Terrella [6] having a magnetic neutral sheet in the simulated nightside magnetosphere, were used to validate Störmer's orbit theory [7] and Alfvén's perturbation theory [8]. Advances in understanding iconic features in the solar-terrestrial environment, such as Birkeland currents, electric fields along B-lines, and electrostatic particle acceleration, were made with these "configuration simulations" and the role of current filamentation and inhomogeneity was debated. Birkeland [9], Danielsson and Lindberg [10], Podgorny [11], Ohyabu and Kawashima [12], and Bratenahl and Yeates [13] were primary laboratory contributors.

A sophisticated physics-orientation effort ("process simulation") was needed, in which suitably designed lab experiments could be performed for clarifying physical processes. Accomplishments in understanding collisionless resistivity, the interaction between a plasma and a magnetic field (magnetic-field penetration), the beam–plasma interaction, parallel electric-field generation in a magnetic mirror geometry, and electric-charge double-layer formation were responsible for widely recognized highlights during this period of investigation. Noteworthy laboratory contributions include the discovery of a double sheath (layer) [14], a high-voltage parallel potential drop [15,16], an instability due to a perpendicular ion beam [17], and fast-electron production by an electrostatic beam-plasma wave [18].

An international workshop, held in 1980, on the relation between laboratory and space plasmas [19] gathered researchers working on the following topics: critical ionization velocity, formation of double layers, active stimulation (by high-power pulse and high-frequency transmitter) of the auroral ionosphere, energetic electron-beam experiments (in a 17 m × 26 m chamber and from rockets), plasma potential in the presence of strong ion-cyclotron turbulence, a magnetic field-line reconnection experiment, space experiments with particle accelerators (SEPAC) on Spacelab, and the EXOS-B/Sipole-station VLF wave–particle interaction experiment. Significant attention was given to plasma instabilities, for example, from electrostatic shocks in the lab and by the S3-3 satellite, in the high-latitude ionospheric F region, in electrojets, and in the auroral acceleration region.

One workshop focus was documenting the advances in explaining and validating phenomena and processes relevant to probing astrophysical events. The critical ionization velocity experiments were shown to contribute to explaining the pattern associated with ionizing discontinuity driven by stellar UV radiation at the heliospheric shock. Establishing the observational signatures of broad, thin, reconnecting neutral sheets, of the dynamics and topology of magnetic field-line reconnection, and of the effects of high-β was emphasized as an important influence of laboratory experiments to interpreting space and astrophysical data. The topic "Laboratory experiments on quantifying resistivity" was considered an excellent example of the benefit of the lab–space collaboration.

A more recent review paper on the interrelationship between laboratory and space plasma experiments [2] outlined the following benefits of lab experiments to the understanding of space plasmas: illustrating what a spacecraft would detect for specific processes, pioneering diagnostic methods, and being a source of citations on controlled-parameter experimental evidence for space researchers. The number and scope of interdisciplinary (i.e., lab–space) activities are expanding as space missions resemble more and more the multi-point data acquisition associated with laboratory experiments and as present-day activities become more overlapped. Many space plasma physicists concur with the ideas that:

- lab experiments are generally complementary to space observations;
- a well-designed lab experiment has the potential to provide measurements in detail far greater than those that can currently be obtained by in situ measurements;
- such detail can provide new insight into the mechanisms involved and can help direct the development of theories to explain the space observations; and
- future collaboration between space and lab communities would be profitable.

3. Heliosphere–Lab Interrelationship

In reviewing laboratory investigating of the physics of space plasmas, Howes [3] highlights key open questions and lab–space physics successes enumerated in solar corona, solar wind, planetary magnetospheres, and outer boundary of the heliosphere. He identifies velocity space as a key new frontier and outlines a strategy for future lab–space physics investigations on the following topics:

- Plasma turbulence
- Magnetic reconnection
- Particle acceleration
- Collisional and collisionless shocks
- Kinetic and fluid instabilities
- Self-organization
- Physics of multi-ion and dusty plasmas
- Astrophysical connections
- Improved diagnostic capabilities
- Novel analysis methods

From this list, turbulence, magnetic reconnection, particle acceleration, and kinetic instabilities are recognized as four grand challenges. Earlier lab experiments suffered from an inability to model the large scales (relative to kinetic length scales) that are characteristic of space plasma processes, whereas present-day intermediate-scale facilities can generate plasma spanning a substantial dynamic range above the typical kinetic length scales. For example, UCLA's Large Plasma Device-Upgrade (LAPD) produces a 2-mm ion-gyroradius, 17-m long, 60-cm diameter, magnetized plasma column able to axially contain magnetohydrodynamics (MHD) waves with frequencies below the ion cyclotron frequency. The plasmoid instability can be studied in collisional and collisionless regimes when the current-sheet length relative to the ion inertial length or ion Larmor radius exceeds 1000, which is within reach of present-day intermediate-scale facilities. Lundquist number $>10^5$ is possible in soon-to-be inaugurated intermediate facilities. Many other space-related lab devices, not quite at the intermediate scale, are mentioned in terms of their contributions to space, heliospheric, and astrophysical plasma research.

The New Frontier Science Experiments (FSE) Campaign [20] on the DIII-D tokamak, launched in 2017, contributes insight from lab and theory. Subproject titles from the DIII-D FSE initiative are listed below. Four FSE experiments were conducted in FY2017 and another four were conducted in FY2018. Four undesignated slots are tentatively scheduled for the FY2020 campaign. The last experiment on the list is positron generation, a fundamental physics challenge, being taken to new higher yields by using the higher density, higher temperature plasmas confined toroidally in a tokamak. Many positron-related space physics questions motivate this work: Why does matter dominate over antimatter in the universe? Where do gamma-ray bursts originate in space? How do black holes form? The study of positrons provides insight.

(1a) Self-organization of unstable flux ropes (1 day of 2017 runtime)

(1b) Magnetic reconnection and self-organization of flux ropes in tokamak sawteeth (1 day in 2018)

(2)   Impact of magnetic perturbations on turbulence (1 day of 2017 runtime)

(3a) Interaction of Alfven/whistler fluctuations and runaway electrons (0.5 day of 2017 runtime)

(3b) Interaction of Alfven/whistler fluctuations and runaway electrons (0.5 day in 2018)

(4)   Field-line chaos: Self-consistent chaos in magnetic field dynamics (0.5 day of 2017 runtime)

(5)   Electromagnetic ion-cyclotron emission (1 day of 2018 runtime)

(6)   Positron generation in tokamaks (0.5 day of 2018 runtime)

4. Astrophysics–Lab Interrelationship

High-energy-density (HED) astrophysics explores a wide range of topics by exploiting the extreme physical conditions achievable through the use of large off-site facilities specially designed for HED physics and inertial confinement fusion research. Laser energy is used to compress capsules filled with fuel material to high density and pressure in order to generate fusion reactions with the goal of self-sustained fusion burn ("ignition") and the generation of energy. Such work is primarily executed using the 30-kilojoule OMEGA laser at the University of Rochester and the 2-megajoule laser at the National Ignition Facility (NIF). Special diagnostic instrumentation has been developed that makes it possible to study spatial and temporal variations in plasma properties and electromagnetic fields through spectral, temporal, and imaging measurements. Researchers actively collaborate and sometimes lead teams in planning experiments at the facilities and in analyzing the results. Major research topics common to the inertial fusion realm fall within the science of extreme astrophysical phenomena:

- Physics of inertial confinement fusion
- Properties of warm dense matter
- Stellar and Big Bang nucleosynthesis
- Basic nuclear physics
- Astrophysical jets
- Magnetic reconnection
- High-energy-density hydrodynamics
- Nonlinear optics
- Relativistic HED plasma and intense beam physics
- Magnetized HED plasma physics
- Radiation-dominated HED plasma physics

We can test hypotheses concerning the physics of an observation that took place millions or even billions of light years away when dimensionless quantities retain their qualitative ordering. Thus, lab experiments develop, via computer simulation, credible scaling of physical processes. New space telescopes permit observations of ultra-high-energy events, helping us probe these spatially and temporally distant events. The goal of laboratory plasma astrophysics is, quoting from the National Academy of Sciences report [21], Connecting Quarks with the Cosmos, to "discern the physical principles that govern extreme astrophysical environments through the laboratory study of HED physics." The challenge here is to develop physically credible scaling relationships that enable laboratory experiments on the centimeter or meter scale to illuminate physical processes taking place in a distant part of the universe over enormous length scales.

Spectacular outbursts from the notoriously steadily glowing Crab Nebula are altering the theories that have long explained charged-particle accelerations to high energies [22]. Recently, the nebula's gamma-ray flares were observed to fluctuate on time scales of only a few days and even shorter, over just one to three hours, indicating that the charged particles were accelerated within a region representing an infinitesimal fraction of the vast Crab. It is proposed that the electron-positron plasma particles are accelerated near the nebula's center where rapid magnetic reconnection unleashes enormous amounts of energy in the presence of a strong electric field [23]. Laboratory experiments, together with theoretical models, are providing evidence that magnetic reconnection can explain the space observation of this rapid extreme particle acceleration and gamma-ray flaring in the Crab Nebula [23,24].

The remarkable discovery by the Chandra X-ray observatory that the Crab Nebula's jet periodically changes direction provides a challenge to our understanding of astrophysical jet dynamics. It has been suggested that this phenomenon may be the consequence of magnetic fields and MHD instabilities, but experimental demonstration in a controlled laboratory environment remained elusive until experiments were reported [25] that use high-power lasers to create a plasma jet that can be directly compared with the Crab jet through well-defined physical scaling laws, as documented in Table 1. The jet generates its own embedded toroidal magnetic fields; as it moves, plasma instabilities result in multiple deflections of the propagation direction, mimicking the kink behavior of the Crab jet, as illustrated in Figure 3. The experiment was modelled with 3-dimensional numerical simulations that show exactly how the instability develops and results in changes of direction of the jet.

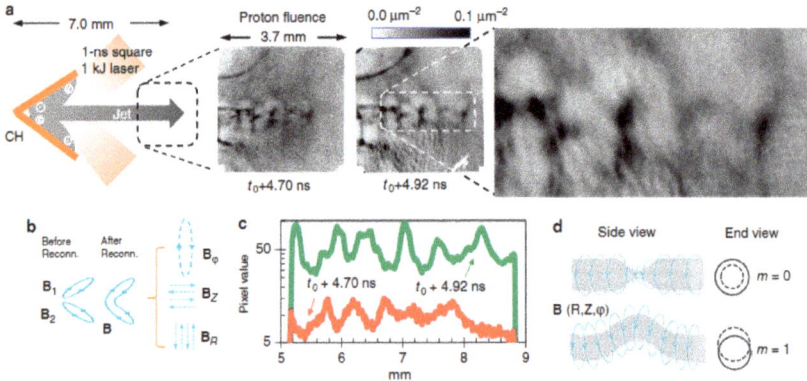

Figure 3. Side-on image [25] shows sequence of clumps and wiggles, resulting from embedded toroidal B fields, indicating MHD current-driven instabilities: mode $m = 0$ (sausage) and $m = 1$ (kink).

Table 1. Scaled lab experiments (refer to Figure 3) explain kink behavior of the Crab Nebula jet [25].

Parameters and Scales		Plasma Jet in OMEGA Experiment *	Scaled to the Crab Nebula †	The Kinked Jet in the Crab Nebula †
Temperature	T_e	~300 eV		~1–130 eV
Ionization state	Z	~3.5		~1
Number density	n_e	~5 × 10^{19} cm^{-3}		~10^{-2} cm^{-3}
Pressure	P	~4 × 10^5 bar		~4 × 10^{-14} bar
Jet radius	r_j	~5 × 10^{-2} cm		~1 pc
Jet velocity	v_j	~400 km s^{-1}	**< 3 × 10^5 km s^{-1}**	**~1.2 × 10^5 km s^{-1}**
Time scale	τ	~10^{-9} s	~1.5 years	~few years
Magnetic field	B	~2 MG	~0.6 mG	~1 mG
Thermal plasma beta	β	~0.1–1		<<1
Magnetization parameter	σ	~1–6		≥1
Mach number	M	~3		>>1
Reynolds number	R_e	~2 × 10^5		~2 × 10^{17}
Peclet number	P_e	~1–5		~4 × 10^{15}
Magnetic Reynolds number	~3 × 10^3		~1 × 10^{22}	
Biermann number	Bi	~6		~6 × 10^8
Radiation number	Π	~3 × 10^5		~1 × 10^{18}

Note: * Near the region of jet launching. † Near the region of the pulsar pole. The bold entries show the physical quantities from the two systems that can be directly compared through the scalings in Equation (3), manifesting how the laboratory experiment parameters scale to match those of the Crab nebula jet.

Understanding the equation of state and chemistry of even more extreme matter stands as a central challenge in validating theoretical models in planetary physics and astrophysics. The interiors of giant planets exist in a density, temperature (n,T) regime where accurately calculating the equation of state is difficult. Molecules, atoms, and ions coexist in a fluid that is coupled by Coulomb interactions and is highly degenerate (free electrons governed by quantum and thermal effects). These strong interactions dominate in the steady-state interiors of giant planets such as Saturn and Jupiter and in brown dwarfs where phase transitions play an important role. Understanding the high-pressure phases of carbon is important since carbon is a major element of giant planets such as Uranus and Neptune. Petawatt-laser-driven shock-wave measurements of diamond's principal Hugoniot curve have been made at pressures between 6 and 9 Mbar using the Laboratory for Laser Energetics (LLE) OMEGA laser. The Hugoniot curve traces the path accessed by the laser-induced shock driven in the material, indicating that, in the solid–liquid coexistence regime in that range between 6 and 10 Mbar, the mixed phase is slightly denser than the one that would be expected from straightforward interpolation between liquid and solid Hugoniot curves.

Near Jupiter's surface (10^{11} Pascal and a fraction of an eV), hydrogen exists in molecular form. However, it dissociates and ionizes deeper into the planet's core (>10^{12} Pascal and a few eV). This transition from insulator to conductor in the convective zone is believed to be responsible for Jupiter's 10 to 15 Gauss magnetic field. An open question is whether there is a sharp plasma phase transition. Experiments performed on the Nova laser at Lawrence Livermore National Laboratory initially suggested that the transition was continuous, and subsequent experiments unambiguously demonstrated that the transition from non-conducting molecular hydrogen to atomic metallic hydrogen at high pressure is a continuous transition. This suggests that the metallic region of Jupiter's interior extends out to 90 percent of the radius of the planet and may explain why the magnetic field of Jupiter is so much stronger than that of the other planets of our solar system.

The study of astrophysically relevant, magnetized high-energy-density (HED) plasmas relies heavily on numerical simulations in limited parameter regimes, where the thermal and magnetic pressures balance ($\beta \sim 1$) and where the magnetic field advects with the plasma (Re$_M$ >> 1), and has had little guidance from controlled laboratory experiments to test underlying principles, even though magnetized plasmas are ubiquitous throughout our universe. Using high-energy lasers, plasma conditions similar to those found in astrophysical systems can be created. Specifically, supersonic plasma flows can arise from irradiating a thin (10 s of μm) solid material with a high-energy laser

Atoms **2019**, 7, 35

pulse in an externally seeded magnetic field. This regime allows us to study the structure of accretion shocks and how the shocks are affected by magnetic fields, which will aid in understanding the spatial structure of hotspots on the surface of a young star [26]. Experimental conditions can be created where a plasma flow encounters a magnetic obstacle, which is similar to a planet's magnetosphere interacting with the stellar wind [27]. Finally, the effects of magnetic fields on collimated outflows, which are observed in young stellar objects [28], can be studied.

The astrophysical Weibel instability has been reproduced for the first time in counter-streaming laser-produced plasmas [29]. The Weibel instability, by generating turbulent electric and magnetic fields in the shock front, is responsible for the requisite interaction mechanism in shock formation in the limit of weakly magnetized shocks. This work confirms its basic features, a significant step toward understanding these shocks. In the experiments, a pair of plasma plumes are generated by irradiating a pair of opposing parallel plastic (CH) targets. The ion–ion interaction between the two plumes is collisionless, so as the plumes interpenetrate, supersonic and counter-streaming ion flow conditions are obtained. Electromagnetic fields formed in the interaction of the two plumes were probed with an ultrafast laser-driven proton beam, and the growth of a highly striated, transverse instability with extended filaments parallel to the flows was observed. The instability is identified as an ion-driven Weibel instability through agreement with analytic theory and particle-in-cell simulations, paving the way for further detailed laboratory study of this instability and its consequences for particle energization and shock formation. Astrophysical shocks, which often manifest as collisionless, typically require collective electromagnetic fields to couple the upstream and downstream plasmas. These shocks can energize cosmic rays in the blast waves of astrophysical explosions and they can generate primordial magnetic fields during the formation of galaxies and clusters [29].

5. Conclusions

An alliance exists between laboratory plasma physicists and space scientists to investigate basic and fusion plasma phenomena relevant to space. Dedicated lab studies (1) probe and elucidate fundamental plasma physical phenomena and processes, (2) provide benchmarks for validating theory and modeling, (3) discover unexpected behavior, and (4) establish observational signatures, all in support of interpreting rocket, satellite, and telescope data.

As concluded in the Plasma 2010 report [5], "progress in understanding the fundamental plasma processes in many space and astrophysical phenomena is greatly leveraged by close communication among space, astrophysical, and laboratory plasma scientists." The connections between the different plasma regimes studied in geospace, heliospheric, and astrophysical plasma and the related fields of laboratory plasma physics have led to significant scientific progress in many research areas. Studies of common plasma processes, rather than comparing the large-scale morphology of observed systems, link the different plasma physics communities. Consequently, maintaining and strengthening the linkages between communities is highly desirable.

Funding: Partial funding from U.S. Department of Energy, contract numbers DE-SC0012515, DE-SC0018036, and DE-NA0003874 is gratefully acknowledged.

Acknowledgments: Useful discussions with Greg Howes, Chikang Li, and others in the IPELS (interrelationship between plasma experiments in the laboratory and in space) community are gratefully acknowledged.

Conflicts of Interest: The author declares no conflict of interest. The funding agencies and their employees had no role in choosing the content of this review; in the collection, analyses, or interpretation of data; in the writing of the manuscript; or in the decision to publish the results.

References

1. Fälthammar, C.-G. Laboratory experiments of magnetospheric interest. *Space Sci. Rev.* **1974**, *15*, 803–825. [CrossRef]
2. Koepke, M.E. Interrelated laboratory and space plasma experiments. *Rev. Geophys.* **2008**, *46*, RG3001. [CrossRef]

3. Howes, G.G. Laboratory space physics: Investigating the physics of space plasmas in the laboratory. *Phys. Plasmas* **2018**, *25*, 055501. [CrossRef]

4. Davidson, R. *National Task Force on High-Energy-Density Physics*; Frontiers for discovery in high-energy-density physics; OSTP and NSTC Interagency Working Group on the Physics of the Universe: Washington, DC, USA, 2004.

5. Plasma Science Committee. *Plasma Science: Advancing Knowledge in the National Interest*; National Academies Press: Washington, DC, USA, 2007.

6. Birkeland, K. *The Norwegian Aurora Polaris Expedition 1902—1903*; H. Aschelhoug: Christiania, Norway, 1908.

7. Störmer, C. On pulsations of terrestrial magnetism and their possible explanation by periodic orbits of corpuscular rays. *J. Geophys. Res.* **1931**, *36*, 133–138. [CrossRef]

8. Alfvén, H.; Fälthammar, C.G. *Cosmical Electrodynamics*, 2nd ed.; Oxford University Press: England, UK, 1963.

9. Birkeland, K. Cause of magnetic storms. *C. R. Acad. Sci.* **1908**, *147*, 539.

10. Danielsson, L.; Lindberg, L. Plasma flow through a magnetic dipole field. *Phys. Fluids* **1964**, *7*, 1878. [CrossRef]

11. Podgorny, I.M. Laboratory experiments directed toward the investigation of the magnetospheric phenomena. *Space Sci. Rev.* **1973**, *15*, 8227. [CrossRef]

12. Ohyabu, N.; Kawashima, N. Neutral point discharge experiments. *J. Phys. Soc. Jap.* **1972**, *33*, 496. [CrossRef]

13. Bratenahl, A.; Yeates, C.M. Experimental study of magnetic flux transfer at the hyperbolic neutral point. *Phys. Fluids* **1970**, *13*, 2696. [CrossRef]

14. Babić, M.; Torvén, S. Current Limiting Space Charge Sheaths in a Low Pressure Arc Plasma. Available online: https://inis.iaea.org/collection/NCLCollectionStore/_Public/05/125/5125776.pdf (accessed on 5 February 2019).

15. Hopfgarten, N.; Johansson, R.B.; Nilsson, B.H.; Persson, H. Collective phenomena in a Penning discharge in a strongly inhomogeneous magnetic mirror field. In Proceedings of the 5th European Conference on Controlled Fusion and Plasma Physics, Grenoble, France, 21–25 August 1972.

16. Geller, R.; Hopfgarten, N.; Jacquot, B.; Jacquot, C. Electric fields parallel to the magnetic field in a laboratory plasma in a magnetic mirror field. *J. Plasma Phys.* **1974**, *12*, 467–486. [CrossRef]

17. Ripin, B.H.; Stenzel, R.L. Electron cyclotron drift instability experiment. *Phys. Rev. Lett.* **1973**, *30*, 45–48. [CrossRef]

18. Conrad, J.R.; Walsh, J.E.; Diaz, C.J.; Freese, K.B. Production of Fast Electrons in the Beam-Plasma Interaction. *Phys. Rev. Lett.* **1973**, *30*, 827. [CrossRef]

19. Kikuchi, H. Workshop Proceedings. In Proceedings of the Relation between Laboratory and Space Plasmas, Tokyo, Japan, 14–15 April 1980; D. Reidel Publishing Co.: Boston, MA, USA, 1980.

20. Koepke, M.E. New Frontier Science Campaign on DIII-D launched in 2017. In Proceedings of the 45th Institute of Physics—Plasma Physics Conference, Queen's University Belfast, Northern Ireland, 9–12 April 2018.

21. NAS/NRC Report. *Connecting Quarks with the Cosmos: Eleven Science Questions for the New Century*; National Academies Press: Washington, DC, USA, 2003.

22. Shea, S.B. Available online: https://www.newswise.com/articles/solving-a-plasmaphysics-mystery%3A-magnetic-reconnection?seeOriginal=solving-a-plasma-physics-mystery:-magneticreconnection (accessed on 5 February 2019).

23. Besshoc, N.; Bhattacharjee, A. Fast collisionless reconnection in electron-positron plasmas. *Phys. Plasmas* **2007**, *14*, 056503. [CrossRef]

24. Lyubarsky, YE. Highly magnetized region in pulsar wind nebulae and origin of the Crab gamma-ray flares. *Mon. Not. Roy. Astron. Soc.* **2012**, *427*, 1497–1502. [CrossRef]

25. Li, C.K.; Tzeferacos, P.; Lamb, D.; Gregori, G.; Norreys, P.A.; Rosenberg, M.J.; Follett, R.K.; Froula, D.H.; Koenig, M.; Seguin, F.H.; et al. Scaled laboratory experiments explain the kink behaviour of the Crab Nebula jet. *Nat. Commun.* **2016**, *7*, 13081. [CrossRef] [PubMed]

26. Young, R.P.; Kuranz, C.C.; Drake, R.P.; Hartigan, P. Accretion shocks in the laboratory: Design of an experiment to study star formation. *High Energy Density Phys.* **2017**, *23*, 1–5. [CrossRef]

27. Liao, A.S.; Shule, S.L.; Hartigan, P.; Graham, P.; Fiksel, G.; Frank, A.; Foster, J.; Kuranz, C. Numerical simulation of an experimental analogue of a planetary magnetosphere. *High Energy Density Phys.* **2015**, *17*, 38–41. [CrossRef]

28. Manuel, M.J.-E.; Kuranz, C.C.; Rasmus, A.M.; Klein, S.; Macdonald, M.J.; Trantham, M.R.; Fein, J.R.; Belancourt, P.X.; Young, R.P.; Keiter, P.A.; et al. Experimental results from magnetized-jet experiments at the Jupiter Laser Facility. *High Energy Density Phys.* **2015**, *17*, 52–62. [CrossRef]

29. Fox, W.; Fiksel, G.; Bhattacharjee, A.; Chang, P.Y.; Germaschewski, K.; Hu, S.X.; Nilson, P.M. Filamentation instability of counterstreaming laser-driven plasmas. *Phys. Rev. Lett.* **2013**, *111*, 225002. [CrossRef] [PubMed]

MDPI

St. Alban-Anlage 66

4052 Basel

Switzerland

Tel. +41 61 683 77 34

Fax +41 61 302 89 18

www.mdpi.com

Atoms Editorial Office

E-mail: atoms@mdpi.com

www.mdpi.com/journal/atoms

www.ingramcontent.com/pod-product-compliance
Lightning Source LLC
Chambersburg PA
CBHW051720210326
41597CB00032B/5547